Tracking the Franklin Expedition of 1845

Tracking the Franklin Expedition of 1845

The Facts and Mysteries of the Failed Northwest Passage Voyage

STEPHEN ZORN

McFarland & Company, Inc., Publishers
Jefferson, North Carolina

ISBN (print) 978-1-4766-9219-7
ISBN (ebook) 978-1-4766-5114-9

Library of Congress and British Library cataloguing data are available

Library of Congress Control Number 2023030207

© 2023 Stephen Zorn. All rights reserved

No part of this book may be reproduced or transmitted in any form or by any means, electronic or mechanical, including photocopying or recording, or by any information storage and retrieval system, without permission in writing from the publisher.

Front cover illustration by Kristina Gehrmann

Printed in the United States of America

*McFarland & Company, Inc., Publishers
Box 611, Jefferson, North Carolina 28640
www.mcfarlandpub.com*

For Jean, Always

Acknowledgments

First, thanks to Tim Berners-Lee, Vint Cerf and all the other fathers and mothers of the internet. Without their creation, the research for this book could not have been done, at least not in one lifetime. And in particular to the New York Public Library, perhaps the greatest brick-and-mortar scholarly resource in the world, and to Larry Page and Sergei Bryn, who created Google, which begat Google Books and Google Scholar, which in turn made accessible a trove of 19th-century original sources, including virtually all the exploration narratives cited in this book, as well as a trove of hard-to-find material—at least for those of us without daily access to a major research university library.

Thanks also to the worldwide community of Franklin enthusiasts, most visible on the Facebook group "Remembering the Franklin Expedition." And to the dedicated bloggers and website administrators who keep the story alive. As just one example, English researcher Alison Freebairn has showed us that there's always more to be found even in well-used archives. Thanks also to the archivists at, among other places, the Smithsonian Institution in Washington, the Royal Geographical Society in London, the Hudson's Bay Company Archives in Winnipeg, the Public Record Office in London, the National Maritime Museum in Greenwich, the Scott Polar Research Institute in Cambridge and the Arctic Institute of North America in Calgary. While we writers may all, in the words of the late Canadian folksinger Stan Rogers, follow the hand of Franklin reaching for the Beaufort Sea, we would have a much harder time finding our way without the work of these librarians and archivists. And to Alan Cooke, whose brief essay "A Bibliographical Introduction to Sir John Franklin's Expeditions and the Franklin Search" (in the invaluable Patricia D. Sutherland, ed., *The Franklin Era in Canadian Arctic History, 1845–1859*), though now dated, is a model of how to point scholars in the right direction.

And thanks to everyone at McFarland, including Beth Foxwell, Lisa Camp, Kristal Hamby, Susan Kilby, Layla Milholen and everyone else involved.

Closer to home, thanks to Jean Zorn, herself a scholar of faraway places, for her indulgence, even if she never did agree to accompany me on a trek north. Jean's conversations and editing have made this a much better work than it otherwise would have been.

Table of Contents

Acknowledgments	vi
Preface	1
1. The Known Knowns	3
2. A Quantum Theory of History	17
3. *Erebus* and *Terror*	22
4. Who Sailed with Franklin?	29
5. Beechey Island	49
6. Westward Ho?	59
7. Which Side of King William Island?	64
8. Winter 1846–47 and Spring 1847	70
9. The Second Winter Trapped in the Ice	76
10. Where Did They Go?	81
11. Return to the Ships? Mutiny?	87
12. Off the Beaten Path—But Where?	93
13. What Killed Them—and When?	98
14. Cannibalism	114
15. Survivors?	121
16. Sir John's Grave	124
17. Franklin's Legacy	129
18. What Do the Recent Discoveries Mean?	133
Appendix I. The Victory Point Record	137
Appendix II. Erebus *and* Terror *Muster Rolls*	138
Appendix III. Sir John Franklin's Sailing Orders	141
Chapter Notes	147
Bibliography	172
Index	197

Preface

> We were waist deep in the big muddy
> And the big fool said to push on.[1]

The genesis of this book probably goes back to my work for the then–brand-new government of Papua New Guinea in the 1970s. There, I was a young, somewhat radical newcomer. My job qualifications to be an economic adviser to the pro–Independence party in the colonial legislature were (1) that I wasn't Australian (the colonial power) and (2) that I was a quick study. In that role I developed an abiding dislike for all things colonial. That naturally focused my attention on Great Britain, the ultimate world colonial power for centuries and one of the three European countries that had divided up the island of New Guinea in the late 19th century. Next, Roland Huntford's *Scott and Amundsen*,[2] first published in 1979, became for me the paradigmatic takedown of the bumbling British explorer. So it was only natural that, when I discovered the Franklin expedition some 20 years ago, I fixated on Sir John as an exemplar *par excellence* of the hapless British naval officer at sea, as it were, in an environment he was incapable of understanding, an attitude much reinforced by Pierre Berton's *Arctic Grail*,[3] published in 1988. Pete Seeger's lyric, above, was aimed at Lyndon Johnson and the Vietnam War, but it might just as well have applied, in my naïve view, to John Franklin.

As often happens, life intervened. When I returned to the Franklin story a few years ago, I found it to be far more complicated. Not only was Berton's—and my—view of Franklin himself a mere caricature of someone who in real life was not so easy to define, but also there are now so many questions about the Expedition that remained unsolved, despite the discoveries in 2014 and 2016 of the sunken wrecks of *Erebus* and *Terror*. An explosion of literature—fictional, speculative and scientific—in the past four decades has raised more questions than it has answered. The recent exploration of the wreck of the *Terror* has caused some to speculate that there are still-legible written records of the Franklin Expedition patiently waiting to be discovered in Captain Crozier's cabin, but, in the absence of any such evidence, we remain free to speculate.

And so this is a story, or rather stories, about what might have been. Unlike many of the books about the Franklin expedition, mine is not an attempt to insist on a single version of what happened. While more discoveries are sure to come, as the *Terror* and *Erebus* wrecks are studied *in situ* and as newly developed scientific techniques are applied to the bones of the expedition's crew and to the relics retrieved over more than a century and a half, we have perhaps reached the point where we know what we know and we know what we don't know (the latter, in former U.S. defense secretary Donald Rumsfeld's memorable phrase, the "known unknowns"). What we know is less than the whole story of the expedition. And so we are left to speculate. Not to create a patently false narrative, no matter how telegenic, like Dan Simmons's 2007 novel *The Terror* and its eponymous television adaptation. And not to include those suggestions and leads, whether from 19th-century spiritualists or poor translations of Inuit oral history, that are clearly false. But rather to follow a quantum of light as it leaps through the various pinholes of history and see where it leads us. The mixture of factual history, scientific analysis and conjecture is deliberate. The factual chapters tell what we know and reach conclusions based on the evidence. But where there is no evidence, or where such evidence as exists is ambiguous, the book points out the multiple pathways that the facts might have followed, without insisting that any one of those paths is the one that was actually taken, unless the evidence is conclusive.

1

The Known Knowns

> Tragic heroes, bumbling fools, imperial villains, hungry qallunaat (white people, in Inuktitut)—different generations have imagined and remembered the Franklin expedition through distinct vantage points.[1]
>
> If you wish to know what men seek in this land, or why men journey thither in so great danger to their lives, then it is the threefold nature of man which draws him thither. One part of him is emulation and desire of fame.... Another part is the desire of knowledge.... The third part is the desire of gain.[2]

As early as 1576, English sailors, reprising the almost-mythical voyages of the Vikings who had reached what is now Newfoundland and Labrador half a millennium before,[3] and perhaps inspired by the legend of the Irish Saint Brendan, who might have reached Iceland—or somewhere—in a leather-covered curragh in the sixth century,[4] were seeking a Northwest Passage from Europe to Asia, a hoped-for faster route to markets that were only dreamed of at the time, and a route that didn't challenge the gauntlet of Spanish and Portuguese control of the mid-latitude Atlantic Ocean. Even in Elizabethan times, the first of those British sailors, Martin Frobisher, had made the quest something more than a mundane commercial venture: "it is still the only thing left undone, whereby a notable mind might be made famous and remarkable."[5] But, after a flurry of exploration by captains whose names now dot the map of the Arctic—Frobisher himself, John Davis, Henry Hudson, Robert Bylot and Luke Foxe—a flurry that ended with Foxe's unsuccessful attempt to find a westward exit from Hudson's Bay in 1631, Britain didn't send out another serious seagoing assault on the Passage for nearly two more centuries.[6] And even on land, the Hudson's Bay Company, busy extending its fur-trading network north and west, reached the Arctic Ocean only at the mouths of the Coppermine and the Mackenzie Rivers late in the 18th century, and didn't chart that coast until well into the 19th. So, for much of the period after the end of the Elizabethan exploration, the map of the North, and specifically the Arctic Archipelago, was still more in the nature of "Here Be Dragons" than anything resembling a useful guide for mariners.

But, by 1817, a coincidence of events made Arctic exploration, and especially the search for the Northwest Passage, possible and even popular once again. The end of the Napoleonic Wars had left England with an oversized navy that had not all that much to do and with a plethora of officers, mostly on half-pay without a ship. And that same year the whaling captain-turned scientist William Scoresby reported that Baffin Bay west of Greenland was sufficiently ice-free to be navigable, perhaps for the first time since the 17th century.[7]

The Royal Navy's Arctic project was foreshadowed and promoted by Admiralty official John Barrow's publication, in London publisher John Murray's[8] *Quarterly Review* and in book form, of a narrative purporting to be a history of searches for the elusive Passage and a marshalling of reasons to resume the search.[9] Barrow, the second secretary of the Admiralty—in effect, the highest permanent civil servant, as the First Secretary and the First Lord of the Admiralty were political appointees whose identity changed with every change of government and often more frequently—was not necessarily the chief architect of Arctic exploration, perhaps just the chief spin-doctor,[10] but he had the ear of the relevant publics in London—the Admiralty, where he served for four decades, outlasting dozens of his supposed political masters; the scientific establishment, represented by the Royal Society and by newer, more specifically focused scientific associations; and the highbrow reading public, for whom Murray's *Quarterly Review* and the same publisher's expensive and lavishly illustrated narratives of explorers' journeys, not only from the Arctic, but also from such heathen lands as Asia, Africa and North America, were virtually required reading.[11]

While there may have been other reasons why various factions of the British elite promoted Arctic exploration after 1817—to chart the earth's magnetic field, to fill in the incomplete map of the north,[12] to discover the fabled, though wholly imaginary open Polar Sea, to extend the Imperial remit, to forestall possible Russian expansion eastward across the Bering Strait—it was the search for the Northwest Passage that captured everyone's imagination. As several astute commentators have pointed out, all segments of British society lapped up all things Arctic—theatrical productions, explorers' memoirs, panoramas, and unrealistic images of ships under full sail navigating swiftly between icebergs.[13] The fascination with the unknown North reached such a peak that it was justly called "Arctic fever."[14]

The first of these post–Napoleonic War Arctic quests was a double attempt on both the North Pole and the Northwest Passage in 1818. The Navy sent *Dorothea*, commanded by Captain David Buchan, and *Trent*, commanded by young lieutenant John Franklin, through the Norwegian and Greenland Seas toward Spitzbergen, the northernmost land between Greenland and Russia. Beyond that, Barrow thought, based on no empirical evidence whatsoever, lay an open "Polar Sea," with smooth sailing to the North Pole. Barrow was not the only one who at the time believed in this chimera; the serene Polar Sea had even made its way into popular literature, featuring in Mary Shelley's *Frankenstein*,[15] though firmly rebutted in later fiction, such as Jules Verne's *Voyages and Adventures of Captain Hatteras*, featuring a "desert of ice" at the pole.[16] But the open Polar Sea theory had a considerable history, and Barrow's support for the theory could not be dismissed as the mere ravings of some untutored crank.[17] Apart from ancient Greek and Roman mythical descriptions, and the actual arrival of the Norsemen on the North American continent in the first millennium A.D., there had been rumors that Columbus had reached as far north as Iceland in 1467,[18] or perhaps it was 1477, and in 1527 the Englishman Robert Thorne had proposed a serious argument for an open Polar Sea.[19] The theory was put in abeyance for a while after Captain John Wood, searching for a Northeast Passage across the Arctic coast of Russia, was turned back by the ice in 1676, but it was revived again in the 18th century, despite the Royal Navy's having sent Captain Constantine Phipps (with future hero of Trafalgar Horatio Nelson aboard as a midshipman) toward the pole in the 1770s, only to be turned back by the ice north of Spitzbergen.[20]

1. The Known Knowns

In their 1818 voyage, Buchan and Franklin found nothing but ice and weather so fierce that the wind and storms pushed the ships south even as the crews were attempting to haul them north through the ice with anchors and ropes.[21] Both ships, rendered unseaworthy by the ice and the weather, limped back to England. Even had Buchan and Franklin managed to go a bit further north, they would have still ended up in the ice. Those mariners who knew something of the Arctic, like Scoresby, had correctly concluded that an open Polar Sea was impossible, an opinion that Scoresby had published as early as 1818.[22]

The same year, relying in part on Scoresby's reports of ice-free sailing west of Greenland—a circumstance perhaps caused by the warming effects of a massive volcanic eruption in Tambure, in what is now Indonesia, the previous year[23]—Barrow sent Commander John Ross off in *Isabella*, accompanied by *Alexander*, under the command of Lieutenant William Edward Parry, to find the Northwest Passage. But by the time they reached Davis Strait and Baffin Bay, west of Greenland, the balmy conditions reported the previous year had vanished, leaving the water between Greenland and the North American archipelago clogged with ice.

Only at the end of August 1818, not long before the ice could be expected to strengthen again for the winter, did Ross's expedition reach Lancaster Sound, which would become the gateway to the Northwest Passage for dozens of attempts over the next half-century. But 30 miles into the Sound, sailing between Devon Island to the north and the northern reaches of Baffin Island to the south, Ross announced that he had seen, dead ahead, mountains, which he named in honor of John Wilson Croker, first secretary to the Admiralty and Barrow's immediate superior.[24] Whereupon Ross turned his ship around and headed back to England, much to the astonishment of Parry, his deputy, trailing Ross in *Alexander*. Ross didn't choose to spend even a single winter in the ice, a possibility that the expedition had been provisioned for, in the hope of better weather the next year. As it happened, the Croker Mountains proved entirely imaginary, never to be seen by any subsequent Arctic explorer, and Barrow, having savaged Ross in a 50-page review of the latter's

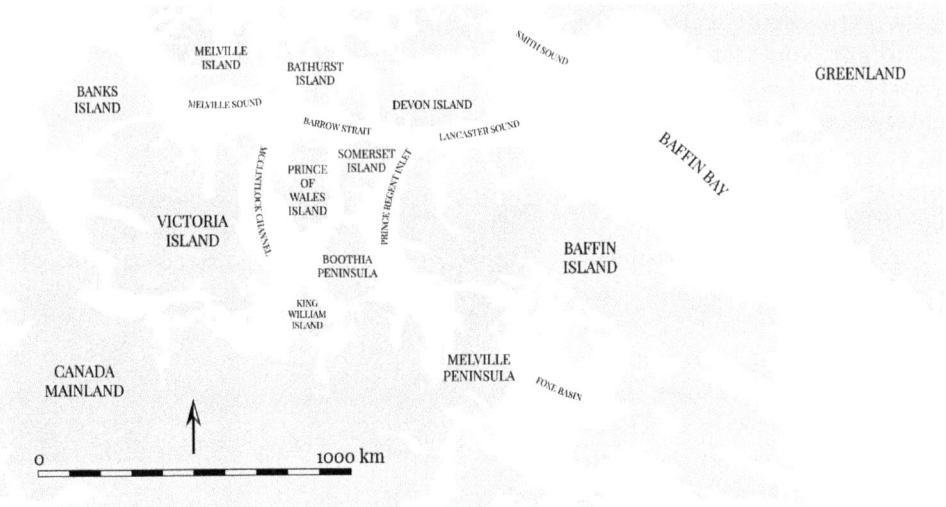

Map of the Canadian Arctic Archipelago (map by David Veller).

exploration narrative, never again entrusted Ross with the command of a Royal Navy Arctic expedition.

The next year, 1819, Barrow sent Parry back to the Arctic, as commander of *Hecla*, accompanied by the not-so-sturdy *Griper*.[25] Leaving early, and benefiting from far better weather than Ross had encountered the previous year, Parry was still slowed by ice in Davis Strait and Baffin Bay, reaching Lancaster Sound only after weeks of hauling the ships through narrow passageways in the ice. But once into the Sound, Parry was able to sail westward several hundred miles down an 80-mile-wide watery highway. After a detour down Prince Regent Inlet, he continued west through the Sound and then through Barrow Strait, reaching the south coast of Melville Island before the ice returned in force in late September and compelled him to find winter quarters, some 1,200 miles west of the nearest settlement in Greenland and at least 700 miles north of the Hudson's Bay Company fur traders at Fort Providence on Great Slave Lake. And then it was 11 months more before Parry's ships were unfrozen and, despairing of ever crossing the formidable ice stream blocking Viscount Melville Sound to the west, he turned back for England, arriving at the end of October 1820.[26] As it happened, no ship made it any further through this northern fork of the Northwest Passage for nearly another century.

Parry tried again in 1821–23, spending two winters in the ice off the east coast of the Melville Peninsula after having failed, as all other explorers had also failed, to find a navigable way to the west from Hudson's Bay.

Meanwhile, then–Lieutenant John Franklin was earning his reputation as "the man who ate his boots"[27] on a four-year odyssey through the northern barren lands and along the Arctic coast of the American mainland, mapping some 550 miles of coastline eastward from the mouth of the Coppermine River. Notwithstanding that nearly half his expedition members died along the way, that Franklin himself survived thanks only to the fortuitous arrival of a band of Indians, and that there were reports of murder, cannibalism and lesser evils among his traveling party, he returned a hero, featured in London panoramas and feted upon his return.[28]

After that, and a subsequent, less dramatic, but arguably more productive mapping voyage—this time heading west from the mouth of the Mackenzie along the coast toward Alaska—Franklin headed off for other assignments (see Chapter 4), but Arctic exploration continued, with John Ross's privately financed epic

The Arctic hero: Sir John Franklin in full dress uniform, late 1820s (Dibner Library Portrait Collection, New York Public Library).

four-year odyssey on the Boothia Peninsula[29] and George Back's near-death experience on an expedition to Hudson's Bay in *Terror*[30] among the most notable of the failed attempts to find a Passage. Despite these repeated failures, however, Barrow's dream never quite died.

In the end, though, Sir John Franklin's final expedition in 1845 may have been less the culmination of Barrow's obsession with the Northwest Passage and the non-existent open Polar Sea than of the British science establishment's parallel obsession with magnetism. Royal Artillery Colonel Edward Sabine's report on the magnetic objects of one more Northwest Passage search in fact preceded John Barrow's formal 1844 proposal to the Admiralty.[31] And before getting the Lords of the Admiralty to approve the new expedition, Barrow also sought the Royal Society's scientific imprimatur, which was duly provided, with an emphasis on completing a series of magnetic observations in 1845 as part of a worldwide scientific effort due to end that year (although the projected was extended, after Franklin had departed, for another three years).[32] In fact, the lure of a navigable Northwest Passage, which was surely known by 1845 to be commercially useless, if indeed such a passage existed at all, would have been inadequate by itself to launch the Franklin Expedition; the promise of magnetic science was almost certainly a necessary, if not alone sufficient, part of the impetus for the voyage.[33] And besides, there were only some 300 or so miles left to fill in on the map of the Arctic coast, thanks to Franklin's own earlier, land-based expeditions, to the trip that Lieutenant George Back had taken down the river that later bore his name and to the Hudson's Bay Company explorers Peter Dease and Thomas Simpson, whose land-based journeys had filled in a good part of the northern continental coastline.[34] Given the scientific cover of the magnetism study, as well as Barrow's need to keep the Royal Navy in the headlines during an era when wartime heroics were unattainable, a final Northwest Passage expedition became almost inevitable.

But who was to command that expedition? Sir John Franklin was certainly no one's first choice. Edward Sabine, the Admiralty's adviser and prominent promoter of magnetic science, had in fact written to James Clark Ross when the latter was in the Antarctic, on an epic voyage of discovery in 1839–1843 in *Erebus* and *Terror*, promising that the decision makers in London were waiting only for Ross and Crozier to return from their southern voyage before sending them north again.[35] But when Ross at last returned to England in 1843, he pronounced himself too tired to undertake yet another polar mission and, to seal his refusal, married a wife who extracted from him a promise not to return to the Arctic.[36] Meanwhile, John Barrow, on the eve of retirement at age 80, pushed for the expedition to be led by polar neophyte James Fitzjames, who had served with some distinction in China and had led a steamship expedition down the Euphrates River in Iraq, but who had no Arctic experience at all,[37] though he had apparently been considered for the post of gunnery lieutenant on James Clark Ross's and Francis Crozier's very successful four-year Antarctic mission of 1839–43 (of which more in Chapter 3).[38] Fitzjames ended up as commander of *Erebus* (though without being endorsed for that role by Franklin[39]) and third-in-command of the expedition overall.[40] While Barrow was still promoting Fitzjames as a possible expedition leader, Franklin's wife Jane was busily engaged in promoting her husband for one last Arctic command, now that Sir John had been recalled from his governorship in Tasmania in, to say the least, questionable circumstances.[41] Lady Jane needed a final act for her husband that would end with applause, rather than snickering. Franklin was confirmed as commander of the new expedition in February 1845, and his list of senior officers with Arctic experience—most of the officers were selected by Fitzjames and had little or no background in polar sailing—included only Franklin himself, Francis Crozier, as commander of *Terror*, and Lt. Graham Gore, who had sailed with George Back on *Terror* a decade earlier. Since the selection of most of the other senior officers was delegated to Fitzjames, most of them were his friends and equally lacking in knowledge of the Arctic. They generally were men who had served with him, either in the Middle East, where Fitzjames had shown courage and resourcefulness on a lunatic mission to build a canal transport route through Iraq, or in China, where Fitzjames had performed heroically in support of British imperial objectives in the Opium Wars.[42] As a result of Crozier's intervention, though, a good many of the junior officers, notably including the two "ice masters," James Reid and Thomas Blanky, both with long histories on whaling ships, did have experience in the north, as did a number of the petty officers.[43]

At the time, there was considerable concern within Admiralty circles about Franklin's age, health and general fitness for the job. An oft-reprinted daguerreotype of him, taken just before the ships sailed from England, shows an aging (he was then approaching 60, a rather advanced age for the era[44]), portly and generally unwell man. Once the expedition was underway, however, there is some evidence that all was not quite as bad as it had looked. For example, Lt. James Fairholme, one of Fitzjames's selections as an officer aboard *Erebus*, sent a letter back from Greenland reporting that "Sir John is a new man since we left. He has quite recovered from his severe cold, looks 10 years younger and takes part in everything that goes on with as much interest as if he had not grown older since his last expedition."[45]

How much of Fairholme's comments represent an objective assessment of his commander and how much is simply an attempt to reassure worried relatives and friends back in England is unclear, though the general sentiment is echoed in other letters sent back in that last batch of missives.[46] And officers who appreciated Franklin's genial manner at dinner but were lacking in relevant experience may not have been the best judges of their commander's fitness for leading them in what turned out to be an impossible task.

Even back at home not everyone with Arctic experience was convinced that the plans for the expedition made sense. John Ross, left on the outside by Barrow as a result of his 1818 failure, had warned that taking large, heavy-draft ships[47] through the Arctic Archipelago was folly.[48] He was, as it turned out, right.[49]

Much later, after the Expedition had disappeared, after the bones and relics of a number of crew members had been recovered by a flotilla of searchers, and after Inuit recollections, rendered more or less accurately by various Franklin searchers, had been absorbed into the standard model of Franklin history, the myth took on a life of its own. As Andrew Lambert states, "Franklin and his men were reborn as paragons of the Victorian cult of service, heroes who had willingly sacrificed their lives for the greater good."[50] Thus was born the cult of the heroic explorer, a model that, a half-century later, Captain Robert Falcon Scott took to its logical extreme as he perished a few miles from a food depot in Antarctica.[51]

Between February and May 1845, the already stout *Erebus* and *Terror* were refitted at Woolwich shipyard, down the Thames from London, in preparation for the Arctic voyage; their hull thickness was doubled, watertight bulkheads installed, iron-reinforced bows added, and retractable propellers, to be linked to the steam-engine locomotives in each of the ships' holds, were installed, as the old Arctic hand Frederick William Beechey had suggested.[52] A test-drive of *Terror* in the Thames in late April 1848 showed that, at least in the absence of ice, the locomotive could propel the ship at the moderate pace of about four knots.[53] Other improvements included heating systems that could pump warm air throughout the ships' living quarters, mirrors and prisms for bringing surface light down into the ships' interiors, and galley modifications for producing fresh water.[54]

On May 12, 1845, *Erebus* and *Terror* were towed down the Thames to the Royal Navy's dock at Greenhithe. There they loaded the last of the ships' supplies, including, at the last minute, canned meat and vegetables from Stephen Goldner's establishment in the East End, and took on the crews, adjusted the compasses and other scientific instruments that Edward Sabine had been schooling the officers on at Woolwich, stowed the extensive libraries,[55] as well as one hand-organ for each ship, each with a repertory of some 50 songs, and otherwise occupied themselves preparing for an expedition that engaged British public opinion to quite an unimagined degree, though there were, even at the time, a few dissenting voices.[56]

Terror's captain Francis Crozier thought the late start a bad omen. In a letter to his friend James Clark Ross, who had commanded the successful four-year Antarctic expedition in *Erebus* and *Terror*, Crozier worried, "We shall have no time to look around and judge for ourselves, but blunder into the ice and make a second 1824 of it," referring to Parry's unsuccessful Arctic voyage in *Hecla* and *Fury*, on which Crozier had been a midshipman. As it happened, Franklin's ships did make it to a safe

Erebus and *Terror* depart Greenhithe, May 1845 (*Illustrated London News* 1845; Wikimedia Commons).

winter harbor in 1845–1846, though not all the way through the Passage, as some perhaps overly optimistic supporters of the Expedition had hoped. But Crozier's pessimism was eventually proved correct.

Finally, on May 19, already at least a week late in the season for the start of an Arctic voyage, *Erebus* and *Terror* were towed out to sea from the Thames estuary by the steam-driven *Rattler*,[57] passing the site of what is now the Sir John Franklin riverside pub in Greenhithe. After stops in Aldeburgh and the Orkneys, Franklin finally shed his escorts and, accompanied only by the supply vessel *Barretto Junior*, set sail for Greenland on June 8, already well into the summer sailing season. They sighted Greenland on June 25, crossed the Arctic Circle on June 30, and reached the Whalefish Islands, off Greenland's west coast, on July 4. There they transferred a year's worth of stores from *Barretto Junior* into *Erebus* and *Terror* before sending the supply ship home with, as it turned out, a few very fortunate men who were too ill to continue, and with letters home from those who were going onward,[58] which would prove to be the last actual first-hand written records of the expedition, except for the one-page Victory Point Record discovered by Franklin searchers in 1859. The last of those letters from Greenland is dated July 12, 1845. After that, silence, except for a brief encounter two weeks later near the entrance to Lancaster Sound, where *Erebus* and *Terror* exchanged signals with two whalers out of Peterhead, *Enterprise* and *Prince of Wales*.

Franklin's sailing orders,[59] reproduced in Appendix III, which were based on what the Admiralty then knew about the geography of the Arctic, and were perhaps partly drafted by Franklin himself, directed him to head across Baffin (or, as it then was, Baffin's) Bay to Lancaster Sound and from there, in a bit of wishful thinking, straight across westward ("you will not stop to examine any openings either to the northward or southward in that Strait but continue to push to the westward without any loss of time"). After passing the known landmark of Cape Walker, just north of the top of Somerset Island, Franklin was told, "From that point we desire that every effort be used to endeavor to penetrate to the southward and the westward in a course as direct towards Behring's Strait as the position and extent of the ice, or the existence of land, at present unknown, may admit." Alas for Franklin and his men, that southwest route would turn out to be almost entirely filled with ice, pouring down through McClintock Channel and then into Victoria Strait, and beyond that with land then largely unknown, namely the very large Victoria Island and its western neighbor, Banks Island. The charts that Franklin carried were mostly, especially along the route that he'd been directed to follow, either completely devoid of information or simply wrong. They depicted an unknown and empty quadrilateral of some 70,000 square miles, bounded by Cape Walker, the southeast corner of "King William Land," assumed to be a peninsula and not, as it in fact is, an island, Wollaston Land (actually, we now know, a part of Victoria Island) and Banks Island. Those same charts suggested that King William Land was not an island, but rather connected to the Boothia Peninsula, foreclosing any notion of sailing around its eastern side—the route that, more than half a century later, Roald Amundsen actually followed when he made the first successful transit of the Passage.[60]

Only if Franklin ran into an ice or land barrier would he have been permitted by the Admiralty's orders to look in a different direction. Building on Barrow's faith in the existence of an open Polar Sea, the orders suggested exploring the channel

1. The Known Knowns   11

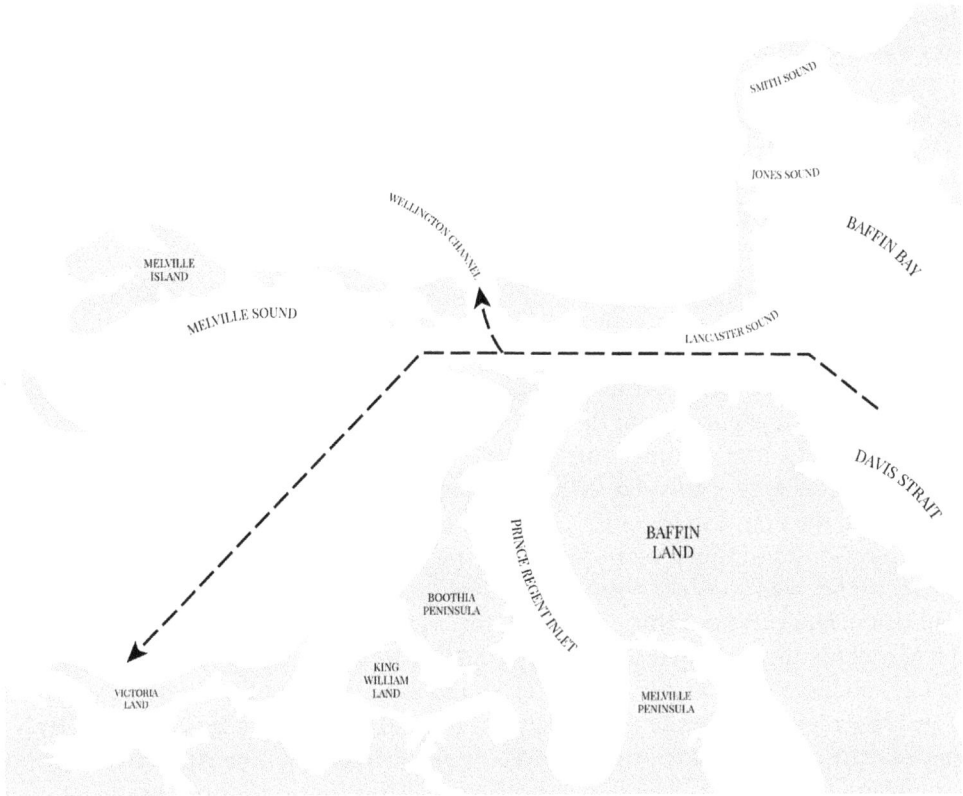

What was known of the Canadian Arctic in 1845 (map by David Veller).

between Devon and Cornwallis Islands, to the north of Lancaster Sound. In fact, as we will see in Chapter 6, Franklin did direct not just an exploration of that channel, but a complete circumnavigation of Cornwallis Island before deciding to winter at Beechey Island in 1845–46 and, if he accomplished nothing else, at least that detour, when its existence was discovered in the Victory Point Record, put an end to the dream of the open Polar Sea, as Franklin saw nothing but ice and more ice further to the north.

Later, when the Admiralty sent out multiple search and rescue expeditions, it told those expeditions' commanders to look where Franklin's instructions had told him to go. In particular, Captain Horatio Austin's multi-ship group in 1850–51 started looking for Franklin, though not finding him, right where Sir John had been ordered to go, in and around the waters south of Devon and Cornwallis Islands.[61] But more often, the large ships that the Admiralty sent out in the early 1850s looked where the ice permitted them to go—generally well north of Franklin's actual route—rather than where he had in fact gone.

While the Admiralty hoped that Franklin would complete the Northwest Passage in a year and by 1846 be well on his way to Hawaii, the orders did recognize the possibility of being stuck in the Arctic for one or more winters, directing Franklin to find "a sheltered and safe harbor" and to find any "Indians or Esquimaux" in the area and "cultivate a friendship with them" as an aid in surviving the winter. Franklin apparently carried out at least the first part of this direction in choosing the sheltered anchorage at Beechey Island in 1845–46, although the location was too far north for there to be any native peoples nearby, especially in the winter, but then the ships spent the next two winters in the open ice between Victoria Island and King William Island. Did they search for a sheltered harbor that next winter (1846–47), but begin their searching too late? Did they think they could outrun the ice by heading south and then realize their mistake only when it was too late? Did Franklin, improbably for such a loyal and rules-conscious officer, willfully disobey that portion of his orders?

Franklin was supposed to keep in touch, to the extent that the technology of the day permitted. Specifically, he was directed that, after passing 65 degrees north latitude, just south of the Arctic Circle, "once every day when you shall be in an ascertained current, [you shall] throw overboard a bottle or copper cylinder closely sealed, and containing a paper stating the date and position at which it is launched."[62] Only a single one of these cylinders or bottles, if, indeed, any more were ever dropped overboard, was ever recovered, either by ships of the numerous search expeditions that looked for Franklin through the 1850s or by whalers who annually sailed into Lancaster Sound in those years. And that one ended up only 35 miles off the coast of Greenland, well south of the 65th parallel, and wasn't retrieved until 1849.[63] Did Franklin in fact carry out this portion of his orders, or was he just unlucky, with dozens of records thrown overboard as instructed but simply never found? With all the whalers off Greenland and in Lancaster Sound each year, not to mention the dozens of search expeditions in the high Arctic from 1850 to 1854, it seems odd that only a single bottle or tube was recovered, and that one dated from June 30, 1845, before the Expedition had even entered Lancaster Sound. Then, nothing. And, despite spending the first winter on Beechey Island and leaving behind abundant debris, he left no written record of what had transpired there, nor any indication of where he was headed when the ships left their winter anchorage. Even that glyph that the folk song imagines, inscribed on one of Devon Island's numerous rocks, with the hand of Franklin pointing toward the Beaufort Sea,[64] would have been helpful.

After leaving Greenland, the two ships were seen by whalers on the western side of the Davis Strait,[65] then nothing, aside from three graves left on Beechey Island in the winter of 1845–46. Professor Janice Cavell has succinctly summed up the little that we can reliably confirm we know about the Expedition from that point on:

> In 1846 he [Franklin] arrived off the north coast of King William Island by an unknown route. The ships wintered there in the ice. On 11 June 1847 Franklin died; on 22 April 1848 the men abandoned the ships, which had drifted slightly southwest into Victoria Strait. Led by Francis Crozier, they trekked down the west coast of King William Island toward Chantrey Inlet, apparently hoping to travel up the Back River to safety. On the way, they passed Dease and Simpson's cairn at Cape Herschel. The rest is conjecture.[66]

Conjecture until, in 1854, Dr. John Rae of the Hudson's Bay Company returned with relics of the expedition and Inuit tales of what might have happened and then, in 1859, when Captain Francis Leopold McClintock discovered the bones, boats and relics of a substantial number of the Expedition's crew on King William Island. Owing primarily to Lady Jane Franklin's insistence that the searches continue until the Expedition was found or its fate revealed, there were, from 1847 through McClintock's voyage that ended in 1859, some 32 different expeditions, involving 47 ships, all searching, or supporting the search, for the missing Franklinites.[67] If there was anything more to be found, one would have expected that the numerous search parties would have found it.

Once freed from the winter anchorage at Beechey Island, Franklin's way was clear; follow the Admiralty's orders to head southwest. The first part of that route almost certainly (almost, because nothing is absolutely certain in the absence of the ships' logs or the sealed records that should have been dropped overboard at regular intervals) led through Peel Sound, a roughly 20-mile-wide opening between Somerset Island on the east and Prince of Wales Island on the west. And, given the magnetic-science focus in Franklin's mission statement, Peel Sound also had the virtue of leading almost directly to the then-location of the north magnetic pole, identified in 1831 by James Clark Ross.

Did the Expedition simply sail southwest until trapped in the ice in September 1846? Were there attempts to round King William Island on its eastern, relatively ice-free side? What were the ice conditions like that limited *Erebus* and *Terror* to only a few hundred miles of progress in the summer of 1846? Why were the ships trapped out in the open ocean, rather than settling into some sort of protected harbor, of which there was no shortage, even in that unexplored part of the Arctic? Known unknowns.

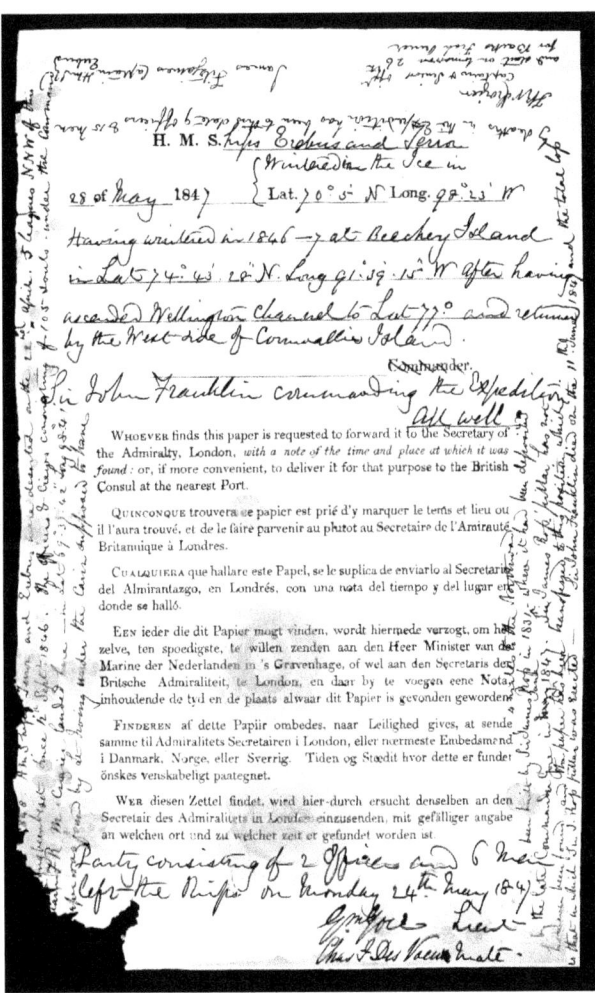

The Victory Point Record (Wikimedia Commons).

In 1859, some 14 years after the Expedition's last contact with other Europeans, Lieutenant (as he then was) Francis Leopold McClintock's search expedition found what's come to be known as the Victory Point Record. The record announced Sir John's death, which had happened in June 1847, and reported a total death toll to April 1848 of nine officers and 15 crew members (including Sir John and the three crewmen buried on Beechey Island). It then noted that the ships had been stuck in the ice for over 18 months and indicated the most minimal of intents: "and start on tomorrow for Back's Fish River."[68] As McClintock himself wrote in the narrative of his search expedition, "a sad tale was never told in fewer words."[69]

The basic outlines of the Expedition crew's journey after leaving the ships in April 1848 are reasonably clear. From Crozier's landing place, a few miles south of Victory Point, the crew, or at least most of it, appears to have headed south and then west, more or less on a line that matches the expressed objective of the Victory Point Record, the mouth of Back's Fish River, from which they would then have had to ascend some hundreds of miles of shallow, fast-flowing streams through dozens of rapids to the nearest Hudson's Bay Company outpost on Great Slave Lake.

Assuming the Victory Point Record is correct in reporting that, as of the time the ships were abandoned, nine officers and 15 men had died would have left only six senior officers to manage some 95 sailors on what many would reasonably have seen as a hopeless death march. Notwithstanding the Royal Navy's stern disciplinary traditions and the leadership qualities of the surviving individual officers, Crozier and Fitzjames among them, it's not hard to imagine that discipline sooner or later broke down, as the already-reduced number of officers dwindled even more and the hopelessness of the situation became ever more apparent. If that's so, it opens up the possibilities for what a lawyer might call frolics and detours, in which some of the men broke off from the main party and went off on their own, whether back to the ships, in a different direction or wherever. Not to mention that only the officers and the ice masters would have had significant navigational skills to guide the group anywhere, other than in aimless meandering.

The string of skeletons and relics found along the west and south coasts of King William Island by McClintock and his second-in-charge Lieutenant William Hobson and later by U.S. Army officer Frederick Schwatka provide the broad outlines of the story: three or more heavy sledges, each carrying a boat for the anticipated travel up Back's River and laden with all manner of supplies, set out from Crozier's Landing, perhaps capable of making between five and 10 miles per day, hauled by men whose strength was diminishing as they walked. McClintock, one of the Navy's sledging

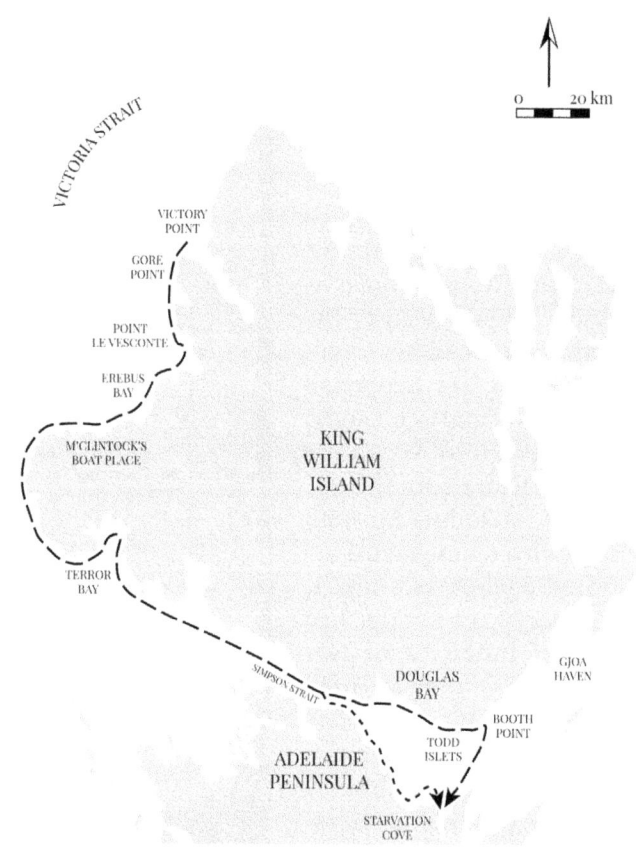

Map of possible King William Island retreat routes (map by David Veller).

gurus, later calculated that the men at their healthiest would have been able to carry food and fuel sufficient for no more than 40 days. In other words, there was no reasonable possibility that the whole party could make it even as far as the mouth of Back's River, much less all the way upriver, past the rapids, to the Hudson's Bay Company post on Great Slave Lake.

In fact, the dead start to appear almost immediately along the eastern shore of King William Island—at Point Le Vesconte, Two Graves Bay and at the "boat place" along the shore of Erebus Bay, where McClintock found an abandoned boat with two headless skeletons inside, plus what turned out to have been at least 14 bodies scattered round. But did all these sailors die on the outbound trek from the abandoned ships, or was that boat, which was found pointing northwest, i.e., back to the ships, a sign that at least some of the crew had second thoughts and were trying to return to the water?

Of those who were left, it appears that a large number, perhaps a majority, eventually came to rest and remained at a camp near Terror Bay, on the south coast. The Terror Bay campsite had the advantage of being the first point on the route that would have provided even a possibility of finding food, as caribou typically crossed near there to King William Island from the mainland in the summer, though perhaps not as early as when the Expedition crew members arrived. In any event, later searchers have confirmed that Terror Bay was a major stopping point, with many fewer than all of the Expedition members continuing any further.[70] Perhaps, though, the encampment at Terror Bay was not just a last resting place on a long, hopeless trek. In 2016, the well-preserved wreck of *Terror* itself was found in the bay. Was it just an accident that she drifted there, washing up many years after her crew had died in the same place? Or had some remnant of the crew returned to *Terror* after April 1848 and sailed her south of the island, either in the summer of that year or perhaps even a year or two later? Parks Canada, in charge of the wrecks, has confirmed that the captain's desk still remains in place aboard *Terror*.[71] Perhaps there are records, still readable, that might answer that question. The possibility that Crozier's desk contains the answers to so many questions—and that those answers are still legible after nearly two centuries under water—continues to animate the online Franklinite community as it settles in for what may be a wait of many years while Parks Canada painstakingly pursues its diving efforts during the brief annual summer seasons.

From the evidence on the ground, it appears that fewer than half of the crew made it beyond Terror Bay. Inuit reported meeting a large party of 40 men, hauling a ship's boat with a sail set (but with that sail apparently providing little assistance) near Washington Bay, further along the southern coast of King William Island.[72] According to the Inuit, all the men appeared sick, showing signs consistent with scurvy, and there was also evidence of cannibalism (of which more in Chapter 14).

The Inuit moved on, and from there the record thins out, with bones to be found heading southwest and eventually across to the mainland, where Schwatka found several sets of remains at the aptly named Starvation Cove, and, five miles inland, a last skeleton. Bones linked to the Expedition have also been found further east, on Montreal Island, along with fragmentary remains on the Todd Islets. Actually in Back's River? Nothing.

So that's what we know. As to the rest, the stories are conflicting, unclear, or nonexistent. There's plenty of evidence. In fact, one can, especially with the availability of so much of the evidence online these days, and with the recent museum exhibition that toured the United Kingdom, Canada and the U.S.,[73] construct a veritable archive of the Franklin Expedition. But, as Kathleen Kasten-Mutkus argues,

> [the] widespread availability online increases access, partially addresses the issue of geography [i.e., substituting for travel to research institutions to examine the Franklin relics] and allows for the recombination of dispersed entities. However, the partial digitization of the Franklin archive nevertheless asks the researcher to combine and recombine disparate and incomplete resources to understand a complicated, fractured whole.... [The available] records constitute a hypothetical Franklin archive which is notable for its ambiguity and inscrutability.[74]

The written portion of that archive is well-known, and almost as inscrutable as the relics: the Victory Point Record, the "Peglar Papers,"[75] and, perhaps, the legends on the gravestones on Beechey Island, discussed in more detail in Chapter 6. For the grandest, best fitted-out discovery expedition of the century, it's a meager, fragmentary lot.[76] Despite all the effort devoted over nearly two centuries, there's much we still don't know. Some of those uncertainties are described in these pages, and some of the questions that are still open may be answered as more material from the wrecks is recovered and analyzed. But, at the same time that we seek answers, we can also recognize that not every question is answerable; that's especially true of the psychological questions: why, for example, did Sir John Franklin and Francis Crozier make the decisions they made? What were they—or any of the officers or crew, for that matter—thinking and feeling, during their three long years in the Arctic? Because we want to know more, we'll keep looking. What I hope to have done here is to lay out what's known and what's still uncertain and to leave the rest to the reader's imagination, knowing that, for the moment, much is still a mystery.

Time to fill in those liminal spaces and illuminate the residue of what can never be known.

2

A Quantum Theory of History

> Everybody's wonderin' what and where they all came from...
>   But no one knows for certain...
>   I think I'll just let the mystery be.[1]

To recapitulate, in 1845, Sir John Franklin and his crew sailed from England in *Erebus* and *Terror*, bound for the Northwest Passage. They sent letters back from a stopover in Greenland and were last seen by whaling ships in Baffin Bay, between Greenland and the Canadian Arctic archipelago. They spent their first winter at Beechey Island, just off the coast of the much larger Devon Island, having traversed Lancaster Sound and headed west into Barrow Strait, all well-worn pathways to the still-elusive Passage. After that, nothing. Until, years later, their bones, some of their possessions, and a single, enigmatic message were discovered. What happened to the expedition? A mystery. Out of perhaps three million wrecked ships[2] at the bottom of the earth's oceans and waterways, none, with the possible exception of RMS *Titanic*, has generated the amount of search and speculation that Franklin's lost expedition has evoked.

As Canadian writer Erika Behrisch Elce recently wrote: "if so much of Canadian cultural history is constructed out of the Franklin mystery, what happens when we take that mystery away … we are far enough from the 'facts' to understand that we're no longer looking for any kind of factual truth. What we continue to want is a good tale well told."[3]

Try telling that to the 3,000-plus members of the Facebook group "Remembering the Franklin Expedition,"[4] whose ongoing discussions, highlighted by exegeses of smudged satellite photos and of nearly indecipherable Victorian-era handwriting, are devoted to the ongoing replacement of the uncertainty or mystery that has surrounded the fate of Franklin and his crew for nearly two centuries with, instead, such "facts" as can be deduced. The discoveries, in 2014 and 2016, respectively, of the wrecks of Franklin's ships, *Erebus* and *Terror*, have, to be sure, eliminated some of the mystery, and increasingly sophisticated scientific analysis continues to shed light on the Franklin sailors' causes of death. And who knows, perhaps there really is a treasure trove of ships' logs and journals hidden away in the desk drawer in Captain Francis Crozier's cabin on *Terror*, and perhaps those papers—if they exist—are, miraculously, still readable after all these years in the water. But at some point, there is still an irreducible core of indeterminacy. When do we just let the mystery be? As one reviewer of a book purporting to reveal "the truth" about polar exploration commented: "I am skeptical of any publication that promises to reveal 'the truth' about any matter relating to polar exploration expeditions. These were complex endeavors, and there can be no single 'truth' about them, nor should we expect one."[5]

And even if, as the evidence increasingly suggests that we should, we give more credence to Inuit accounts of the Franklin Expedition, there is still the problem of determining exactly what those accounts say. They were, after all, filtered through the not necessarily accurate interpreters for John Rae, Charles Francis Hall and other early explorers, and then through the culture-bound preconceptions of the explorers themselves. Even if the stories were transcribed correctly, we now know that not all eye-witness testimony, much less second- or third-hand stories, can be fully believed; what the witnesses "saw"—the images imprinted on their retinas—is not necessarily what they remembered—the processed images stored in their neurons.[6] And more recent exegeses of the Inuit narratives,[7] although illuminating, depend on the presumed accuracy of tales handed down orally for a minimum of three or four generations. While, as David Woodman, Dorothy Eber and others have shown, the basic elements of these oral narratives remain remarkably stable, there is still some uncertainty as to the timing and precise location of various events described in the stories.

This chapter begins the story, or rather stories, about what might have been. Unlike many of the books about the Franklin expedition, this is not an attempt to determine precisely what happened. While more discoveries are sure to come, as the *Terror* and *Erebu*s wrecks are studied *in situ*, as objects are retrieved from the wrecks, and as modern scientific techniques are applied to the bones of the expedition's crew and to the relics retrieved over more than a century and a half, we have perhaps reached the point where we know what we know, and we know what we don't know. Numerous scholarly papers in the past two decades have tested the limits of what can be reasonably inferred from the physical evidence—the bones and relics recovered on and near King William Island, the scraps of paper recovered and then abandoned in various archives, those smudges on the satellite photos. What we know—and what we will know, no matter how much more is revealed—is less than the whole story of the expedition. And so we are left to speculate. We need to reject the proven-false narratives, like, to take just one example, Adam Beck's story, told to Franklin searchers in the 1850s, that Franklin's ships had been wrecked off the Greenland coast and some of the crew murdered by Inuit,[8] or even the less vivid version, proposing that Franklin gave up in 1846 after wintering on Beechey Island and that his ships were wrecked on the way home, possibly in Baffin Bay. And we need to avoid creating a patently false narrative, like Dan Simmons's novel, *The Terror* and its televised adaptation[9] or even that book's better-written comrades, like Mordecai Richler's *Solomon Gursky Was Here*[10] or William Vollmann's *The Rifles*.[11] And we shouldn't include those suggestions and leads, whether from 19th-century spiritualists or poor translations of Inuit oral history, that can easily be shown to be false. But let us follow Schrödinger's cat as it emerges from the sealed box of its history and see where it leads us.

Regarding that cat: the Austrian physicist Erwin Schrödinger described his famous thought experiment as follows:

> One can even set up quite ridiculous cases. A cat is penned up in a steel chamber, along with the following device (which must be secured against direct interference by the cat): in a Geiger Counter there is a tiny bit of radioactive substance, so small, that perhaps in the course of the hour one of the atoms decays, but also, with equal probability, perhaps none; if it happens, the counter tube discharges and through a relay releases a hammer

that shatters a small flask of hydrocyanic acid. If one has left this entire system to itself for an hour, one would say that the cat still lives if meanwhile no atom has decayed. The first atomic decay would have poisoned it. The psi-function of the entire system would express this by having in it the living and dead cat (pardon the expression) mixed or smeared out in equal parts.

It is typical of these cases that an indeterminacy originally restricted to the atomic domain becomes transformed into macroscopic indeterminacy, which can then be resolved [only] by direct observation. That prevents us from so naively accepting as valid a "blurred model" for representing reality. In itself, it would not embody anything unclear or contradictory. There is a difference between a shaky or out-of-focus photograph and a snapshot of clouds and fog banks.[12]

In other words, until that hour has passed, and we open the box and look at the cat, it is both alive and dead at the same time, and we can spin out wonderful tales well told of either its burial or its further adventures with equal veracity, namely, "maybe it was like this."

A similar physics problem that reflects my approach to the Franklin story is the so-called double-slit experiment, in which a beam of light or electrons is sent at a plate pierced by two parallel slits, and then the light passing through is observed or recorded on a screen behind the slits. When you shoot the light at the plate, you don't know where each photon is going to end up on the screen, though you can statistically predict, with a high degree of accuracy, the overall result of shooting trillions of photons or electrons.[13] All the light will end up somewhere on the screen, but any particular quantum of light could go either way. But until it gets there, you don't know where "there" is.

Or, to try another less than intuitively satisfying explanation, "particles were always nowhere and everywhere at the same time. They arrived at their destination with a spooky knowledge of every available path that they could have, or might have, taken, with no certainty as to which path they actually did take."[14]

These experiments and explanations illuminate the importance of uncertainty and the limits of knowledge. We all understand that the future is uncertain: even a "sure thing" can lose the next horse race. But it requires a bit more imagination to see that the past is also uncertain; something—perhaps something small and of little importance, but perhaps something large and fundamental to the story—will always be missing when we try to recreate or describe what happened before. So how does that uncertainty apply to the Franklin Expedition? If we just keep on rummaging through the new evidence that continually appears, might we not someday know it all?[15]

We can be pretty sure that all the 129 men who sailed west from Greenland in *Erebus* and *Terror* in 1845 are, by now, quite dead, unlike Schrödinger's hypothetical cat.[16] But many questions about the Franklin expedition are not so easily answered. Did Sir John follow his orders to the letter or, once those orders proved mistaken, did he try other routes? Where, exactly, did the crew go when they left the ships in 1848? Did they all go in the same direction? Did some return to one or both of the ships? Were the ships manned when they arrived at their final resting places? What were the causes of death for the crews? Did anyone survive beyond the summer of 1848? What was the extent of cannibalism? Was the crew disenchanted, angry, even mutinous or did they remain loyal Royal Navy sailors to the end? Was Sir John Franklin buried at sea or on land and, if the latter, where? As long as we don't "know" the answers to these questions, the history of the Franklin Expedition remains a mystery. The rest of this book sends the Expedition through a kind of double-slit experiment and imagines some of the possible results. Or, in other words, we look at both the dead cat and the one that's still alive and consider them both equally valid. In Elce's terms, all the possible answers are available in the hope of providing good tales well told.

So, what don't we know? In a sense, we don't know anything at all beyond the (very limited) written and physical artifacts: the Victory Point Record retrieved by the McClintock expedition in 1859, the bones and relics recovered over the years by an army of Franklin searchers and, most recently, the sunken wrecks of *Erebus* and *Terror*, under examination now by Parks Canada's underwater archaeologists. We don't know what the men (if indeed, they were all men)[17] felt, said or thought. If indeed there were no women, and with no port visits, we don't know to what extent they obeyed the Admiralty's strict prohibition of homosexual activity.[18] We don't know very much about where they went, given the vast possibilities of a then-uncharted Arctic. We don't know, though we can speculate, how they spent the three long winters aboard their ships.

Even though we have, in Schrödinger's terms, opened the box and found the cat quite dead—or have we?—much remains uncertain. After all, some writers still argue that there may have been one or more survivors, and there are still many smaller uncertainties within the larger narrative. What happened that first winter on Beechey Island, when the first three of the 129 expedition members died? How far west did they sail the next summer before turning south? When they did turn south, which side of King William Island did they try, and how did they end up trapped in the open ice of Victoria Strait, rather than in a safe winter harbor? What did they do that winter in the ice, when they still had expectations of moving on the next year? What sledging journeys did they undertake, to what end? How did Franklin die, and what happened in the wake of his death? Was he buried on land, and, if so, where? Then, when the ships remained trapped in the ice into a second winter, why did the crew not then abandon them and seek to escape while still relatively healthy and with adequate supplies? Was there any reconnaissance or planning in advance of the abandonment of the ships in April 1848? When they did leave the ships, did everyone head southwest, toward what became known as Starvation Cove, or did some head elsewhere, perhaps toward the well-known food cache at Fury Beach on Somerset Island? In the first months of the trek toward safety, did any of the crew return to the ships, either temporarily or permanently? Was there a mutiny, or several? Did the ice break up in the summer of 1847 or 1848, releasing the ships and perhaps permitting *Terror* to be sailed to its final resting place in the eponymous Terror Bay? Was anyone aboard *Erebus* as it drifted toward O'Reilly Island just off the North American mainland? Where else might the ever-weaker crew have headed? And then, there is the much-debated issue of what killed the expedition members. Shoddy canned provisions leading to botulism? Lead poisoning? Scurvy? Exposure? Starvation? Tuberculosis? Zinc deprivation? Something else entirely—Dan Simmons's monster, perhaps? And did they, or, rather, to what extent did they, engage in cannibalism? Did any of the expedition survive beyond the summer of 1848, and, if so, where and for how long?

2. A Quantum Theory of History

Chapter 1 summarized what we know for sure, the known knowns. The next few chapters add some detail, covering what we know about the Expedition ships, *Erebus* and *Terror*, and some of what we know about Sir John and those who sailed with him. From there on, we'll propose some possible paths to answering the remaining questions in the succeeding chapters and look at the extent, if any, to which the recent discoveries of *Terror* and *Erebus* have narrowed the range of possibilities or even answered some of our questions.[19] What's left unanswered will be the enduring mystery of the Franklin Expedition.

3

Erebus[1] and *Terror*[2]

Erebus was, oddly but perhaps presciently, named for the dark place under the earth that, in Greek mythology, the dead passed through on their way to Hades, as well as for the primordial god of darkness. She was launched in 1826, one of a class of "bomb vessels" based on the design of her predecessor in Arctic exploration, *Hecla*, which James Clark Ross had sailed three times in search of the Northwest Passage (and once, also abortively, toward the North Pole). Bomb vessels were originally designed as floating platforms for mortars, to be used in bombarding fortresses and other targets on shore.[3] They were not built for speed but for strength, to withstand the downward recoil of the two heavy mortars they carried on deck. It was supposed that the same dogged qualities would stand them in good stead when their job was to push the polar ice out of the way.

One of the first recorded uses of something that might be described as a bomb vessel in war was in 1347, when Edward III of England used a single-decked ship to launch round stone projectiles at the French in Calais. (Yes, comparisons to the siege scene in the 1975 film *Monty Python and the Holy Grail* are entirely in order here, though there is no record of the French defenders at Calais hurling barnyard animals back at the English vessel.) More to the point of our narrative, during the late 17th century and then throughout the 18th century, the French navy refined the design of short, wide-beamed vessels that carried two forward-facing mortars. Once Huguenot exiles had secretly brought the plans to England, the Royal Navy continued the process of modifying the design, eventually moving the mortars toward the middle of the ship, one between the foremast and the main, and another between the main and the mizzen (rear) masts. Some bomb vessels actually did launch bombs in anger; *Hecla*, the first of her class, had been part of the British victory at the Battle of Algiers in 1816, *Terror* had been one of the ships launching the bombs bursting in air over Fort McHenry during the Anglo-American War of 1812, and earlier versions of the design, including a previous incarnation of *Hecla*, had been used against Napoleon and his allies as early as the Battle of Copenhagen in 1801. But *Erebus* herself, built after the end of the Napoleonic wars, never saw action for her intended purpose.

By modern standards, *Erebus* was not a large ship: 105 feet long,[4] 29 feet wide at the beam, 372 tons, as measured at the time.[5] Her full complement was only 67 sailors, compared to the hundreds that served aboard the two- and three-decker ships of the line. Her barque rigging—square sails on the fore and main masts and fore-and-aft sails on the mizzen (rear) mast—contrasted with the stately three-masted square-sail look of larger men of war, as did her size—Horatio Nelson's flagship, *Victory*, was 2,141 tons, compared to *Erebus*'s 372. With the sails fully

3. Erebus *and* Terror

"Erebus in the Ice" (Painting by François Étienne Musin, 1846; Royal Museums Greenwich).

spread, and a fair wind at her back, she might make as much as eight or nine knots, a pace more aptly described as stately than sprightly.[6] And she really did not have room for the three years' worth of supplies that the Navy sent her off with in 1845. On her previous voyage, a three-years-plus sojourn to the shores of Antarctica, she had not needed so much in the way of provisions, spending two of the three southern winters in the somewhat balmier waters of Hobart, capital of Van Diemen's Land (shortly to be renamed Tasmania). Perhaps coincidentally, Hobart was then home for the time being of Lieutenant-Governor Sir John Franklin, who entertained the officers and men of the ships and who spent quite a lot of time carrying out magnetic observations with them.[7] Provisions were plentiful in the temperate climate of Hobart. The third southern winter found her in the much less hospitable venue of the Falkland Islands, where provisions were hard to come by, but the ships were able to replenish their supplies at Cape Town and Rio de Janeiro on the way home to England in 1843. So, no lack of food, and, despite the nearly four years away from home, virtually no evidence of scurvy, probably thanks to the fresh food available during those southern winters in Hobart.

Built in the Pembroke, Wales, shipyard in 1824–26 at a cost of £14,603—very roughly, $145 million in today's money—*Erebus* was then outfitted with masts, sails and weapons—the two mortars that bomb vessels carried, plus 10 or so cannon—at the Royal Navy's yard at Devonport, near Plymouth, and then promptly stood down "in ordinary," or, in plain English, left to sit in the harbor without a crew and without any work to do.[8] The ship didn't head out on an actual naval mission until February 1828, when she left on what turned out to be a two-year tour of the Mediterranean, showing the flag and reminding potential adversaries that Brittania ruled the waves.

Erebus returned to England in June 1830, without having actually fired a gun at an enemy; nor would she at any point from then on.

After another spell in ordinary, *Erebus* found a new career, exploring the ends of the earth. She had first been reinforced for travel to the polar regions back in 1836, when it was intended that she be used in a mission to rescue whalers thought to be in danger off the coast of Greenland. That mission never happened, as the whalers were reported safe before *Erebus* even set sail. Early in 1839, the Admiralty approved a discovery voyage to the barely-known Southern Ocean that circled Antarctica, with polar veteran James Clark Ross in command. *Erebus* was accompanied by *Terror*, with Francis Crozier in charge. For this expedition, *Erebus* was completely refitted at the Chatham shipyard: the raised quarterdeck and forecastle were removed, leaving a single, flush upper deck; the mortars and eight of the 10 cannon were taken away; the hull and decks were reinforced, in anticipation of potential damage from ice; and space was made so that the ship could carry as many as nine small boats, for use in ice-crowded seas, much as today's eco-tourism cruise ships carry a fleet of Zodiacs to ferry their well-heeled customers in search of seals, penguins and the occasional whale.[9]

Even with the guns removed, *Erebus* wasn't exactly a luxurious, commodious place for the crew to spend years together. Across the stern of the ship was a large captain's cabin, with five double-glazed rear-facing windows. But no one else got much space. To port, leading forward from the captain's space, were four small cabins for the lieutenants and the sailing master, and on the starboard side, two cabins for the surgeon and the purser. Further forward were even smaller cabins for the warrant officers, and then the tightly packed communal area for the rest of the crew, where each sailor might have a 17-inch space to hang a hammock. And last, even further toward the bow, the galley, with its Fraser patent stove and water-melting apparatus, and the sick bay.

The Antarctic voyage was a triumph. *Erebus* and *Terror* left England in September 1839, celebrated Christmas at sea, and arrived in Simonstown, the British naval station, captured from the Dutch in the 1790s, near Cape Town, South Africa, in March 1840. It wasn't until August, after battling the typically awful weather in the Southern Ocean and making brief stops at the isolated and unwelcoming Possession and

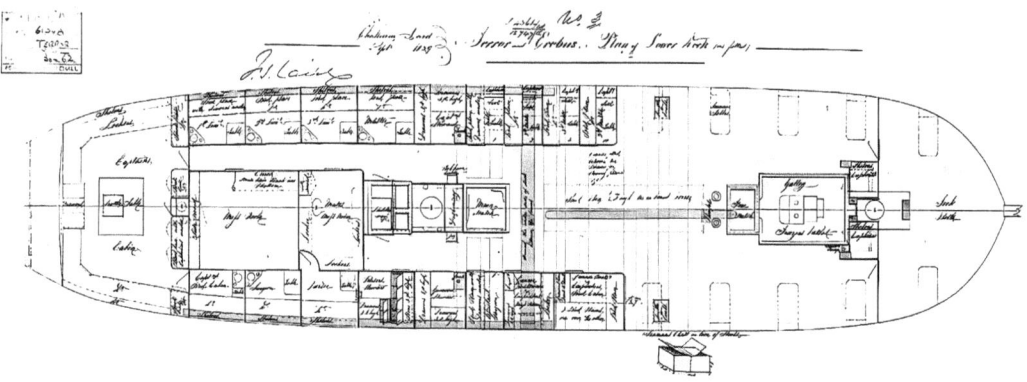

Lower deck layout of *Erebus* and *Terror* as of 1839 (Francis J. Laire, Master Shipwright, Chatham Dockyard; Wikimedia Commons).

Kerguelen Islands, that the ships reached Hobart, where they were hosted by Sir John Franklin, recently posted there as colonial Lieutenant-Governor and a very willing participant in the magnetic observation program that the Admiralty had assigned to the expedition. Franklin went so far as to build a more-or-less permanent magnetic observatory at Rossbank, close to Government House. Then, just three months later, *Erebus* and *Terror*, newly caulked and repainted after their voyage from England, sailed south. Early in January 1841 they crossed through pack ice and reached 71°14' south, surpassing James Cook's furthest south of more than half a century earlier. A few days later, in sight of the Antarctic continent, they reached 74°23' south, matching the record set by British captain James Weddell, who had been in the area hunting for seals in 1823.[10] Then, passing 76° south, the ships came in sight of the 300-foot-high frozen barrier of the (subsequently named) Ross Ice Shelf. Finally, after passing 78° south and then sailing along the ice barrier for more than 100 miles, Ross and Crozier decided it was time to escape before the southern summer waned and, dodging icebergs on the way, made it back to Hobart in early April, with no loss of life among the 128 on board the two ships—virtually the same size crew as that which sailed with Franklin a few years later into the Arctic—and with only a few spars and one small boat damaged.

The highlight or *Erebus*'s second winter in Hobart was undoubtedly the June 1, 1841, fancy-dress ball for some 300 of Hobart's elite, hosted by the officers of the two ships, for which the deck of *Erebus* served as the ballroom and that of *Terror* as the dining room.[11] (There was also an apparently dreadful play celebrating the voyage performed in a Hobart theater in May 1841.)[12] This may have been the time that Francis Crozier began his ultimately unsuccessful quest to convince Franklin's niece, Sophia Cracroft, to marry him. Then it was back into the ice, this time by way of Sydney Australia, where they set up yet another magnetic observatory, and New Zealand's Bay of Islands, making a hasty departure on November 23, 1841, in the wake of a Maori attack on some English settlers.[13] They reached the Ross Ice Shelf in February 1842 and setting, just barely, a new record for furthest south, at 78°9'30", a record never surpassed by any sailing ship, nor by any ship at all for another 60 years.

On the way back, this time by way of the Falkland Islands in the South Atlantic, *Erebus* and *Terror* survived narrow escapes from icebergs and storms. The most perilous of these occurred in the early morning hours of March 12, 1842. *Terror*, leading *Erebus*, veered to port at the same time that *Erebus*, right behind her, veered to starboard, each attempting to avoid huge icebergs. The two ships collided, with *Erebus* carrying away *Terror*'s anchor and much of the booms and rigging on her foremast. Then they smashed together repeatedly until Crozier was able to steer *Terror* through a fast-closing gap, no more than twice the width of the ship, between two massive bergs. *Erebus* was even more severely damaged, with most of her rigging gone. James Clark Ross was able to avoid losing the ship by loosing the mainsail and using the wind to push the ship backward, away from the icebergs, and then turning to sail through the same gap that *Terror* had just found. Despite the damage, which was recorded on several pages of the ships' logs, both vessels somehow emerged seaworthy and were able to limp north to the nearest place they could be repaired, the windswept Falkland Islands in the South Atlantic.

Reaching Port Louis in the Falklands in April 1842, Crozier learned of his recently announced promotion from Commander to Captain. As at Hobart, the expedition built an observatory, repaired and recaulked the ships and, in December 1842, headed south for a third Antarctic summer, though this time the effort met with rather less success. Heading for the Weddell Sea, to the east of the Antarctic Peninsula, the ships were repelled by ice at much more northerly latitudes than in the previous years, finally sailing into Simonstown in South Africa in April 1843 and then, after a stop in Rio de Janeiro for yet more magnetic observations, returning to England that September. Then, until 1845, *Erebus* was once again stripped down and placed in ordinary, waiting for work, though this time the wait was a good deal shorter.

For the 1845 Expedition, *Erebus* and her sister-ship *Terror* were refitted at Woolwich, a few miles downriver (east) of London on the south bank of the Thames. Even after the successful Antarctic mission, there were still improvements that could be made before sending the ships into Arctic waters. To start with, the already massive bows were covered in sheet-iron, extending back some 20 feet from the bows, rather than the copper used in the Antarctic voyage, which had "crumple[d] like brown paper"[14] in the ice.

Both *Erebus* and *Terror* were equipped with railway locomotives for (very limited, because coal supplies were sufficient only for a few hours' steaming) powered cruising, or, more likely, for pushing the ships through the ice when wind and manpower would not suffice. In *Erebus*'s case, the locomotive was a second-hand 25-horsepower unit obtained from a local railroad line.[15] The dockyard's master shipwright, Oliver Wright, designed a mechanism for lifting the screw propeller that was attached to the locomotive out of the water when necessary, so that the propeller wouldn't be trapped, motionless, by the ice. Other improvements included the addition of a piping system for circulating hot water to warm the sleeping areas of the ship and for melting snow to supply fresh water.

One result of the strengthening work, which added a substantial amount of timber to the interior of *Erebus*, and of the massive tonnage of stores that she was required to carry, especially after transferring supplies onto the Expedition ships and sending the supply vessel *Barretto Junior* home from Greenland, was that the ship had very little elbow room below decks. Even Commander (as he then was) Fitzjames had a cabin that was barely six feet across, just large enough for a bed, a small table, a chair and a few shelves. Sir John, of course, occupied the large captain's cabin in the stern, where the daguerreotype apparatus, specifically requested by Franklin, probably resided, though thus far there has been no evidence that either the camera or any images recorded by it survived. It's probably a good thing that *Erebus* and *Terror* never had to clear the decks for action, as warships did, since there was nowhere to clear anything into once that final year's worth of supplies, carried to Greenland on *Barretto Junior*, had been loaded aboard in early July 1845.[16]

HMS *Terror*

Like *Erebus*, *Terror* now has her own biography.[17] *Terror* was older than *Erebus*, part of the *Vesuvius* class of bomb vessels—along with her sisters *Vesuvius* and *Beelzebub*—developed during the Napoleonic Wars. The design, by noted naval architect Sir Henry Peake, combined the stoutness of the traditional bomb design with a lesser draft and sleeker lines of the *Cherokee* class of frigates—of which the most famous is HMS *Beagle*, which Charles Darwin sailed on later in the 19th century—that had done some important exploring in non–Arctic climes. *Terror* was launched in 1813 at Robert Davy's Topham shipyard in South Devon, just in time to sail on April 7, 1814, as part of a convoy joining the West Indies and North America station of the Royal Navy, that was charged with blockading the fledgling United States during the War of 1812. She cost nearly £15,000, including the rigging and fitting-out work done at the Royal Navy's own yard at Portsmouth after she was launched—(very) approximately equivalent to about US$150 million today. During her initial campaign, *Terror* famously participated in the bombardment of Fort McHenry at Baltimore (along with an earlier incarnation of *Erebus*—not the same ship that sailed with Franklin into the Northwest Passage), accounting for the "rockets' red glare" immortalized by Francis Scott Key in the U.S. national anthem, all as part of the 1814 Chesapeake campaign by the Royal Navy and the Army that also resulted in the 26-hour British occupation of Washington, D.C., and the burning of the White House and other U.S. government buildings. For good measure, *Terror* also bombarded Stonington, Connecticut, and St. Mary's, Georgia, before returning to England when the war ended in 1815. John Franklin, meanwhile, at the time a lieutenant on HMS *Bedford*, was also involved in the War of 1812, wounded at the battle of Lake Borgne in December 1814 and then serving as part of the naval support for the British army at the Battle of New Orleans, a battle most famous for having been fought after the treaty concluding the war had been signed but before anyone in the area knew about the peace deal.

Terror was very similar in size and design to, though just slightly smaller than, the later-built *Erebus*: 102 feet long with a beam of 27 feet and a displacement of 325 tonnes. Interior arrangements were much the same as well, although on the 1845 Northwest Passage expedition, the ship's commander, Francis Crozier, had the large stern captain's cabin to himself, while over on *Erebus* the ship's commander, James Fitzjames, occupied the small cabin typically used by the ship's first lieutenant, while John Franklin, the overall expedition leader, shared the large stern cabin with surgeon Stanley's stuffed birds, assistant surgeon (and naturalist) Harry Goodsir's biological samples, the daguerreotype apparatus and some of the scientific instruments.

Like most warships, *Terror* didn't find much work after returning from North America. While still in ordinary, laid up without a crew or a mission, she underwent extensive repairs in 1821–22, at a cost of some £12,487 (perhaps $125 million today). In 1824, she was sent on a brief anti-piracy mission to Algiers, returning to Portsmouth the same year. Finally, in 1828, she was recommissioned and sent to join the British fleet in the Mediterranean, but that assignment was cut short when she was caught in a hurricane off the coast of Spain and badly damaged, though miraculously blown onto a beach, rather than being slammed into the rocks that made up most of the coast, and had to limp back to England. James Fitzjames, at the time stationed with the Royal Navy at Lisbon, was sent to help with the repairs needed to make *Terror* sufficiently seaworthy to return to England, his first (brief) encounter with either of the vessels that went to the Arctic in 1845. Repaired, this time at a cost of £7,839 (about $75 million today), *Terror* was then refitted for polar exploration (another £6,440, for work that included major strengthening of her timbers to withstand work in the ice as well as reconfiguring the lower deck to turn the space previously occupied by the guns and their supporting structures into additional cabins for warrant officers and more room for the crew). For a month or so, the newly remodeled vessel served as a tender to HMS *Howe*, a British ship of the line. Then she was

HMS *Terror* wrecked on an iceberg, July 1837 (watercolor by Owen Stanley; Library and Archives Canada, Acc. No. 1960-109-36).

sent out in May 1836 under the command of polar veteran George Back, who had traveled with Franklin on one of the latter's overland journeys, with the goal of reaching Repulse Bay or Wager Bay, in the northwest corner of Hudson's Bay, and determining whether Boothia, recently explored by John Ross and his nephew James Clark Ross, was a peninsula or an island. If the latter, then perhaps a Northwest Passage lay to the south of the Lancaster Sound-Barrow Strait "highway" that Parry had traversed two decades earlier.

Back and *Terror* never reached Repulse Bay on that 1836–37 trip. *Terror*, it appeared, was accident-prone. Once through Hudson Strait, she was trapped in the ice off Southampton Island, further east, for 10 months; at one point she was pushed almost 40 feet up against the face of a cliff, and then further damaged in a collision with an iceberg as she attempted to return home the next summer. George Back barely nursed her across the Atlantic, beaching her on the west coast of Ireland in September 1837. From there, she was partially repaired, just enough so she could sail to the Royal Navy yard in Chatham for more complete repairs.

And from there, her story runs in parallel with that of *Erebus*. To Antarctica with James Clark Ross and Francis Crozier in 1839–43, yet another refitting and strengthening, then, in 1845, to the Arctic. Until, sometime after April 1848, the two ships separated once more, *Terror* to come to rest in the eponymous Terror Bay, off the south coast of King William Island, and *Erebus* to drift southward and eventually sink just north of the continental land mass. Perhaps, as we'll explore in subsequent chapters, one or both ships were actually remanned and sailed from the point at which they'd been abandoned.

4

Who Sailed with Franklin?

We know a lot about a few of those who sailed on the Franklin Expedition—mostly the senior officers—and we know a little about a lot of the rest of the crew. Sir John himself is the subject of a dozen or more full biographies,[1] not to mention a good many epic poems, of perhaps questionable literary quality.[2] Perhaps the most compelling portrait of the Expedition's commander is the wholly fictional, but compelling, depiction in Sten Nadolny's novel *The Discovery of Slowness*[3] of a man who took his time thinking things through in a manner perhaps not entirely appropriate for a naval commander required to make quick and clear decisions. Franklin's second in command, Francis Crozier, is the subject of at least two published lives,[4] and the third-ranking officer, Commander James Fitzjames, also has his own literary testaments, both factual and fictional.[5] But none of them leap from the pages of their chroniclers as fully developed characters; rather, for the most part, they seem to be cardboard cutouts or caricatures, exemplars of the Victorian archetype of the hero-explorer, bravely carrying on with his mission against all odds, for the greater glory of, well, who knows?

So that we can have at least some faith that our projections of what might have been comport with what we know of the real people who sailed with Franklin, this chapter summarizes what we do know about as many of the expedition members as possible. As with much about the Franklin Expedition, there's only a limited amount that we actually know, and a much larger area for speculation. The background, and to some extent, the personalities, of many of the officers are reasonably well known, both from the extant biographies and from the letters they wrote. The lower ranks—a good many of them functionally illiterate—left much less of an historical record in the archives. Here's what we know.

Sir John[6]

A thoroughly middle-class background. Born in 1786, the fourth son in a family of nine children. His father, Willingham Franklin, was a merchant in Spilsbury, Lincolnshire, though descended from minor country gentry who had lost most of their estates through the profligacy of Sir John's grandfather and great-grandfather.[7] Despite some evidence that Willingham wanted John to pursue a career as an Anglican clergyman (among his brothers were Willingham Franklin, Jr., a Supreme Court judge in Madras and, like John, eventually knighted, and James Franklin, a Fellow of the Royal Society and a major in the 1st Bengal Cavalry), off young John ran to sea, first on a merchant ship and then, at age 14, joining the Royal Navy on the 64-gun ship of the line HMS *Polyphemus*, seeing his first action in the Battle of Copenhagen in 1801, then accompanying his relative Captain Matthew Flinders as a midshipman on *Investigator*, circumnavigating Australia, learning the basics of navigation and, foreshadowing his later misadventures, being shipwrecked near the Great Barrier Reef and having to wait several months for rescue while Flinders sailed off in a ship's boat to seek help.

Franklin was back in England by 1804, and his career looked like that of many other young men in the Royal Navy. With the Napoleonic Wars creating a constant challenge for Britain, young officers with any talent at all were sure of employment. Franklin served as a midshipman and then master's mate on HMS *Bellerophon* at the Battle of Trafalgar in 1805, then on HMS *Bedford* from 1807 to 1815, most notably in the absurd Battle of New Orleans. Along the way, he was promoted to lieutenant, then served briefly at the end of the Napoleonic Wars as first lieutenant on HMS *Forth*, a smaller 40-gun warship.[8]

The end of the European wars in 1815 meant the end of hope of rapid promotion for Royal Navy officers, as most officers were sent ashore on half-pay and most ships laid up in ordinary, sold to private interest or scuttled. Franklin was among the fortunate few who found continuing employment without lengthy stints ashore. His polar adventures began in 1818, when he was rescued from peacetime half-pay and given command of the hired brig *Trent*, accompanying Captain David Buchan on a mission to explore the Arctic Ocean north of Spitzbergen, and, in the dreamworld of John Barrow, to sail through the open Polar Sea to the North Pole. In reality, the two little ships spent only a couple of months in the ice fields near Spitzbergen, pushed back by the ice faster than they could make progress by sail, suffering repeated injuries from ice and storms and limping back to England, barely afloat, that same year.

The next year, 1819, the Admiralty sent Franklin on another ill-fated mission, this time to head via Hudson's Bay and then across rivers and tundra of what is now Manitoba and Saskatchewan to the north coast of the North American continent, where he was supposed to map the shore of what would become the southern branch of the Northwest Passage. That three-year journey was another disaster; roughly half of those who traveled with Franklin died along the way, the expedition endured repeated bouts of near-starvation, during one of which Franklin reportedly cooked and ate his boots. One of the *voyageurs* in the party killed one of his companions—and probably ate and served the dead man to others in the guise of "wolf meat." But Franklin emerged a hero nonetheless. He had managed to survey more than 500 miles of the Arctic coastline, and on his return to England in 1822 (and especially after John Murray's publication of the expedition narrative in 1823,[9]) was feted as "the man who ate his boots."[10] Apparently enough of a success that he was promoted to Captain in November 1822, elected a Fellow of the Royal Society in February 1823, became a founding member of the literary and scientific Athenaeum Club, and was thought, or at least thought himself, sufficiently established to marry Eleanor Anne Porden, daughter of the architect of the King's stables and herself a well-regarded poet, in August 1823; their daughter, Eleanor, was born the following June. In yet another of those odd coincidences that dot the Franklin Expedition story, Porden had already won acclaim for her 1818 poem *The Arctic Expeditions*, written in response to Sir John Barrow's influential *Quarterly Review* essay and, as much as the more famous

Frankenstein by Mary Shelley, representing the early 19th-century English romanticization of—and ecological concern with—the polar regions.[11]

Marrying Porden, whose own religion was a very conventional Church of England practice, also steered Franklin away from the dangerous—at least in terms of professional advancement—shoals of "radical" Calvinist and Methodist beliefs that he had been exposed to through the publications distributed by Lady Lucy Barry and her circle[12] and that he carried with him on the Coppermine expedition in 1819–21. Some of that more muscular Protestantism, however, such as the very British belief in the necessity of enduring adversity with patience and an optimistic, cheerful outlook,[13] seems to have stuck and carried on throughout Franklin's subsequent career.

In 1825, even though Eleanor was dying of tuberculosis, Franklin departed for the Arctic once again—Eleanor died soon after he left—on a second cross-country expedition from Hudson's Bay to the Arctic coast, mapping the area from the mouth of the Mackenzie River westward to Point Beechey in what is now Alaska, at the same time that another exploring party mapped eastward from the Mackenzie to the Coppermine, the point from which Franklin had set off on his previous surveying journey. This time the expedition proceeded with somewhat less *sturm und drang* than on the previous trip; Franklin returned to England in 1827, duly published his second memoir,[14] was knighted in 1829, received an honorary degree from Oxford and, in November 1828, married his second wife, the energetic and ambitious Jane Griffin.[15] While waiting for his next command, Franklin became a Fellow and Councilor of the Royal Geological Society, Vice President of the Royal Geographical Society and traveled to Russia where he met Tsar Nicholas I and some of the leading Russian geographers. Not bad for the son of a country merchant who had himself left school for the sea at age 14 and had never attended university.

Eventually, in 1830, Franklin was given command of the 28-gun sloop *Rainbow*—a very small ship for a full Post Captain, but in the shrunken peacetime Navy, perhaps all that was available—for what became more than three years of diplomatic duty in the Mediterranean, principally displaying the British flag in support of the ineffectual teenage Prince Otho of Bavaria, whom the allied powers in Europe had decided to install as King of Greece after the combined navies of Britain, France and Russia had routed the Ottoman Empire fleet at the Battle of Navarino in 1827 and detached Greece from that Empire. Not much in the way of naval action, but lots of time for seeing the classical ruins and exploring the countryside while Lady Jane, who accompanied Franklin, did her own tours around the Mediterranean. Back in England at the end of 1833, Franklin racked up one more honor, becoming a Knight Commander of the Guelphic Order of Hanover, the personal order of King William IV.[16]

But, once again, there were no ships available for command. Franklin was offered the governorship of Antigua, but turned it down, because he would have had to report through a superior in the Caribbean rather than directly to London. Then along came an opportunity sufficient for his and Jane's ambitions, the governorship of Van Diemen's Land (to become, in 1855, Tasmania), the last of Britain's penal colonies and a place that Franklin had in fact visited many years before, on his voyage with Matthew Flinders.

Like its predecessor as a dumping ground for convicts, New South Wales, Van

Diemen's Land was peopled by free settlers, convicts, and an ever-diminishing number of its Aboriginal inhabitants.[17] Franklin probably arrived with all the zeal of an evangelical reformer, but was ultimately mired in settler politics and, as others have described,[18] forced out ignominiously in 1843, returning to London determined to vindicate his name and, if possible, undertake one more exploit to redeem his reputation (or at least Lady Jane was so determined). His sins, such as they were, appeared to have been opposing the more greedy of free settlers and colonial officials in their quest for more land ownership and profits.

So the Sir John who embarked at Greenhithe in May 1845 was very much a known quantity. He was reportedly popular among his crews, possibly because he was known as a very light disciplinarian, and he was well thought of in England's scientific circles, thanks to his dogged diligence in pursuing magnetic observations around the world. Not, perhaps, the sharpest knife in the drawer—his own niece Sophia Cracroft commented that John Franklin's "outer man was none of the smartest"[19]—but known to the Admiralty as someone who would carry out his orders and who could always be counted on to lead a good Sunday service aboard ship.

Lady Jane Franklin (Dibner Library Portrait Collection, New York Public Library).

Although he had by then been quite dead for more than five years, the inexorability of the Admiralty's seniority system kept grinding away, with the result that Franklin was promoted to Rear Admiral in October 1852, before finally being declared dead, along with the rest of his crew, as of March 31, 1854.

Francis Crozier[20]

For someone so central to the story of the Franklin Expedition, Francis Rawdon Moira Crozier remains a bit of a mystery. Even Crozier's date of birth is a matter of some dispute. Before 2009, most accounts stated that he'd been born in September 1796, probably on the 17th. That's based on a memorial tablet in Crozier's parish church in Banbridge, County Down in what's now Northern Ireland and a subsequent statue erected in the town, both of which listed "September 1796" with no specific date, and also on the first full-scale biography of Crozier, May Fluhmann's *Second in Command*, which in turn relies on a somewhat fuzzy note of Crozier's found in the Scott Polar Research Institute archives. In 2009, R.J. Campbell suggested, based on sailors' recollections of the "Captain's birthday," that the actual date might have been August 16.[21] And most recently, researcher Alison Long, digging in the Downshire archives, found correspondence from Crozier's father suggesting the actual date of birth was October 17.[22] No matter, a difference of a month or even two won't change the reality that, in 1848, tasked with leading the Expedition's survivors, Crozier was a seasoned naval and Polar veteran and considerably older than most of the remaining officer corps and the remainder of the crew.

Crozier was a younger son of a prosperous County Down family; his father, George Crozier, was a lawyer for some of the largest landowners in the area—Francis's middle names are an homage to local gentry—as well as for the leading linen merchants in the area around the family home in Banbridge. Two of Francis's brothers, like their father, also became solicitors, while a third became a vicar in the Church of Ireland, the Irish version of the establishment Church of England, despite the family's history as part of the emigration of Presbyterian Scots to Ireland under King James in the early 17th century. While Francis's older brothers eschewed the family's centuries-long history of military service, Francis went—or was pushed—into the Royal Navy in June 1810 at the age of 13, in the midst of the Napoleonic Wars,[23] joining HMS *Hamadryad* at Cork.

Perhaps because of his lack of "interest" at the Admiralty, a euphemism for powerful sponsors, Crozier made relatively slow progress in the Navy. He was appointed Midshipman—the lowest officer rank—at the end of 1812 and shipped out on HMS *Briton*; during the ship's Pacific cruise, he met with the Pitcairn Island descendants of the famed HMS *Bounty* mutiny.[24] An odd coincidence, in view of the likely mutiny of at least some of the Expedition's crew during the retreat across King William Island (see Chapter 11 below). In the wake of the *Bounty* mutiny of 1789, the Navy had slightly improved conditions for those serving at sea, and eased the harshest aspects of discipline, but throughout Crozier's career, most commanders still authorized the occasional use of lash for punishment.

In 1816, Crozier served on HMS *Meander*, on guard duty at the mouth of the Thames. Then the following year he passed the exams for mate and was assigned to *Queen Charlotte*, a first-rate ship of the line, patrolling the English Channel. And the next year he was appointed as mate on *Dotterel*, a three-year posting that included voyages to St. Helena in the south Atlantic.

Of all those on the Franklin Expedition, Crozier had by far the most experience in the ice. When he returned to England from his assignment on *Dotterel*, he faced the same issues as most young Royal Navy officers now that the major wars were over: too many candidates chasing too few ships. So he chose an alternate path that he would follow for most of the next quarter-century, the discovery service, planting the British flag in the furthest corners of the world.

Crozier volunteered to join Edward Parry's second Arctic expedition, which left England in 1821. Parry had gained attention in his first Arctic trip, sailing through an ice-free Lancaster Sound all the way to Melville Island, the furthest west that any British explorer reached for many years. His subsequent voyages, the ones that Crozier served on, were less successful. On the 1821–23 trip, Parry sailed north through Hudson's Bay and the Foxe Basin, looking for a route to the west and finding only the ice-choked Fury and Hecla Strait. Parry and his crew spent one winter at the large Inuit settlement of Igloolik, on the northwest corner of the Melville Peninsula, so Crozier would have had some opportunity to compare British and Inuit ways of dealing with the environment: the Inuit diet of seal meat and blubber versus the scurvy-inducing Navy fare of salt meat and hard tack; the light, flexible Inuit dog-hauled sleds versus the heavy British man-hauled sledges, etc. How much of that knowledge he was able to apply to his situation in 1848 is a matter for future chapters.

Back in England after Parry's failure to penetrate Fury and Hecla Strait—which remained an impassable barrier until 1948, when an icebreaker finally managed to bull its way through—Crozier, still officially a midshipman, promptly signed up for Parry's next attempt. This time Parry, again taking the bomb vessels *Fury* and *Hecla*, was sent in 1824 to probe Prince Regent Inlet, the southward-trending opening to the west of the Boothia Peninsula, and to see whether that offered a route through to the west. Even less success this time; after a winter in the ice, *Fury* was wrecked on the eponymous Fury Beach, at the southeast corner of Somerset Island, and her supplies offloaded and left there for the use of future expeditions. Crozier would have seen first-hand where this potential supply depot was set up and undoubtedly would have taken its existence into account as he led the crew off *Erebus* and *Terror* in 1848.

After that unsuccessful voyage, Crozier was finally promoted to lieutenant in 1826, at age 29. In 1827 he sailed once more with Parry on *Hecla*, this time in an attempt to reach the North Pole via Spitzbergen, due north from the northern coast of Norway. While Parry and James Clark Ross attempted to man-haul their sledges toward the pole, Crozier stayed with the ship, setting up supply depots. Eventually the pole-bound sledges returned, having barely made headway against the drifting ice. One more lesson in the futility of using heavy man-hauled sledges, but a lesson that the Royal Navy was very slow to learn.

Crozier had served with James Clark Ross on two of the three *Hecla* and *Fury* voyages, as well as on *Hecla* when Parry had tried for the North Pole. This connection would reassert itself some years later, but, when he returned to England in late 1827, Crozier once again faced the problem of being in a navy that had too many officers for the number of ships that it needed. He finally found a ship—HMS *Stag*, patrolling off Portugal—in 1831 and, still a lieutenant

Captain Francis Crozier (Unknown artist. Wikimedia Commons).

as he approached age 40, remained with it until 1835. Then, finally, back to the Arctic. Crozier was recruited by James Clark Ross at the end of 1835 to be second in command on HMS *Cove* on a mission to rescue stranded whalers off Greenland. While the mission eventually proved fairly pointless—most of the whalers freed themselves and returned to England—it did cement the relationship between Ross and Crozier and led to Crozier's promotion, in January 1837, to the rank of commander, one step short of full captain.

And then, perhaps after a failed attempt to find a wife,[25] Crozier sailed once more as second-in-command to Ross on the 1839–1843 Antarctic voyage, commanding *Terror* with Ross, on *Erebus*, who was again in charge of the expedition as a whole.[26] The two ships were away for almost four years, wintering twice in Hobart, where Sir John Franklin was the colonial governor, reaching a new furthest south, charting much of what's now known as the Ross Ice Shelf and giving Crozier yet another chance to be disappointed in love, this time with Franklin's coquettish niece Sophia Cracroft. The exertions of that voyage, perhaps especially the narrow escape when Crozier steered *Terror* between two fast-closing icebergs, had turned his hair gray and his "hands shook so much they could hardly hold a glass or a cup"[27] by the time he returned to England in 1843, even though he was still in his 40s.[28]

Back in England and finally promoted to captain, as he had learned when the ships had docked in the Falkland Islands in 1842, but once again without a command, Crozier by all accounts became seriously depressed, took a year's leave from the Navy and wandered off to Europe, occasionally sending letters back to Ross and other friends that reflected his gloomy outlook.[29] As one relative of his reminisced, Crozier "went out [to Antarctica] a young man and came home broken down and rather old-looking for his years."[30]

Despite his 30 years' experience in the ice, Crozier never got quite the same adulation and official recognition as other British polar explorers. The London publisher John Murray never asked him to write a narrative, and unlike all the other senior figures of the time—Parry, James Clark Ross, George Back and Franklin—he never received a knighthood, let alone of the rarified status of Sir John's Guelphic Order of Hanover (though Crozier is now commemorated by an eponymous crater on the moon as well as various terrestrial geographical features). Although Crozier was nominally Protestant—his family was Presbyterian rather than the official Church of Ireland—as were much of the gentry in Northern Ireland, and thus not subject to the more blatant prejudice—and sometimes legal restrictions—directed against Catholics in British institutions of the time, and although some officers with Irish background had advanced to high levels in the Navy, including the Navy Hydrographer Francis Beaufort and two would-be Franklin rescuers, Robert McClure and Leopold Francis McClintock, of whom we'll hear more later, Crozier was still an Irishman with few sources of influence in the Admiralty, at a time when "interest" was often all that mattered.[31]

By the time he sailed with Franklin, Crozier's hair had turned gray and his hands had developed a shakiness or trembling. Perhaps those were just the results of a hard life at sea, or perhaps related to the trauma of the near-fatal collision with *Erebus* in the Antarctic—James Clark Ross also exhibited similar trembling—but in any event neither Crozier nor Franklin were, as they left Greenhithe, in peak physical, or even, possibly, mental, condition. Crozier was also subject to something that

Senior officers of the Franklin Expedition. Top row, from left: Edward Couch (mate, *Erebus*); James Fairholme (Lieutenant, *Erebus*); Charles Osmer (Purser, *Erebus*); Charles Des Voeux (mate, *Erebus*). Second row, from left: Francis Crozier (captain, *Terror*); Sir John Franklin (Expedition commander); James Fitzjames (commander, *Erebus*). Third row, from left: Graham Gore (Lieutenant, *Erebus*); Stephen Stanley (surgeon, *Erebus*); H.T.D. Le Vesconte (Lieutenant, *Erebus*). Fourth row, from left: Robert Sergeant (mate, *Erebus*); James Reid (ice master, *Erebus*); Harry Goodsir (assistant surgeon, *Erebus*); Henry Collins (2nd master, *Erebus*) (*Illustrated London News*, 1851, based on 1845 daguerreotypes by Richard Beard; Wikimedia Commons).

looks, in retrospect, a lot like chronic depression. One shipmate commented that "the ship was very uncomfortable, owing to the captain being very much out of temper."[32] And Franklin, in a letter to his wife, commented that Crozier, on the months-long trip to Greenland, had come across to dinner with Sir John only twice and that "I do not think he has had his former flow of spirits since we sailed,"[33] a polite, or perhaps merely uniformed, way of suggesting that Sir John's second-in-command was deeply depressed. Not a psychological profile that would have appealed to an Admiralty looking for heroes who could be immortalized in best-selling narratives of conquering the polar vastness.

What follows are some brief sketches of a few of those—primarily the officers—who sailed with Franklin. We know a fair amount about most of the officers; they were literate, and they generally wrote lots of letters. About the petty officers and able seamen, not so much, with a few happy exceptions.

On *Erebus*

James Fitzjames

In addition to Franklin and Crozier, one other officer, third-in-command Commander James Fitzjames, has his own book-length biography.[34] As Crozier's biographer Michael Smith describes him, Fitzjames was an "engaging and able officer with … friends in the right places but no Arctic experience."[35] So, how did someone with no history of work in the ice, and not even any real history of lengthy exploration voyages, end up as a senior figure on the Franklin Expedition? The answer has more to do with the personal politics of "interest" at the Admiralty, and in particular the advantage of having key bureaucrat John Barrow on one's side, than it does with ability or relevant experience.

For a long time, based on the work of the 19th-century naval historians Albert and Clements Markham, Fitzjames was thought to have had some sort of minor aristocratic background. In fact, as William Battersby demonstrates in his biography, Fitzjames was the illegitimate son of a British diplomat, Sir James Gambier, and a woman who may have had some position in the Rio de Janeiro court of the Portuguese royal family, then in exile in Brazil as the Peninsular War raged in Europe. James was born in Rio in July 1813, and, when Gambier returned to London in 1814, young James was baptized, listing wholly imaginary "Fitzjames" parents on the parish records, and then, accompanied by the Portuguese-speaking nurse who had traveled with him on the voyage from Brazil, promptly fostered off to the Coningham family, country gentry with an estate at Rose Hill, Abbotts Langley, Hertfordshire, where Fitzjames lived until going off to sea at age 12.

Fitzjames's first posting was as a volunteer (i.e., a young boy) on *Pyramus*, which just happened to be commanded by his older natural second cousin Robert Gambier and which sailed in 1835 on a diplomatic mission carrying three English consuls to Latin America. On the ship's return, Fitzjames stayed with it, serving in the Experimental Squadron, a group of ships trying out different designs and techniques. Eventually, with a bit of deception on his part about whether he actually had the required service time for the promotion, he ended up as a midshipman on HMS *St. Vincent* from 1831 to 1834 in the Mediterranean, including duty as part of the young King Otho's installation by the European powers as the nominal ruler of Greece, as was John Franklin on *Rainbow*.

Fitzjames's next two assignments earned him a reputation as a brave and daring officer. First, in 1835–37, he was part of an expedition to establish a link from the eastern end of the Mediterranean across the Syrian desert and then down the Euphrates River by steamboat to the East India Company port at Basra on the Persian Gulf, in what is now Iraq. The mission was quixotic; one of the two steamboats was wrecked, the costs were far over budget, and in the end the route didn't save all that much time compared to sending large ships across the Indian Ocean, around the Cape of Good Hope and up through the Atlantic to England. But Fitzjames, despite several bouts of malaria, conveyed, nearly single-handedly, an important collection of mail from India for 1,000 miles up the river in a small boat to Baghdad and then across the desert, mostly by camel, to Beirut.

Promoted to Lieutenant after his return, Fitzjames served on the gunnery training ship *Excellent* in 1838, along with Henry Le Vesconte and Edward Couch, who would both join the Franklin Expedition some years later. Because of his newfound expertise—Fitzjames graduated at the top of his class from the gunnery school—the Admiralty refused him permission to sail on James Clark Ross's Antarctic expedition of 1839–43, sending him instead as gunnery lieutenant on HMS *Ganges* to the eastern Mediterranean and then, in 1841, off to China in a similar role on HMS *Cornwallis*, part of the British task force sent in the First Opium War to force the Chinese emperor to buy opium from (British) India, with the profits then invested in Chinese tea, silk and ceramics, with the customs duties on those products back in England largely financing the Royal Navy and other British entities.[36] Fitzjames, with his typical daring—actually recklessness would be more like it—was prominent in the battles leading up to the signing of the Treaty of Nanking on August 29, 1842.

Battersby, on the basis of a letter written by Fitzjames from Singapore to John Barrow, Jr., in London, concludes that Fitzjames must have gotten another of Sir John Barrow's sons, George, then a colonial official, out of some sort of scandal, thus placing the senior Barrow in his debt.[37]

And then, off into the Arctic, and into the imaginings of so many writers.

Graham Gore[38]

Unlike many of the senior officers on the Expedition, Graham Gore, first lieutenant on *Erebus*, came from a long line of distinguished naval officers. Gore's grandfather, Captain John Gore, had accompanied Captain James Cook to the Pacific, taking command of the expedition after Cook was killed in Hawaii and the second-in-command, Charles Clerke, died. John Gore eventually circumnavigated the globe four times. Graham Gore's father, also named John, rose to Captain before retiring shortly after the Franklin Expedition sailed from England and, in retirement, was promoted to Rear Admiral. Graham went to sea at age 11, joining his father and older brother (yet another John, who also became a Navy lieutenant but died at sea) on *Dotterel*; coincidentally, Francis Crozier served as mate on *Dotterel* at the same time, before leaving in 1821 to join Parry's Northwest Passage expedition.

After some time at the Royal Naval College at Portsmouth, Gore served on HMS *Albion* at the Battle of Navarino, the last great sea battle between Nelson-era sailing ships. This was the battle in which the combined forces of the British, French and Russian navies defeated the Ottoman Empire, leading eventually to Greece's independence. The European powers insisted that Greece become a monarchy, not a republic, and conveniently found a king: the 17-year-old Otho, son of King Ludwig of Bavaria, to replace the Ottoman viceroy in Athens. It was the major powers' interest in propping up the new monarchy that eventually drew both Franklin and Fitzjames to the eastern Mediterranean in the following years.

Another of those with Arctic experience, Gore had been the mate on George Back's 1836–37 expedition to attempt the Northwest Passage in *Terror* via Repulse Bay. After the perilous return from that voyage, where the damaged ship barely made it to the west coast of Ireland, Gore, newly promoted to lieutenant, served in a British campaign to secure a coaling station in Aden, ousting the traditional Arab ruler, and then, like Fitzjames, went off to China to fight in the First Opium War. He returned from that assignment on Charles Darwin's former ship HMS *Beagle* by way of Australia, where he served under Captain John Lort Stokes, who himself would be considered to lead the Northwest Passage expedition before Franklin was named and who painted some quite professional scenery pictures. (Stokes's landscape of the Flinders River in Queensland—named, in one of the many coincidences that permeate the Franklin story, for John Franklin's cousin Mathew Flinders—is in the collection of the National Library of Australia.) Gore, now leap-frogged by Fitzjames on the promotion list, had a brief assignment aboard *Cyclops* before finding a place in the Franklin Expedition.

Thanks to the Victory Point Record, we know a little bit about what Gore accomplished on the Expedition. We know that, in May 1847, the lieutenant led a sledge party to the western shore of King William Island, leaving records in cairns on the northwest coast of the island. And we know from Fitzjames's and Crozier's additions to one of those records, the one thought to have been left at Victory Point, that by April 1848 Gore himself had died, though probably after the death of Sir John. Fitzjames's 1848 addition to the Victory Point Record refers to "the late commander Gore," which indicates that Gore may have received a "battlefield" promotion to fill the gaps in the seniority order after Franklin had died and Fitzjames had moved up to Captain.

Graham Gore's father eventually emigrated to Australia, where his descendants still live, so DNA analysis from there may yet connect some of the bones found on King William Island with the *Erebus*'s first lieutenant.

Henry Le Vesconte

Another of Fitzjames's old China hands. Le Vesconte was, like Gore, from a naval family, originally from the Channel Islands (hence the French surname). His father had fought in the Napoleonic Wars at the battle of Copenhagen and with Admiral Nelson at Trafalgar. Henry, born in 1813, went to sea at age 16, eventually serving as mate on the *Calliope* in the First Opium War in 1841 and then as a lieutenant on *Clio*, when it was captained by James Fitzjames. When Fitzjames was choosing officers for the Franklin Expedition, Le Vesconte was undoubtedly near the top of the list. Le Vesconte, in a letter to his father, comments that Fitzjames "worked hard to get me appointed first [lieutenant] on one of the ships [*Erebus* or *Terror*] but it was no use."[39] Le Vesconte ended up as second lieutenant on *Erebus*, junior to Gore and to Fitzjames himself.

Le Vesconte was expected to be in charge of the surveying work once the Expedition reached the unknown portion of the map west and south of Lancaster Sound and Barrow Strait. And, like many of the other officers, he was looking forward to using the Passage as a route to promotion, so, as he wrote to his mother, he would have the wherewithal to marry the daughter of another prominent Channel Islands family. He thought, with youthful enthusiasm and without experience in the Arctic, that the expedition would be far less dangerous than voyages he had already been on, in the Opium War and, on *Clio* with Fitzjames, suppressing the slave trade off the coast of Africa.[40] By the time the Expedition departed England, Le Vesconte's immediate family had moved to Canada, so he, somewhat jokingly, talked of leaving the ship once they had made it through the Passage and coming home by way of North America to see his parents and siblings. A skeleton, found by Charles Francis Hall in 1869, was thought to be that of Le Vesconte and dutifully buried back in England. But recent analysis[41] indicates that the skeleton is actually that of *Erebus*'s assistant surgeon Harry Goodsir. Given the re-identification of the remains originally thought to have been his, Le Vesconte is probably still in North America, if not anywhere very near the descendants of his family.

James Fairholme

Though not one of Fitzjames's comrades in the First Opium War, James Fairholme, third lieutenant on *Erebus*, had served with Fitzjames before. Fairholme, then at the age of 19, was aboard *Ganges* in the eastern Mediterranean at the same time as Fitzjames. Moreover, Fairholme also had some exploration experience, albeit not in the polar regions. He had been mate on *Albert*, one of the three steam vessels that were sent to explore the Niger River in West Africa in 1841, making it some 350 miles up the river before heading back. That expedition earned Fairholme a promotion to lieutenant in 1843, as well as a trip home to England to recover from a tropical disease he had acquired on the expedition.

So, Fairholme's connection to the Franklin Expedition was most likely through Fitzjames. Unlike Gore and Le Vesconte, he did not come from a naval family. He did have some minor aristocratic connections, however; his maternal grandfather appears to have been the Scottish Lord Walter Forbes,[42] and his father was a landowner and banker in Perth, on the eastern edge of the Scottish Highlands.

Fairholme is the source of one of the quotations most often used by those writers who differ from the bumbling-idiot view of Franklin; in a letter to his father sent back from Greenland, Fairholme comments that "Sir John is a new man since we left. He has quite recovered from his severe cold, looks 10 years younger & takes part in everything that goes on with as much interest as if he had not grown older since his first Expedition. We are all delighted to find how decided he is in all that he resolves on, & he has such experience and judgement that we all look on his decisions with the greatest respect."[43]

Because the Admiralty did not declare the Franklin Expedition crew dead until March 1854, Fairholme figures in a Scottish trusts and estates legal case. His uncle, Adam Fairholme, had died in 1853, leaving James a significant sum of money, which would have gone to his parents had James survived Adam. But the court, in 1858, ruled that, notwithstanding the Admiralty declaration, and relying on evidence presented by John Rae among others, James Fairholme must have died before 1853, and so the money that Adam left him reverted to Adam's estate, from which it was distributed to Adam's heirs, a category that did not include James's parents.[44]

Harry Goodsir

Harry Goodsir, assistant surgeon and naturalist on *Erebus*, is one of the voices from the Franklin Expedition that speaks to us most clearly from the past. In his letters, collected by Russell Potter and colleagues,[45] he enthuses about the opportunities for a serious naturalist, as he considered himself, in the Arctic, while at the same time obsessing over the cost of his clothes (four pounds for the hat he would need if he were invited to a reception for Queen Victoria!), the silverware required for each officer (monogrammed forks and spoons which, recovered from the Inuit years later, became important evidence of the fate of the Expedition), the dredges for scooping up samples from the sea floor, etc. Only once the ships have left Stromness and potential creditors behind do Goodsir's letters begin to focus primarily on his shipmates (including his superior, surgeon Stanley) who, according to Goodsir "appears to spend the greater part of his time reading novels in bed"[46] and on his biological specimens, laid out in Sir John's capacious stern cabin.

Along with his father and three of his brothers, Goodsir had qualified as a physician, graduating from the Royal College of Surgeons in Edinburgh in 1840 and being appointed in 1843 as Conservator of the Surgeons' Hall Museum there, a post that went to his younger brother Archibald when Harry left to join the Expedition. But, if Harry expected to return to Scotland when he first left to join *Erebus*, he seems to have had a change of heart as he joined the ship's company in Woolwich and as he learned that the Admiralty had approved a plan for appointing a full-time naturalist to all future discovery missions. Even as he left England, Harry was looking forward to signing up for more exploration as soon as he returned. His own work up to that time was much more about scientific research than it was about treating human patients; with his older brother Robert, he had authored a pioneering work on cell theory.[47] And it didn't detract from Goodsir's new plans for his career that the Admiralty was offering £300 a year for naturalists.

In the television series made from Dan Simmons's novel *The Terror*, Harry Goodsir became the man forced to carve up dead shipmates as cannibalism emerged. Did that happen to the real Assistant Surgeon, and, if so, was he, like his onscreen counterpart, sickened by the experience? Well, that's just one of the mysteries that remains. One can't deny, though, that Harry would have had the necessary knife skills.

On *Terror*

John Irving

United States Army Lieutenant Frederick Schwatka's 1878–80 search expedition did find an above-ground grave near the western shore of King William Island. The bones in and near the grave, along with Lieutenant John Irving's maths medal from the Royal Naval College, found next to the grave, were returned to Edinburgh, where the body that might have been Irving's was given a military funeral in 1881. If the grave was indeed that of Lt. Irving, the third lieutenant on *Terror*, then perhaps, as David Woodman speculates, he died at what subsequent searchers labeled the "Boat Place" on Erebus Bay while leading a sledge party back to the ships at some point after splitting from the main group of the retreat.[48] More recently, however, some commentators have cast doubt on whether the grave was actually Irving's.[49] Perhaps, as Russell Potter hypothesizes, the grave that Schwatka found was actually that of Sir John himself, constructed well before the ships were deserted, while the crew was still healthy enough for hard work.[50] If so, the centuries-long quest for Sir John's grave has already ended, and the commander's bones are already interred in Britain, just under the wrong headstone. More about that in Chapter 16.

Whether or not that grave was Irving's, we do know a little bit about the lieutenant.[51] Born in Edinburgh in 1815 and, like Crozier, a younger son of a lawyer, Irving entered the Royal Navy at 13, but rather than going directly to sea, he attended the Royal Naval College in Portsmouth, where he was awarded the maths medal that identified his putative grave. At age 15, after two years at Portsmouth, Irving served briefly on the brig *Cordelia* and then for three years in the Mediterranean on the frigate *Belvidere*, and then on HMS *Edinburgh*, serving, as did Franklin himself, in the show of European powers' support for the coronation of the new King Otho. He also, in 1836, suffered frostbite while climbing Mt. Etna in Sicily, which apparently resulted in some lasting disfigurement—though that's not apparent in the daguerreotypes taken just before the Expedition set sail. In any event, something—the frostbite or the death of his 24-year-old sister that same year—seems to have caused some sort of mental upheaval, with the result that Irving resigned his commission in the Navy in 1837 and went to Australia with his younger brother David, hoping to run a sheep farm. Not a happy period in Irving's life: he was robbed shortly after arrival, the price of sheep dropped from 30s to 5s a head, the farm failed, and he came down with a serious case of dysentery. Somewhat unusually, in a peacetime Navy with far too many officers already on half-pay, Irving was able to be reinstated in the Navy and assigned to several ships between 1843 and 1845, before being tapped for the Franklin Expedition.

Aside from these bare facts, the one aspect of Irving that stands out is his unashamed religiosity. Much like Sir John himself—at least early in Franklin's life when he was under the influence of Lady Lucy Barry[52]—Irving was a firm believer in the evangelical, missionary sort of Christianity favored by those who belonged to Presbyterian, Puritan, Methodist or other churches outside the orbit of the official

Church of England. His letters often refer to his essentially Puritan belief in a direct relation to God and to his distress when shipmates, including in one instance a ship's chaplain, uttered even the mildest of oaths. We don't know whether this aversion to blasphemy persisted right up until his death in the spring of 1848.

The bones identified as Irving's were recovered from a shallow grave on King William Island by Lieutenant Schwatka on his 1878–80 search expedition. Schwatka had the bones sent to Edinburgh, relying on the maths medal as sufficient identification, and those bones were interred in Dean Cemetery after a large public funeral. If Irving died, as is most likely, in 1848, he would have been just 33 years old.

Alexander McDonald

On *Terror*, assistant surgeon Alexander McDonald was among those with previous polar experience, even though he was only 28 when the ship sailed from Greenhithe. McDonald, like Harry Goodsir, had graduated from the Royal College of Surgeons in Edinburgh, then for three years had sailed with Scottish whalers into the waters off Greenland, thanks to his acquaintance with whaling captain William Penny, who later led two of the search journeys,[53] first on the whaler *Advice* in 1849, when he was accompanied by Harry Goodsir's brother Robert and then the next year with HMS *Lady Franklin* and *Sophia*.

McDonald also may merit the title of the first anthropologist to report on the customs of the Inuit—though there was no such academic discipline at the time. In April 1840, Penny had transported the Inuk Enoolooapik, whom Penny had brought to Scotland on a previous trip, back to his home in Baffin Island. The next year, McDonald published his monograph *A Narrative of Some Passages in the History of Enoolooapik*,[54] documenting Inuit customs and belief systems. It would remain the only full-length account of Inuit customs to be published until the 20th century.

Before embarking with Franklin, McDonald served as surgeon aboard HMS *Belvidere* in the Mediterranean from 1841 until early 1845. He returned to England to discover that Franklin had read his monograph and was promptly recruited for the Expedition, signing on as assistant surgeon to John Smart Peddie on *Terror*. In the letters sent back to England from the Whalefish Islands in July 1845, McDonald repeated, without specifically endorsing, the optimistic view of some that they might make it through to the Pacific without stopping for as much as a single winter.[55]

Eventually, six items that can reasonably be connected with McDonald were recovered by various searchers—five from Inuit who bartered with the searchers and one, a silver teaspoon, that was found in the debris surrounding the two skeletons found at McClintock's "boat place" near Erebus Bay. But whether McDonald died there or somewhere else remains unknown, at least until DNA evidence from the boat place site turns up a match to some living relative.

Thomas Blanky

Mordecai Richler was right.[56] There was indeed at least one Jew on the Expedition. But his name wasn't, as Richler imagined, Ephraim Gursky. And he probably wasn't the sole, improbable survivor. Thomas Blanky, a veteran sailor from Yorkshire, had changed his name—as so many Jews did in so many places—to hide his origins.[57] Blanky hailed from Whitby, as did his final commander John Franklin, though they were of very different social classes—Blanky's father was probably a laborer. As befits someone appointed ice master—in effect, an adviser to those charged with navigating the ship, usually stationed, when in the ice, high up on the mainmast and looking ahead for possible ways through—Blanky was, by 1845, not only a veteran sailor, but also someone experienced in Arctic travel. He had gone to sea at age 11 and worked on board merchant ships and whalers, traveled as far as Riga, Québec, Alexandria and St. Petersburg, and had been on three prior Arctic discovery voyages: as an able seaman on George Lyon's 1824 trip in *Griper* that accompanied Edward Parry's unsuccessful third Northwest Passage attempt, next on Parry's abortive North Pole voyage in 1827, and then as mate on John Ross's four-year odyssey to Boothia, Fury Beach and back in 1829–33, during which he accompanied James Clark Ross, John Ross's nephew, on a sledging trip to find the north magnetic pole and impressed his commander sufficiently that, on the expedition's return, Blanky was able, on the strength of Ross's recommendation, to get the command of a merchant vessel.

Compared to most of the officers and crew, Blanky, perhaps on the basis of his experience with John Ross's expedition, was more realistic about the likelihood of quick success; he wrote his wife from the Whalefish Islands in Greenland not to despair even if the ships were gone as long as six or seven years.[58] Though by 1854, when the crew was officially declared dead by the Admiralty, even that forecast was too optimistic.

Perhaps a few of the subordinate officers were less suited for the Arctic and for the daily shipboard routine. Fitzjames, who selected many of the junior officers, said in one letter to Barrow that a candidate "is just the man for the expedition—being active & energetic—a capital shot & a pleasant fellow."[59] And some were simply unsuitable to a long voyage in cramped quarters that demanded at least a modicum of team spirit; of Engineer James Thompson on *Terror*, fresh from a career with land-based steam engines, Crozier remarked that he was "a dead or alive wretch full of difficulties and is now quite dissatisfied because he has not the leading stokers to assist him in doing nothing."[60]

Petty Officers and Crew Members

For more than a century and a half, little was known about most of those on the Expedition. But thanks in large part to the tireless efforts of researcher Ralph Lloyd-Jones, we now have at least some knowledge of the Royal Marines, warrant officers, petty officers and crew members who sailed into Lancaster Sound with Franklin. Lloyd-Jones's four journal articles,[61] based on painstaking analysis of the Muster Books of *Erebus* and *Terror*—copies of which were conveniently sent back to England on the supply ship *Barretto Junior* when the Expedition departed Greenland in July 1845—and the Description Books from the crews' previous voyages, as well as church baptism and marriage records, the Allotment Books showing who among the crew assigned part of their wages to family members, and the 1841 British census, together give us some information on quite a number of the crew members and a good picture in the aggregate—how old they were, what their past experience was and where they came from.

If nothing else, the Royal Navy was very good at record-keeping. Perhaps because their lordships' view was that ordinary sailors were not to be trusted and might desert at the first opportunity, the Description Books, which would be consulted when searchers looked for a deserter, recorded such things as height, complexion, smallpox status—vaccinated or scarred—and even described the crew members' tattoos. So, what can we learn from Lloyd-Jones's heroic dives into the minutiae? Here are some of the facts, in no particular order:

First, while most Navy ships carried very young recruits—remember, Franklin, Crozier and many of the other officers had signed on to a ship while in their early teens—the average age of the sailors on the Expedition was 28, and there were only four "boys" listed on the rolls of the two ships: George Chambers and David Young on *Erebus* and Robert Golding and Thomas Evans on *Terror*, and all were listed as being at least 18 years old.[62] Similarly, there were no Ordinary Seamen, the most junior, least-skilled, lowest-paid rank; all the crew were listed either as Able Seamen, which implied at least some previous shipboard experience, or as Petty Officers with a defined specialty. In fact, at least a few who had been listed as Ordinary Seamen on earlier ships signed on to the Expedition as stewards, or domestic servants, a lower-rated specialty, because the men working on deck were expected to be highly skilled, or at least highly experienced.[63]

And while 14 crew members on each ship were described as "first entry," meaning they had no prior Royal Navy service, that didn't mean they were lacking in experience on ships. A number of the crew likely had worked on whalers, perhaps in the same polar waters toward which they were headed in 1845, and many of the others were from Deptford, Woolwich or other port and dockyard towns, suggesting they had at least some acquaintance with ships. Most of the sailors hailed from the southeast counties of Middlesex, Surrey and Kent, rather than from the industrializing Midlands, and there were 15 Scots (including Royal Marines), of whom two were Orkneymen whom Fitzjames allowed to go into port to see their families just before the ships left for Greenland.[64] Lloyd-Jones concludes that, excluding the officers and the first-entry men—some of whom may well have had whaling experience in the Arctic—there were 62 sailors on the two ships who, among them, had experience on some 41 different Royal Navy ships.[65] We know for certain that nine of the crew, including the two cooks, John Diggle on *Terror* and Richard Wall on *Erebus*, had been on Crozier and Ross's 1839–43 Antarctic journey, five of them, including the Boatswain's Mate Thomas Johnson and Captain's Steward Thomas Jopson, sailing on *Terror* both times.

Only a few of the crew, apart from Crozier himself, were Irish; the Allotment Books list William Pilkington, Cornelius Hickey and Marine James Daly as being Irish.[66] Those three may or may not have been Catholics, who were a distinct minority in the Navy at the time, but William Hardy, the bosun's mate on *Erebus*, certainly was, having married his wife Mary in the Roman Catholic chapel in St. George's Fields, Lambeth, just a month before the ships sailed.[67]

While the officers, especially those chosen from among Fitzjames's pals, may have been lacking in relevant experience, the crew as a whole was not. Lloyd-Jones concludes from his analysis that "the [crew] were very well aware of the dangers that they faced, very well qualified to undertake them, and tragically unfortunate in their lack of success. Failure in no way diminished their heroism."[68]

Maybe. But, whatever their previous experience and accomplishments, nothing could have prepared them for the trials of three successive winters in the darkness and the ice, two of those trapped out in the ever-threatening pack ice, and for the travails of a near-hopeless march to an imaginary salvation.

In addition to the naval officers and sailors, each ship carried a detachment of seven Royal Marines—a sergeant, a corporal and five privates. The Marines had been created as a separate branch of the service only in the mid–18th century. While nominally part of the Army, and armed, like the regular infantry, with "Brown Bess" muzzle-loading muskets, they served under the orders of the naval officers on their ships. On the Franklin Expedition, the 14 Marines were on average a bit older than the sailors, as befits a force intended to enforce discipline and prevent mutiny. How well they performed that task, once the men had left the ships and started trekking across King William Island, is another of the things that will never be known.

And that is about all we know of the lives and character of the men who served under Franklin. We know a little more, but not much, about what happened to them after they died. We know for certain where three of the crew members ended up. John Hartnell, an able seaman on *Erebus*, and John Torrington, the leading stoker on *Terror*, whose illness was apparently a cause for complaints from his boss, James Thompson, died in January 1846, and Marine Private William Braine died in April that same year. All three were buried in marked graves on Beechey Island, and their bodies have since been examined and analyzed for hints as to their causes of death.[69] Apart from those three, few of the men who sailed with Franklin have been identified in death.

Of the 105 officers and men who quit the ships in April 1848, the bones of perhaps as many as 92 have been found or reported, if, that is, one includes remains found by Inuit (many of which have since disappeared), bones found by early searchers like McClintock and Schwatka, and more recent finds by the many archaeologists and other searchers who have reexamined King William Island and adjacent areas since 1980.[70] Perhaps 30 to 40 individuals at the Terror Bay camp; it has long since been obliterated by the sea, but Inuits reported finding, in the 1840s and 1850s, many bones with knife marks indicating cannibalism. There are, in addition, the remains still extant of perhaps as many as a dozen at Starvation Cove across Simpson Strait from King William Island. Other bones were found in locations scattered all the way from the April landing site near Victory Point, sometimes called "Crozier's Landing," to as far east as the Todd Islets in Chantrey Inlet, almost directly across King William Island from the original landing site, and on the mainland as far south as five miles south of Starvation Cove.

But only a few of those bones can be reliably linked with specific individuals. There's the grave long identified as that of John Irving, but, as noted above, that grave may or may not really be Irving's, despite the presence of his Royal Naval College maths medal. Then there are the two graves found at the site called by the Inuit Setumenin, toward the southeast corner of King William Island. One of those skeletons was recovered by Charles Francis Hall in the 1860s and eventually sent back to England, tentatively identified as Lieutenant Le Vesconte, on the basis of its prominent nose and massive lower jaw. It was buried under the Franklin Expedition monument in Greenwich and thought to be Le Vesconte's until 2009, when renovations to the monument allowed the remains to be exhumed, leading to an archaeological study and facial reconstruction that tentatively re-identified the body as that of Harry Goodsir.[71]

Another of the skeletons tentatively identified over the years as Harry Peglar, captain of the foretop on *Terror*, was found by McClintock in 1859. Along with the skeleton was a wallet containing a collection of papers, including the seaman's certificate for Peglar. These aptly named "Peglar Papers" have provided ample fuel for complex exegeses and theorizing by scholars and Franklin aficionados over the years; the papers themselves are a confusing mix of partial sentences and apparently random thoughts, written both forward and backward.[72] But whether the skeleton was actually that of Peglar is a different question. The scraps of uniform found with the skeleton were those of a personal steward or servant, not of a crewman who had actual sailing duties. That led Richard Cyriax and A.G.E. Jones, among others, to theorize that the bones might have been those of Thomas Armitage, the gunroom steward on *Terror*.[73] In a remarkable piece of detective work, based on the records of prior shipboard service of Peglar and Armitage, Glenn Stein, an expert on polar medals and their history, argues that the bones could well have been those of William Gibson, the subordinate officers' steward on *Terror*.[74]

Even more recently, scientists have attempted to identify the remains by using samples from currently living descendants of the Expedition crew for DNA analysis. The first of these projects to bear fruit, in 2021, found that a jawbone recovered from near Erebus Bay was that of John Gregory, the engineer on *Erebus* responsible for the locomotive.[75] Gregory was identified based on a DNA swab from his great-great-great grandson, living in South Africa. Presumably, as more relatives of Expedition members contribute DNA samples, there will be more such identifications and perhaps, as in Gregory's case, even facial reconstructions by the archaeologists to bring, if not to life, at least to memory, more of the crew. Perhaps Peglar/Armitage/Gibson and other as-yet unidentified members of the Expedition's crew will be, if not resurrected, at least given some modicum of recognition as this work continues.

And what of those left behind? Thanks to her own copious correspondence, a number of biographies, and even a few novels based on her character, we know a lot about Lady Jane Franklin, and her tireless efforts to send out rescue missions to search for her husband's ships and, later, of her focus on having Sir John appropriately memorialized as the discoverer of the Northwest Passage,[76] but we know very little about the wives and families of most of the rest of the crew.

The hopes and fears of some spouses, siblings and parents come to life in the letters to the missing explorers collected by Russell Potter and colleagues,[77] and, thanks to Erika Behrisch Elce's illuminating research, we now know a little bit about some of the crew members' widows, their fight for recognition and (generally meager) pensions, and the unexpected benevolence of some bureaucrats within the Admiralty who undertook considerable effort to secure a measure of financial relief for those widows when the letter of the law might have denied it.[78] As Elce points out, the Admiralty regulations of the time were exceedingly strict, but, where they could, some civil servants there pushed against the letter of the regulations as far as possible, ignoring the rule that a fixed date of death had to be established and lengthening out some officers' terms so as to satisfy the requirement that an officer must have been commissioned for 10 years prior to death before his widow would be entitled to a pension. The widows of the Expedition's ice masters, Thomas Blanky on *Terror* and James Reid on *Erebus*, were particular beneficiaries of this unexpected largesse, since the ice masters held civilian, not naval, positions and would, under a strict interpretation of the regulations, not have qualified for pensions.

So, this chapter ends where the men ended, dead. But there's more to their story. Let's return to the beginning of the Expedition, sailing off, seeking fame, glory, promotion and double pay.

5

Beechey Island

Beechey Island, where *Erebus* and *Terror* spent the winter of 1845–46, isn't anyone's idea of a vacation spot. At that latitude—nearly 75° North—the winter includes almost three months of unbroken darkness.[1] So what did the Franklin Expedition crew do during their winter at Beechey Island? Here's what we know:

Three of the crew members—*Terror* stoker John Torrington, *Erebus* seaman John Hartnell and *Erebus* Marine William Braine—died. Their graves on Beechey Island have become pilgrimage sites for the ever-increasing number of disaster tourists who, now that global climate change has pretty much eliminated ice in the higher latitudes during the summer, have been arriving by the cruise ship-full, singing Stan Rogers's "Northwest Passage"[2] in the ship's lounge as they go.[3] Fortunately for modern science and for the worldwide coterie of Franklin aficionados, the bodies were buried in permafrost, which is much better than mere ice for preserving human remains.[4] The dead men's hair and fingernails have been put through the machinery of modern science to determine what killed them. (Anti-climax: at least for these first few deaths, it was more likely tuberculosis, or pneumonia brought on and aggravated by tuberculosis, than anything else, not that that conclusion goes unchallenged by the advocates for lead poisoning, zinc deficiency or botulism, among many other possible causes. But more of that later in Chapter 13.)

They left a lot of paper behind. As recently as mid–2019, researcher Alison Freebairn uncovered a trove of paper that the many search expeditions of 1850–51 had found on Beechey Island and that had apparently lain untouched in the Admiralty archives for more than a century and a half.[5] But, while historians are ever hopeful that the very next scrap of paper will be the one that reveals all, the reality is that, while we now have more scraps of papers in Fitzjames's and Alexander McDonald's hands, some lists of numbers that are clearly magnetic observations, and an old chocolate wrapper—all recovered by the search expeditions in the early 1850s and left untouched in a London archive for 150 years—nothing in this latest find appears likely to add very much to what we know. Nor does it subtract very much from what we don't know.

What those papers, at least so far, do not give us is any kind of a record of what the 126 sailors who survived the winter at Beechey Island did over those long dark months. We can only presume that they did the things that we know British sailors always did when wintering over on the Arctic, following well-established routines that had begun with Parry's furthest-west voyage in 1819–1820. They used some of the vessels' sails to create a kind of tent over the decks of the ships, furling and covering the rest of the sails on the yards for the winter or storing the sails below, and

Beechey Island, looking across the ice to North Devon Island. Graves in the foreground (photograph by Ansgar Walk, August 1997; Creative Commons Attribution-Share Alike License 2.5. Wikimedia Commons).

they packed the decks in snow, topped with a mixture of gravel and ashes to provide a safe footing for those up on deck.[6] They piled snow up all the way round both ships, providing some insulation against the winter wind as well as making the ships themselves a bit more stable. An incidental benefit of the snow-packing exercises was that it provided at least some weeks of useful employment and physical exercise for the crew, who otherwise didn't have a whole lot to do through the interminable winter.

In the nearly three months of total darkness, with the sun below the horizon from November into February, there were two principal dangers: outdoors, it was easy to get lost. Frequent signposts, rope guides, and, when those failed, rocket flares, would have been used to guide wayward crew members back to the ships. Inside, the danger was fire, always a threat on ships, but even more when lights and flames were probably in use much more often than on a voyage through the tropics, and when a source of water to put out fires was not so easily accessible. Most Arctic expeditions dug a four-foot square fire hole through the ice near each ship to reach liquid water, and chiseled it open as frequently as every half hour.[7] Presumably, Franklin's crew did the same while they were anchored at Beechey Island.

Another hazard inside the ships was the incessant condensation; moisture from cooking, drying damp clothing, or merely from breathing would collect on the frozen hard surfaces and then drip down. To limit, though not eliminate, the humidity, Franklin's ships used a modified version of the Fraser patent stove and hot-air piping system, pumping warm air below deck and, incidentally, providing a system for melting snow to provide fresh water for the crew.[8] That system, some argue, may have been a source of the lead poisoning (discussed in more detail in Chapter 13) that was detected in so many of the crew members' remains.

While we don't have records from Franklin's ships to show how the crews spent their days, but we do have a record of the daily routine from Captain Horatio Austin's winter orders for HMS *Assistance*, part of the 1850–51 Franklin search flotilla. It's likely that Franklin's and Crozier's crews followed a nearly identical schedule:

06:00: wake up, lash up hammocks
06:15: all hands on deck; ventilate lower deck
07:45: public prayers
08:00: breakfast, then chores and exercise on the ice
11:30: issue lemon juice
12:00: lunch
12:30: grog, then chores/exercise
16:00: all in off the ice
17:00: dinner
17:45: exercise on deck
18:30: school or entertainment
21:00: lights out[9]

Officers, naturally, had somewhat more flexible schedules, though most had well-defined scientific responsibilities, including conducting the magnetic and climate observations that were a major focus of the Franklin Expedition, as well as leading hunting parties. We believe, for example, that the *Erebus*'s assistant surgeon and

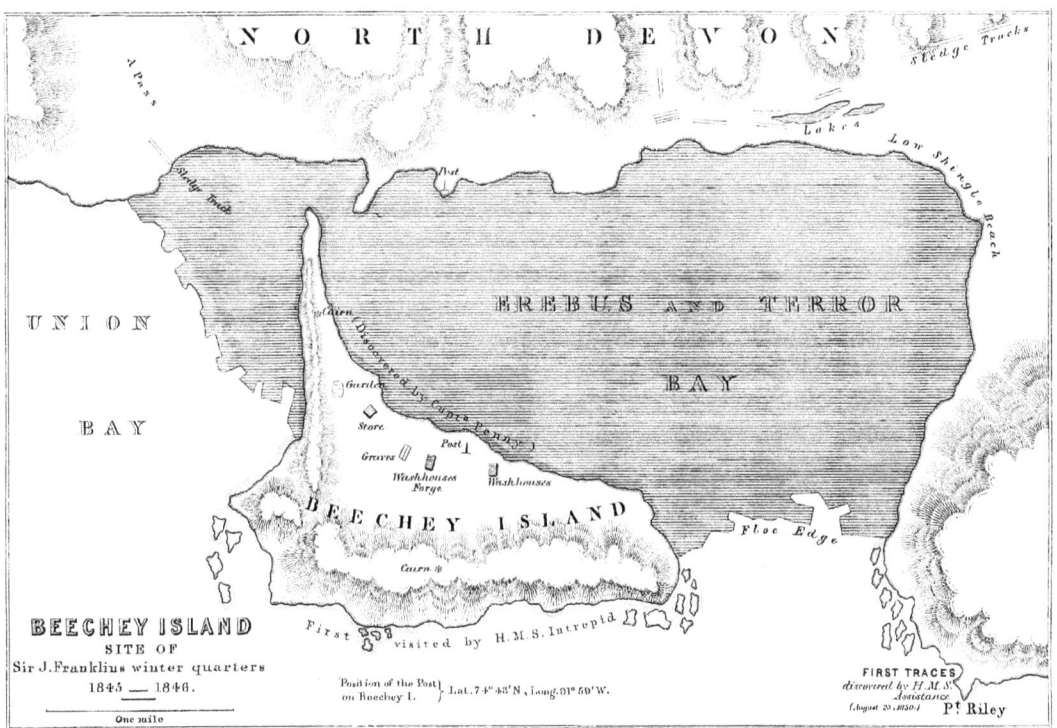

Map of Beechey Island Anchorage, 1845–46 (Franklin Expedition Wiki).

naturalist, Harry Goodsir, commandeered a good portion of Franklin's spacious cabin for his plant and animal specimens and drawings.[10]

The crews set up storage depots on land, offloading the supplies that had been perched precariously on deck since Franklin had sent the *Barretto Junior* home from Greenland. They created a rubbish tip that eventually contained some 700 to 1,000 empty provision tins, making yet another pilgrimage station for future visitors. They almost certainly made thousands of meteorological and magnetic observations, some of which, as well as the instructions for setting up their observatories, apparently survived to form part of the trove recently discovered by Alison Freebairn. As Andrew Lambert observed, "[Northwest Passage] voyagers spent more time recording data than sailing, their long dark winters occupied with endless observations of terrestrial magnetism, astronomy and meteorological phenomena, taking geological specimens while subjecting the flora and fauna to the deadly categorization of the preserved specimen."[11]

In fact, without the magnetic observations, there might well not have been any expedition at all. While discovery of the still-uncompleted Northwest Passage might excite the British public, the commercial irrelevance of the Passage had long since been proven, and British governments were not generally inclined to fund irrelevant projects, despite John Barrow's obsession with the Passage and with the possibility of an open Polar Sea. What made the expedition viable was the likelihood of improving the map of the North American Arctic archipelago and the promise of completing a scientific project, the study of terrestrial magnetism, that had been a focus of the gentlemen-scientists of the Royal Society for the past 25 years. Franklin himself was very much a part of this project, having built a magnetic observatory in Tasmania when he was lieutenant-governor there and having personally participated in taking magnetic measurements of the "term days" when scientists all over the world made simultaneous observations.[12] With Royal Artillery Captain Edward Sabine, an old Arctic hand and adviser to the Admiralty on scientific matters—and later president of the Royal Society—spearheading the case for the Arctic expedition that Franklin eventually commanded; with Franklin's orders specifically urging magnetic observation; and with Sir John's personal involvement in the magnetism effort, we can be certain that there was a magnetic observatory and that quite a lot of the officers' time that first winter was spent sitting there in the cold making the intended observations. Fitzjames, in particular among the officers, had the responsibility for coordinating the magnetism studies and for overseeing the scientific equipment dispatched with the Expedition. But, who knows, perhaps Sir John was himself a regular participant in the magnetic observations, hiking to the observatory and spending time sitting in the cold like the rest of the observers, manipulating the "dipping needle" to obtain—or at least attempt to obtain—a measure of the earth's magnetic field that could then be used to correct the compass readings that, so near the North Magnetic Pole, were all but useless as navigation guides in the Arctic.[13] Crozier, the one person on the Expedition with the greatest knowledge of magnetism—and the greatest recognition by the scientific world for his contributions—had been passed over when Fitzjames was anointed chief magnetologist; it's likely that he sat out a lot of the observation work, sulking, if you will, like Achilles in his cabin.

Did they publish the shipboard newspapers and stage the amateur theatrics that were staples of so many other polar expeditions? Most likely; Sir John was never one to do things differently. In particular, they almost certainly staged a major Christmas celebration, with separate dinners for officers and crew (the officers' dinner was perhaps enlivened by their own stores of wine or even, as Captain Henry Kellett of *Resolute* provided in 1852, a complete Christmas dinner from Fortnum & Mason in London).[14] In any event, there were certainly plum puddings and mince pies aplenty in that first Arctic winter. And elaborate planning for a Christmas pageant would have been a useful way of occupying the crew during the dark days of November and December.[15] As for publications, a shipboard printing press was a common feature on multi-year Arctic voyages, with the occasional broadsheet a source of both useful occupation and often-amusing reading.[16]

We know they put up some buildings on the island.[17] The largest of them, approximately 60 by 70 feet, was probably the storehouse for the extra supplies that had been taken on from *Barretto Junior* in Greenland and had nowhere to go on the winterized *Erebus* and *Terror*.[18] They also built smaller structures for those weather and magnetic observations, a laundry, and apparently a forge for the Expedition's armorer/blacksmith. They erected a mysterious post that was, before the weather knocked part of it down, topped by a black hand with a pointing index finger. But where had that been pointing? Was it just a handy guide back to the ships? Or was it, as Stan Rogers surmised in "Northwest Passage," "the hand of Franklin reaching for the Beaufort Sea,"[19] pointing out the way the Expedition had gone? (For those who think Rogers's song is, well, banal, a better alternative might be Iron Maiden's "Stranger in a Strange Land,"[20] based on the archaeologist Owen Beattie's exhumation of John Torrington, the first of the Expedition crew to die and one of three crewmen buried on Beechey Island.)

Did they follow the typical Arctic practice of launching sledging expeditions to explore their surroundings in the early spring, when sledging conditions are good and before the ice breaks up to release the ships? They must have, for, again, Sir John was a great one for doing what those who went before him had done. But where did they go, what did they find? The evidence, so beloved of historians, is missing, and recent efforts to find traces of such expeditions have come up empty.

So, let's assume that in April and May 1846, as daylight returned and as the ice became more friendly for sledge travel, some of the officers and men of *Erebus* and *Terror* did indeed venture forth, hauling their heavy wooden sledges and seeking to know their environment a bit better. Where might they have gone and how did they get there—and back? And what, if anything, did Sir John do with the information the sledging parties brought back? Did he factor their findings into his plans for the 1846 sailing season, or did he just do his duty, follow his orders and, when the ice broke up in the summer, head west and then southwest, following his orders, into the unknown?

A Digression on Sledging[21]

Travel options in the high Arctic were limited in the mid–19th century. If—a big if—the ice receded, ships might have some sailing room in July, August and September, before the waterways froze over for the next winter. Summer—the period when the ice had receded enough to allow ships to make some progress—rarely lasted more than 10 weeks and was often much shorter. From October through March, in conditions of total or near-total darkness, depending on exactly how far north they were, sailors would have been hard-pressed to venture much farther than from their ship to the depots on the nearby shore or to the meteorological and magnetic observatory on the nearest hill, and then only along well-worn paths, perhaps with stanchions and rope guidelines to lead them from one known location to another in the darkness and the snow. Only in April, May and June were conditions even moderately acceptable for land travel, across both the still-frozen sea ice and on land, which might or might not still be covered with ice and snow.

While a few explorers made these journeys carrying their belongings and food in knapsacks, and a very few attempted to use wheeled carts—with, generally, a distinct lack of success[22]—the vehicle of choice for these springtime excursions over the ice was most often the sledge, a predominantly wooden contraption ranging in size from something resembling the Flexible Flyers of our childhood, perhaps a foot and a half wide and no more than six feet long, to 1,500-pound behemoths of up to 30 feet long and a dozen feet wide, which could then be loaded with ships' boats and up to a ton of supplies.

Sledge-like devices had been used for millennia by the indigenous people of the Arctic. Whether pulled by dogs or by reindeer, or on occasion even by their human makers, small sleds were ubiquitous around Inuit and Indian settlements, carrying slain seals back from the breathing holes where hunters had dispatched them or ferrying an entire family's belongings on the household's seasonal migrations. But Inuit sleds had a few characteristics that, despite their obvious utility, weren't always imitated by the newly arrived Englishmen. The indigenous sleds were mostly made from bone—whale or caribou, generally—which was much lighter than the wooden contraptions nailed together by the English. Moreover, the Inuit sleds were lashed together with animal sinew, making then a good deal more flexible than the rectangular, nail-fastened English constructions.

Sledges, especially when heavily loaded, are hardest to move when starting from a dead stop or when moving uphill, even at a relatively slight angle. Naval historian Richard Cyriax tells us that large, heavily loaded sledges sometimes required that the dozen or so men on the haul ropes be pulling as much as 268 pounds apiece.[23] One commander in the Franklin search effort concluded that man-hauled sledging was the hardest work that free men have been put to in modern times.[24] Not an easy task for men who were already likely to be undernourished and certainly under-vitamined at the end of a sunless Arctic winter. A popular prayer when a lengthy sledge journey was about to be launched was "O Lord of life and death ... have mercy on those that are appointed to die."[25]

Some British explorers had followed the Inuit example and used indigenous dogs for pulling their sleds. As early as 1818, William Edward Parry had seen these Inuit sleds in action, when he accompanied John Ross. On Parry's own first voyage as commander, in 1819–20, he made much use of small sleds, perhaps seven feet long and no more than two feet wide, that could be either pushed or pulled by one man. But they were mostly used for short trips to move supplies and for hunting parties, and dogs were not used to pull them.

Parry himself undertook one major exploring journey, in the late spring of 1820, from Winter Harbor to the north coast of Melville Island and back. But rather than sleds, he used a wheeled wooden cart, which, as one might expect, broke down during

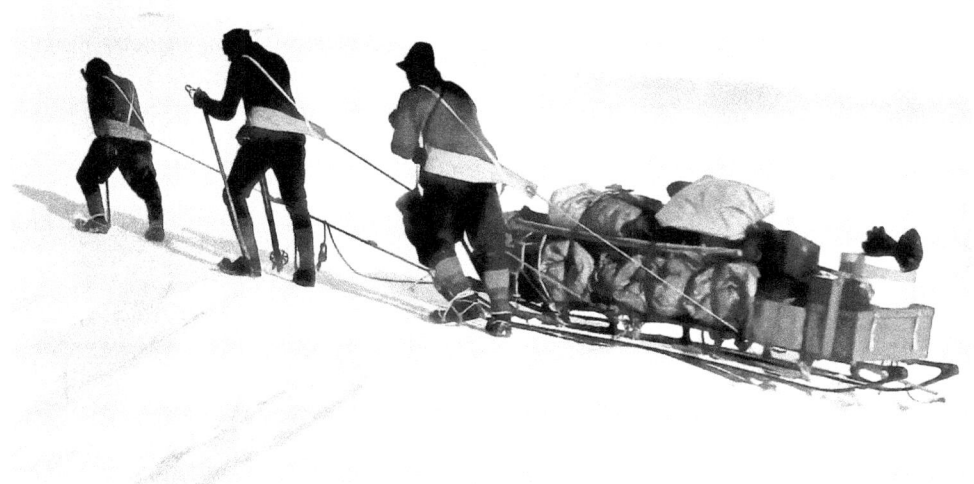

Men hauling a sledge during Scott's Terra Nova Expedition (photograph possibly by Lawrence Oates; Wikimedia Commons).

the trip, requiring the men to carry what was essential on their backs and to leave a portion of their supplies behind.[26]

On Parry's second voyage, in 1821–23, he sent his second-in-command, George Lyon, along with seven other crewmen on one extended trip, with each man pulling his own personal sledge, loaded, in the case of officers, with 90–95 pounds, and, for the "other ranks," with roughly 120 pounds; the British class system is a hardy perennial. Traveling once again toward Melville Island, Lyon intended to be gone for three weeks, but returned after just 14 days, nonetheless recording the longest sled journey (180 miles) up to that time by a Royal Navy officer.

Finally, during the winter of 1822–23, while wintering over at Igloolik, near the Melville Peninsula, Parry and Lyon bought sled dogs from the local Iglulingmiut, learned to drive the dog teams and made a number of short trips. But the one long journey, an attempt to reach the open sea that the Inuit said was on the western side of the Peninsula, was cut short because of the weather. It's not clear from Parry's sanitized and edited official narrative of the trip[27] whether the dog-drawn sleds proved unsuitable or whether the trip was simply bedeviled by bad weather. Parry tended to send out sled journeys late, usually not leaving the base camp until June, when meltwater made traveling over the ice messier and slower than earlier in the spring, when the ice would have been firmer, so perhaps his relatively unsuccessful trips were just the result of bad decision-making.

Although not part of the search for the Northwest Passage, Parry's subsequent expedition in search of the North Pole solidified the Royal Navy's predilection for using heavy, man-hauled sleds. On that trip, beginning in June 1827, Parry first tried wheeled and reindeer-drawn sleds, then, when those efforts resulted in no movement at all, switched to heavy man-hauled sledges carrying wooden boats; fully laden with supplies as well as the boats, each sledge likely weighed more than a ton.

Even after adopting James Ross's recommendations for lighter, more maneuverable sleds, the Navy still pushed the limits. For example, the multiple sledging expeditions in the spring of 1851, launched by the large official Royal Navy Franklin search flotilla, averaged 1400 pounds per sledge, including a tent (good Englishmen wouldn't just make snow houses on the model of Inuit dwellings), sleeping bags, cooking gear, food and fuel.[28] Even without carrying a boat on the sledge, that cargo list added up to 200 pounds or more per sailor, still far more than made sense for fast, safe travel. Despite those deficiencies, the 1851 trips did add some 1,200 miles of coastline to the maps and, by concentrating on destinations to the north and west of Beechey Island, and finding nothing, effectively ruled out those areas as possible Franklin destinations or stopovers, although Lady Jane and others in England kept the northern search concept alive for some years more.[29]

Despite the proven advantages of using dogs to pull lightweight sledges, the Royal Navy generally stuck with larger, man-hauled equipment. Pierre Berton suggests that "it was a form of cheating to use animals for transport ... to the English there was something noble, something romantic, about strong young men marching in harness through the Arctic wastes, enduring incredible hardships with a smile on their lips and a song in their hearts."[30]

Franklin would have used only man-hauled sledges; he didn't have sled dogs with him when he was last seen in Baffin Bay, and there are no Inuit reports of his having acquired any after that. The many search expeditions that followed in Franklin's wake often did use dogsleds, following Parry's example, although as many as 70 percent of their sledge journeys were carried out in man-hauled vehicles,[31] but Franklin himself could not have. And most of those dogsled trips undertaken by the search expeditions seem to have been for message-bearing communications trips, rather than for the longer search-and-explore journeys that the Franklin Expedition may have carried out in the spring of 1846, when perhaps Sir John sent out sledge parties to scout the best route for the ships to take once they were freed from the ice around their winter harbor.[32] In Franklin's defense, these large sledges fit well into the Royal Navy's model of operations: each sledge could function as an independent command, almost like a small ship run by a lieutenant, flying its own pennant and crewed by six or eight hearty seamen. And sailors, of course, were well accustomed to hauling on ropes.

In contrast to the Royal Navy's approach, which one commentator described as "[r]ather stupid but faintly admirable,"[33] Hudson's Bay Company emphasized traveling light and living off the land:

> The ability to live off the country, depending on a gun, a trap, and a fishing net, allowed the Hudson's Bay men to preserve their pemmican for emergencies, to lighten their loads and to travel without the cumbersome equipment that other expeditions required. The techniques and hardihood of the traders-turned-explorers was exemplified in the voyages of Dr. John Rae.... Twice during his career Rae walked 1000 miles within two months, and he had the reputation of being able to walk 100 miles in two days.... Thomas Simpson's capacity to cover a steady 20 miles a day on foot along the Arctic shore greatly enlarged the range of the Dease-Simpson expedition.[34]

Back to the Facts

About those 700—or was it 1,000?—empty tin cans that were found piled up in the rubbish tip on Beechey Island, discovered by the first of the search expeditions. Several writers have suggested that the number of empty tin cans found on

Beechey Island implies that Franklin's men had found problems with their food supply and emptied out much more than what they actually would have needed as food.[35] But more realistic historians of the Expedition have done the calculations and have decided that 700, or even 1,000 or more, tins would not be an unreasonable amount to have been used up feeding the large Expedition crew over the winter.[36] And no clear evidence has emerged from the discarded tins on Beechey Island that would suggest that either the tinned food itself or even the lead from the soldering used on the tins had put the Expedition in peril before *Erebus* and *Terror* sailed away from their winter harbor in the summer of 1846. Insofar as ships' crews on extended discovery expeditions were ever healthy, we can safely assume that, when the sun emerged in March 1846, Franklin's crew was still in adequate condition to go on with its quest.

And we know, thanks to Alison Freebairn's discoveries in the archives, that they did indeed make those meteorological and magnetic observations, as they had been commanded to. One can imagine Fitzjames and other officers spending frozen nights and days (though how would one tell the difference in the three months of Polar darkness?) with their dip circles, trying to measure the distance from the magnetic pole,[37] perhaps dazzled by the aurora borealis, perhaps just sitting in the cold, waiting for their shift to be over so they could return to the relative warmth of the ships.

The Royal Navy had already established that spring was for sledging. So, let's assume that the spring of 1846 did see sledge expeditions. Where might they have gone and what might they have found?

The objective was the Northwest Passage. So they must have explored to the west of their winter home on Beechey Island. Behind them, to the east, lay well-known sea routes and adequately mapped points of land—no point in exploring what they already knew. And to the northwest lay Cornwallis and Bathurst Islands, stepping-stones to Parry's furthest west on Melville Island, and from there only a short, if inconveniently ice-clogged, at least until the global-warmed 21st century, sail to the Beaufort Sea, the Bering Strait and the riches of the Orient. As the Victory Point Record states, in late 1845, just before going into winter quarters, they had sailed north through the Wellington Channel, then completely around Cornwallis Island, establishing both that the latter was indeed an island and that, apart from a narrow band of navigable water near its northern edge, it was bounded further north by an impassible ice barrier. Returning that previous summer around the western and southern shores of Cornwallis, they had satisfied at least one portion of the Admiralty's instructions and had found no reason to think that the Northwest Passage would lie in that direction unless, like Fitzjames, one was an evidence-averse fanatical believer in the chimera of the open Polar Sea.[38] So the logical direction to look would have been to the south or southwest, following their orders.

But remember the chart that the Expedition carried? To the west of their winter anchorage, it showed a watery path leading at least as far as Parry's furthest west on Melville Island. To the southeast, it showed Prince Regent Inlet, leading down to Fury Beach and beyond to John Ross's winter harbors, though the southern edges of the inlet were not known. Directly south was North Somerset Island, and a bit west of that, the last landmark identified in the Admiralty's sailing instructions, Cape Walker. As it turned out, Cape Walker was itself on a relatively small bit of land, Russell Island, just north of the much larger Prince of Wales Island. But, as far as Sir John and the Admiralty knew at the time, it could well have marked the entrance to a lane

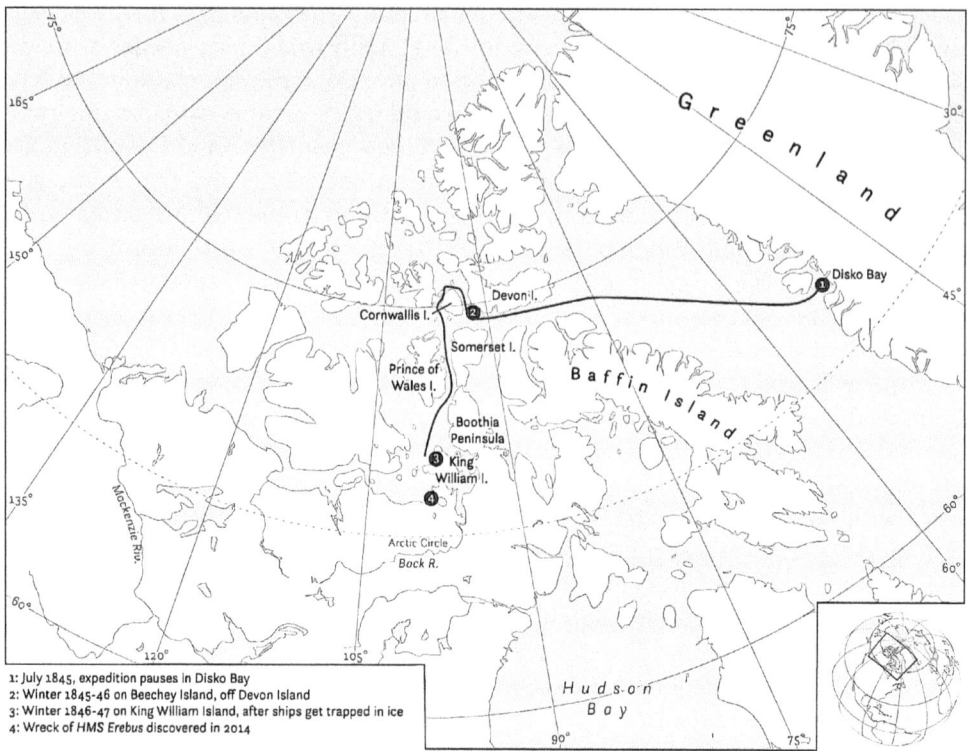

1: July 1845, expedition pauses in Disko Bay
2: Winter 1845-46 on Beechey Island, off Devon Island
3: Winter 1846-47 on King William Island, after ships get trapped in ice
4: Wreck of HMS Erebus discovered in 2014

The Franklin Expedition's most likely route (map by "Smurftrooper," Creative Commons Attribution 4.0 International License; Wikimedia Commons).

of open water leading to the Pacific. Such sledge journeys as the Franklin Expedition mounted in the spring of 1846, though unrecorded in the meager documentation that has thus far emerged, must have been undertaken with the aim of discovering a possible route toward the southwest.

Who ventured out with those sledges? And did those exploring parties, each likely headed by a lieutenant, explore the entrance to Peel Sound, between North Somerset and Prince of Wales Islands? Did they go even further, seeking to update Ross's old map that suggested King William Island was connected to the Boothia Peninsula, and thus there would be no route all the way down the sound to the (presumably) open water along the continental coast? Did they look for a possible winter harbor for 1846–47, in the lee of an island and protected from the surging highway of ice in the McClintock Channel and Victoria Strait? How close to the edges of that black, unknown quadrilateral on the Admiralty charts did they reach? What did they find?

6

Westward Ho?

So where did Franklin go when, sometime in the summer of 1846, he left the Expedition's winter quarters on Beechey Island? His orders, as we've seen, told him to head generally southwest, through what Barrow and the Admiralty erroneously believed to be open water pretty much all the way to the North American mainland.[1] And Franklin hoped to find a navigable strait somewhere between the known land masses of Banks Land (now known to be an island) and Wollaston Land and Victoria Land, now known to be the southwestern and southeastern appendages, respectively, of the much larger Victoria Island.[2] As Robert McClure discovered, there is in fact such a channel, now named Prince of Wales Strait, but it wasn't all that easy to get to it from the east, through the often ice-choked Viscount Melville Sound, and McClure, having sailed into the strait from the west after rounding Alaska, was forced to lead his crew on a walk across that channel over the ice in 1853, rather than sailing through it.[3] For his pains, McClure was awarded a share of the British government's £10,000 prize for "discovering" the Northwest Passage, an honor much resented by Lady Jane Franklin. The wreck of McClure's ship, the *Investigator*, was finally discovered in 2010 by Parks Canada. Like Franklin, McClure lost his ship to the ice; unlike Franklin, he managed, along with most of his men, to walk away from the disaster, the beneficiary of both somewhat better planning—he gave up and started looking for rescue a bit earlier than Franklin, even so losing a number of men to scurvy, malnutrition and the cold—and simple good luck; McClure and his men were found by a sledge party from HMS *Resolute*, itself one component of the numerous missions that had by that time been sent out in the early 1850s to rescue the Franklin crew. *Resolute* herself was trapped in the ice and didn't make it out of the Arctic right away—McClure and his remaining crew members traveled back to England in other vessels, long before the ice-bound *Resolute* was recovered by an American whaler in 1856 and returned to England. Eventually its wood was used to make the "Resolute Desk" that was presented to U.S. President Rutherford B. Hayes in 1880 and has been in prominent use in the White House, serving as the official desk for most Presidents in the past 60 years.

Franklin's orders were to go as far west as Cape Walker, on the southern margin of Barrow Strait just north of Prince of Wales Island, and then head southwest as best he could. Most histories of the Expedition assume that *Erebus* and *Terror* sailed south through Peel Sound, which actually lies to the east of Prince of Wales Island, and only then headed in a more westerly direction, emerging from the somewhat protected waters of the Sound in time to be locked in the murderous ice stream flowing down from the northwest through the much larger McClintock Channel into Victoria

Strait. There's no documentary or physical evidence to support the choice of the Peel Sound route—no cairns along the way, no references to the route in the Victory Point Record—and Franklin wasn't one to disregard his orders (especially if, as seems likely, he personally had a hand in drafting them). What if he did what he'd been told to do and sailed west, at least as far as Cape Walker? In fact, there was no *a priori* reason to suspect that the McClintock Channel was a worse route than Peel Sound, though in actuality that turned out to be the case. Another Franklin searcher, William Browne, part of Captain Horatio Austin's squadron in 1850–51, also concluded that the best approach to a Passage would be the direct southwest route and not Peel Sound.[4] And one scholar writing not long after the Expedition was lost hypothesized that Franklin had, in fact, sailed west toward Viscount Melville Sound, that the Expedition ships had been abandoned somewhere in that vicinity and that the ships, by then unmanned, had been pushed back eastward all the way through Lancaster Sound and out towards Newfoundland's Grand Banks.[5] Interesting theory, but rather conclusively disproven by the discovery of the actual wrecks of *Erebus* and *Terror* off the coast of King William Island a few years ago. But at least so far nothing, including the recent discovery of Franklin's ships, has either proved or disproved the Peel Sound hypothesis. The Victory Point Record merely identifies the location where the ships were beset in the ice in September 1846, without stating how they got there. And if any location records were, as ordered, thrown overboard during the 1846 sailing season, none have been recovered.

Could Franklin have headed west into Viscount Melville Sound, aiming for the channel originally spied by Parry's expedition 25 years earlier, to the east of Banks Island (the channel, explored by McClure in 1850–51, is now known as the Prince of Wales Strait), and then been forced back by the ice drift down McClintock Channel? Certainly not impossible. In fact, for several reasons, that would have been a more likely route for Franklin to have followed. And if, in hindsight, we now know that heading vaguely south, along Peel Sound, protected from the massive ice floes by the bulk of Prince of Wales Island, would have been a better

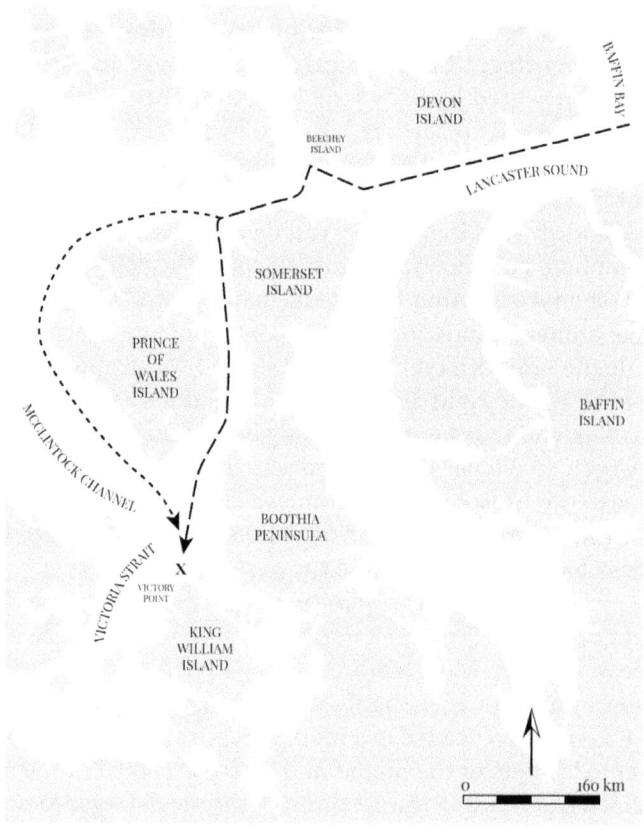

Peel Sound and alternative routes (map by David Veller).

route to the southwest than the open water/ice of McClintock Channel, well, Sir John didn't have the benefit of that hindsight.

Instead, Sir John likely would have relied on the wisdom of earlier Arctic explorers, preserved in their published narratives, which he and his officers had undoubtedly read and which were surely among the many volumes carried on board *Erebus* and *Terror*. A quarter-century before Franklin set sail, Edward Parry, in his 1819–20 journey in *Hecla* and *Griper*, had sailed through Lancaster Sound and Barrow Strait, as far west as Melville Island. He was stopped there by solid pack ice, which he believed might be a permanent obstacle to reaching the Beaufort Sea via a northern approach. Drawing a lesson for future seekers of the Northwest Passage, Parry suggested that they stay away from open water and look for navigable channels near coastlines, where the water might be deep enough for ships to navigate but shallow enough that the heavy multi-year pack ice would be grounded some distance offshore, leaving a lane, at least in the summer, of relatively open water.[6]

Had Franklin followed Parry's advice about staying close to the available coasts, there would have been five possible routes: (1) south through the relatively narrow Peel Sound, then along the west coast of Boothia and, at Cape Felix on King William Island, choosing either the east or the west coast of that island; (2) past Cape Walker and then down the west side of Prince of Wales Island and across a bit of open water as far as Cape Felix, then around King William Island; (3) west of Cape Walker into the McClintock Channel and then down the east coast of Victoria Island; (4) across McClintock Channel and then along the north coast of Victoria Island. That route would have eventually led into what would later be McClure's path at Prince of Wales Strait and the route that Henry Larsen followed in 1944 when he completed the Passage. Larsen, it should be noted, made the crossing—in a climate that was likely much warmer than that which Franklin faced—in a mere 18 days[7]; or (5) following Parry's 1819 route past Melville Island and then out through McClure Strait north of Banks Island into the Beaufort Sea. And, remembering that the 1840–1860 period was likely one of the coldest eras of the millennium,[8] it would be well to keep in mind Larsen's own caution that "[o]ur voyage showed that the Northwest Passage can be traversed in a single year, but does not prove that this could be accomplished every year."[9] We don't know which of these routes, if any, might have been open in the summer of 1846, and thus we don't know whether Franklin made the wrong choice or was just unlucky, making the attempt in a year when none of the choices would have been the right one. Would some of them have allowed Franklin to find a safe winter harbor in 1846–47, before the ice locked the ships in again? Unless Francis Crozier's captain's logs miraculously appear in readable form from their supposed resting place in the captain's cabin of the sunken HMS *Terror*, there's no way of knowing.

That *Erebus* and *Terror* ended up in none of those more likely places, but rather in the middle of an ice-choked Victoria Strait, suggests either that Parry's advice was based on his observations in what would turn out to be an abnormally warm year (1819), or that Franklin, ignoring the wisdom of prior explorers, but heeding the intuition of John Barrow[10] and his own belief that any Passage to the southwest would lie far to the west and could be reached by following a mostly coastal route west of Cape Walker,[11] blundered into a deadly trap. As it turned out, the hoped-for land west of Cape Walker didn't exist, and so no coastal route was available. Did that mean Franklin turned down Peel Sound, to the east of Cape Walker, instead? His own orders, which he had helped to draft, told him not to explore any openings either to the north or the south until he had passed westward of Cape Walker, and then, if blocked to the southwest, to look north, instead, to the Wellington Channel and Melville Island. But, as the Victory Point Record tells us, the Expedition had already done its northward foray, circumnavigating Cornwallis Island in 1845 before returning to winter quarters at Beechey Island. So, no particular reason to look north another time.

If Franklin did sail down Peel Sound and emerge into open water north of King William Island, he must have been very lucky. Various of the Franklin search expeditions tried the Peel Sound route in 1849, 1851, 1852, 1858 and 1859, and in each of those years found it thoroughly blocked by ice.[12] Given the research suggesting that 1845–50 was a particularly cold moment in the Arctic, it's puzzling that the modern consensus so thoroughly embraces the Peel Sound route. Even McClintock, whose search expedition was the first to find *in situ* relics of the Franklin Expedition, eventually concluded that Peel Sound, which McClintock himself had tried and failed to navigate the previous year, was likely to be navigable only once every four or five years,[13] even though McClintock had just a few years earlier confidently stated that Franklin had, in 1846, sailed down a largely ice-free Peel Sound and then, "leaving clear water behind him," sailed on into the pack in Victoria Strait.[14] But a sledging trip down part of Peel Sound on James Clark Ross's Franklin search expedition in 1848–49 changed McClintock's mind, leading him to conclude that any attempt to take a ship through that route "would not only fail, but lead to an almost inevitable risk of destruction."[15]

As for Victoria Strait itself, in the wake of the Franklin disaster, mariners stayed as far away as they could from the site for most of the next century.[16] With the impact of climate change and the advent of larger ice-breaking ships, however, by the 1960s Victoria Strait had become the most-used route through the Passage, for all but the very largest ships.[17]

Some four decades ago, three researchers tested—or at least attempted to test—the hypothesis that Franklin had sailed west of Prince of Wales Island into the McClintock Channel, rather than heading south through Peel Sound, as postulated by what we might call the Standard Model of the Franklin Expedition.[18] A field trip by Clifford Hickey, James Savelle and George Hobson in 1982 sought cairns or other evidence that Sir John's ships had tried the western route, near the east coast of Victoria Island. The western option was certainly not unimaginable. Indeed, Parry had gone even further west, although to the north of Victoria Island, way back in 1819, passing 110° West before settling in at Winter Harbor on Melville Island. And, despite near-consensus support among Franklin researchers for the Peel Sound route, there's no tangible evidence for that direction either. All we know, from the Victory Point Record, is that the ships were locked in the ice off the northwest coast of King William Island by September 1846. How they got there is unknown, no matter how certain some of the experts may be that the Expedition took the Peel Sound route.

But as often happens in scientific work, Hickey, Savelle and Hobson's 1982 trip to test their hypothesis was inconclusive. No cairns, no Expedition relics washed up on the shore of Victoria Island, or at least not on those portions of the shore where the researchers were looking, no graves. In a word, nothing to prove the existence of Sir John's putative western route. But absence of evidence is not, as we all know, evidence of absence. Might it have happened?

6. Westward Ho?

The lack of physical evidence of a Franklin expedition visit to Victoria Island doesn't mean it didn't happen. By way of example, Robert McClure's *Investigator* was abandoned in 1853 in Mercy Bay, on the north shore of Banks Island, but before leaving the ship, the crew cached thousands of pounds of supplies on the shore for the use of future explorers in that exceedingly remote place. And a subsequent expedition sent from Henry Kellett's *Intrepid* the following year found the *Investigator* still (barely) afloat and removed even more thousands of pounds of supplies, adding them to the onshore cache. At some point after that, the Copper Inuit from the mainland discovered the cache and, according to their oral history, began regular visits there while on their annual hunting trips to the north, retrieving whatever of the food and supplies seemed usable. By the time the next non–Inuit visits the locale—Second Officer O.J. Morin of the Canadian Coast Guard's ship *Arctic*, on a sledge journey in 1909—there was nothing left of *Investigator* or of the large cairn left to mark the site of the offloaded supplies. Just some loose coal, a few barrel staves and fragments of sail and rope.[19] A century on from that find, the Arctic will have taken away all signs of what might have been. So, a late 20th-century search for the remains of a mid-- 19th-century expedition might well have come up empty, especially when the search could, inevitably, cover only a small portion of a very large and mostly uninhabited area.

So, what if Franklin, like McClintock and Browne in subsequent years, had taken a look down Peel Sound, perhaps even sailed some distance into it, but had become convinced that further progress down the Sound would lead only to premature entrapment in the ice? Might Franklin have decided to turn around, head back into Barrow Strait and then southwest, perhaps hoping to outrun the stream of ice that blocked the way into Viscount Melville Sound? And how, from there, did the ships end up beset in Victoria Strait that September? If they had managed to sail further west, within range of the Victoria Island shore, might they have found a safe winter anchorage? What if?

Ideally, Sir John would have sent out exploratory sledging missions from the Expedition's winter quarters on Beechey Island in the spring of 1846, perhaps as far as the southern end of Peel Sound and the southwest corner of Prince of Wales Island. If they had made it that far, the officers in charge of the sledges might have returned with actual geographic knowledge to supplement the imaginings recorded on the Admiralty charts. But did the sledges go that far? Did they return with useful information? Or was Franklin left to sail off to the southwest holding little more first-hand knowledge than he had when he left Greenhithe in May 1845?

7

Which Side of King William Island?

When in 1846, and in which direction did Franklin's Expedition leave Beechey Island? How late in the season did the ice break up, and where did the breakup allow *Erebus* and *Terror* to go? We have already referred to evidence that the mid–19th century was a bit colder in the high Arctic than was the 20th century—and quite a bit colder than the Arctic, as recently affected by global warming, is today.[1] One meta-survey estimates that winter temperatures at the time Franklin sailed were anywhere from 1–4 degrees Celsius (roughly 2–7 degrees Fahrenheit) colder than in the 20th century.[2] More specifically, it seems to have been a lot colder in the winter months then than it is now, and perhaps a bit warmer then than the 20th-century average in the spring, when the crew would have been making sledge journeys and would possibly have been impeded by slushy conditions. The best sledging was done when the ice was still firm, but late enough into the spring that there would have been abundant daylight and that the temperatures would have been warm enough to permit extended outdoor activity, that is, in most years, from late April through early June.

Judging from the lack of records left behind at Beechey Island to guide future search missions—not even a finger-post pointing in the general direction that Franklin had traveled—and from the odd bits of personal gear that were left behind on the island, including some officer's cashmere gloves, it seems that perhaps the ships left their first winter's anchorage in something of a hurry, taking advantage of a sudden breakup of the ice. If so, had the Expedition cut its sledging expeditions short that spring, limiting the amount of new geographical information that the officers could incorporate into their planning? Or maybe Franklin and his crew were just bad at record-keeping and personal organization. Certainly, the winter-quarters location was somewhere that would have been a logical place to leave not merely a record that the Expedition had been there, but some hint of their plans for the coming summer sailing season. But the Beechey Island site, visited by hundreds of would-be Franklin rescuers just a few years later, yielded nothing that gave the slightest suggestion of where the Expedition had headed or which route they planned to take.

John Rae reported that he saw Canada geese migrating north, a harbinger of summer, as early as April 30, 1846, but that was at York Factory, midway down the western shore of Hudson's Bay, well south of the latitudes where Franklin's crew was sailing. Even as far south as York Factory, however, the ice did not break up enough that year to allow Rae's two small (22-foot) boats to set out into the bay until June 13.[3]

At Beechey Island, some 2,000 kilometers (roughly 1,200 miles) north, the breakup of the ice would have been much later, and Franklin's two large ships would have required a good deal more open water in which to navigate than did Rae's small boats. Late June? Unlikely. Late July or even August? More probable, leaving a sailing season of less than two months at most, more likely as little as a month. Remember that the Victory Point Record reported that the ships had been beset in the ice since September 12, 1846; did a sailing season of perhaps six weeks leave time for the Expedition to explore the available alternatives?

Since the time of Franklin and McClure, by which point the British had completely given up on the idea of a useful trade route through the Arctic, the world has warmed, the ice has receded, and the Passage has, for better or worse, opened up. Through the end of the 2022 navigation season, there have been 351 successful transits of the Passage, 200 heading west, as Franklin did, and 151 heading east, as McClure attempted.[4] And that doesn't include submarine voyages, mostly through the deep-water route that parallels Edward Parry's successful first voyage, via Melville Island and then under the ice into the Beaufort Sea, that the U.S., British, Russian and presumably Canadian navies prefer to keep secret (though the Canadians may not have nuclear submarines capable of the long under-ice journey).[5] From the time of the Franklin Expedition in 1845 until 1950, there were only three successful voyages: Roald Amundsen's historic trip in 1903–1906 and two by Royal Canadian Mounted Police officer Henry Larson in the 1940s From 1950 to 1975, there were 13 transits of the Passage, almost all by icebreakers or their support vessels, including the largely symbolic voyage of the oil tanker *Manhattan,* carrying a single barrel of oil from Alaska's North Slope. Then from 1976 through 2000, the Passage began to be a regular route: some 58 ships made the journey, including small yachts, some of them sailed single-handed—Kenichi Horie of Japan was the first, taking three years in 1979–82 and wintering in Resolute on Cornwallis Island and Tuktoyaktuk in Canada's Northwest Territories—passenger vessels, beginning with the *Lindblad Explorer* in 1984, first in a long line of ships carrying adventure-bound cruise passengers, and the start of freight traffic through the Passage, when Russia's icebreaker *Admiral Makarov* and the tug *Irbis* brought a steel floating dock from Korea to the Caribbean in 1999. Then, with global warming, came the deluge. Two hundred seventy-seven of the 351 successful transits of the passage have been made after the year 2000, 79 percent of the total, with a record 27 in 2022 alone, when traffic rebounded after two COVID-affected years of few attempts. By the end of 2022, a total of 72 passenger ships had made it all the way through, in addition to hundreds of adventure cruises that went part of the way and returned. And, much to the chagrin of John Barrow, if only he had known about it, the Netherlands-based Wagenborg shipping line has established regular freight service through the Passage, to and from Asia to the East Coast of the United States and Europe. In 2022 alone, Wagenborg sent eight ships through the route; 175 or so years late, from Barrow's point of view, but still, the Northwest passage appears, finally, to have become a viable sea route.

The route through the Passage that, most scholars agree, is the one that Franklin himself tried to follow passes through Davis Strait, Lancaster Sound, Barrow Strait, Peel Sound, Franklin Strait (as it's now known), Victoria Strait, Coronation Gulf, Amundsen Gulf and on into the Beaufort Sea. Roughly one-quarter of all Northwest

Passage trips have followed this route, which is suitable for vessels of up to 14 meters (46 feet) draft. The nearby route, which involves going around the east side of King William Island, is much shallower; with a minimum depth of 6.4 meters (21 feet), many shoals and complex currents. The latter is the route that Amundsen and many subsequent smaller vessels have taken, avoiding the danger of an ice-filled Victoria Strait. Taking the two Peel Sound routes together, they account for 154 of the 351 Passages, or 43 percent of the total.

There are at least two questions as to how Franklin and his crew got to where the ships ended up. First, did he sail down Peel Sound, to the east of Prince of Wales Island, as most Franklin scholars believe, or to the west, directly into McClintock Channel and Victoria Strait? Second, however he got there, at some point Franklin must have been within range of Cape Felix, at the northern tip of King William Island. From there, he once again had a choice, stay out in the ice-clogged Victoria Strait or try to go around King William Island to the east, as, more than half a century later, Roald Amundsen eventually did when he made the first successful transit by sea of the Northwest Passage in the much smaller *Gjoa*.

There's no evidence—nothing in the Victory Point Record, no cairns or other indicators of the Expedition's presence—that Franklin even tried sailing east of the island, where he would have been shielded from the river of ice pouring out of the northwest. So we probably must assume that, following orders, he headed straight out into Victoria Strait and an icy entombment?

We've already seen that the Admiralty had instructed Franklin to head southwest from Cape Walker, which would have carried him out into the McClintock Channel and then the Victoria Strait ice stream. Perhaps, as we've said, he was just following orders. And he would have been encouraged to do that by a map prepared by James Clark Ross, who, we remember, had traveled by sledge to Victory Point, on the northwest shore of King William Island back in 1822. Ross's map presumed, on the basis of his observation of some foggy shapes in the far distance, that what we now know as the James Ross Strait was instead only a bay, terminating in some sort of land connection between the Boothia Peninsula and what we now know to be King William Island. That would have made the island a peninsula of a peninsula and precluded traveling round it from the east, as Amundsen eventually did in 1903. The available Admiralty charts of the Arctic that Franklin had with him would have shown a dotted line connecting Boothia with King William Land, at the bottom of an imagined "Poctes's Bay."[6] Franklin would likely have believed both James Ross and his charts, and, avoiding the imaginary dead end to the east, instead sailing west, into the ice.

Of course, had Franklin actually tried to sail *Erebus* and *Terror* to the east of King William Island, it's exceedingly likely that his two large—for the time—ships would have run aground. When Amundsen went that way in 1903, he was sailing the much smaller, shallower-draft *Gjoa*, and even then had to jettison cargo to lighten his ship and make it through. *Gjoa* drew only 10 feet, compared to at least half again as much for Franklin's two ships.[7] While it's true that vessels with a draft similar to that of *Erebus* and *Terror* might, with superior seamanship and a good deal of luck, have been able to wend their way through the James Ross and Rae Straits and their connected waterways, it is also true that the many shoals and hazards on the way are of such difficulty that the geography has caused a recognized authority on Arctic navigation to conclude that this route is suitable only for very small ships.[8] Since Amundsen's time, another 60 boats have completed the Passage using the eastern route, but most of them have been very small, with a small or even one-man crew.[9] It's almost certain that Franklin's ships could not have managed getting all the way through Rae Strait, James Ross Strait and Simpson Strait, the waterways that lead around the eastern and southern shores of King William Island toward the open water near the continental shore, at least not in a single season. As late as 1996, the *Hanseatic*, with a draft only slightly greater than that of *Erebus* and *Terror*, ran aground in Simpson Strait and was stranded for three weeks.[10] And Amundsen himself, in the tiny *Gjoa*, took several years to navigate the passage.[11] Writer Ken McGoogan, however, argues that the eastern, Rae Strait route would actually have been passable if only Franklin had tried it. McGoogan cites the example of the adventure cruise ship *Clipper Adventurer*, with a draft of 4.72 meters (15 feet, six inches), which made the trip through Rae Strait in 2012. *Terror*, by way of contrast, had a draft of 4.47 meters (14 feet, eight inches), and so, with good, careful seamanship, would have been able to make it round King William Island. As to why Amundsen, whose *Gjoa* drew only 10 feet, repeatedly ran aground, well, McGoogan just blames that on poor seamanship, sticking too close to the shore, where the water wasn't deep enough.[12] Had the same happened to Franklin, late in the brief summer sailing season as he was, he might still have been trapped in the ice in the winter of 1846–47, as indeed he was trapped out in Victoria Strait. But he would have been trapped in a far more favorable environment, shielded from the worst of the pressure from multi-year ice sheets and with perhaps a better chance for Sir John and his officers to consider their circumstances calmly and arrive at plans that might have permitted the eventual survival of at least some of the crew. With safe winter quarters on the eastern side of King William Island, it's possible that the senior officers could have reached a consensus in 1847 that it was time to abandon the mission and head for safety. At that time, at least, the men would have been a lot healthier than they were when the ships were finally deserted in April 1848.

Had Franklin tried the eastern route around what we now know to be an island, he would at least have secured a protected winter anchorage for his ships, rather than leaving them to the not-so-tender mercies of the river of ice in Victoria Strait. And, if the ice had not broken up in the summer of 1847, permitting him to sail on toward the continental coast, his crew would have been much closer to a potential rescue, with the option of heading overland toward Fury Beach, with its additional food and other supplies, on Somerset Island, or north toward the more frequented sailing grounds of Lancaster Sound and Barrow Strait—the very area where the Franklin search concentrated when that search was finally launched some years later.

But—there is almost always a but—*Erebus* and *Terror*, even if they had made it through the Rae Strait on the eastern side of King William Island—not to mention the equally, if not more, difficult James Ross Strait at the northern end of the passage around King William Island—and even if the ice broke up sufficiently for them to leave an imaginary winter harbor in the summer of 1847, would still have faced the problem of navigating Simpson Strait, the narrow passage between the south side of the island and the continental shore. That's where the *Hanseatic*, a vessel that has made multiple transits of the Passage and had a draft of 4.8 meters, just inches deeper than *Erebus* and *Terror*, ran aground and was stuck for three weeks in 1996.

Franklin, as we have seen, was not one to flout his orders. And those orders had been based on the charts that showed no way of connecting to the already identified waterways along the north coast of the continent by sailing to the east of King William Island. But what if? What if he had sailed down James Ross and Rae Straits (or into what he thought, based on James Clark Ross's 1818 observations some two decades earlier and reflected in the charts, prepared by Ross and carried by Franklin, was "Poctes Bay")? Would Sir John have turned around because he saw something that looked like land blocking his way, reminiscent of John Ross's sighting of the "Croker Mountains" supposedly blocking the western exit from Lancaster Sound? Would he have determined that King William Island was in fact an island, but then have been unable to maneuver his ships through the treacherous, at least for ships of his *Erebus*'s and *Terror*'s dimensions, waters? Would such a frolic and detour have put the Expedition into Victoria Strait too late in the season to escape the oncoming ice? Would the horseshoe-shaped Matty Island, in the middle of the James Ross Strait, have appeared to be an impenetrable barrier? The Victorians of the mid–19th-century Royal Navy weren't much for the ironic, preferring instead the heroic tales of bold adventure,[13] but such an outcome—discovering what in fact was the correct route but turning back from it only to be frozen in on the wrong side of King William Island—would certainly satisfy our 21st-century taste for delicious irony.

However they got there, whether through Peel Sound or around the western side of Prince of Wales Island, at some point in August or September 1846, the ships ended up within sight of Cape Felix. Once again, as at Cape Walker, a choice presented itself. But now, nearly half the year's sailing season, if such it could be called in the ice-choked waters of Victoria Strait, was gone. It had surely become evident, even to the ever-optimistic James Fitzjames and to Sir John himself, that the expedition would be spending another winter in the Arctic. The question became: should they head to the east of Cape Felix and then south into what might well be a dead end, requiring sailing out of it again the next summer, or stay out in the ice of Victoria Strait and hope the ships and crew would survive the winter and then be close enough to the continental shore so that next year's breakup of the ice would allow them to reach the already charted open water heading west near the coast? None of the choices were good, but which would have been least bad? Perhaps, if the Expedition had headed southeast, into "Poctes Bay," they might have found a winter harbor near Matty Island, a location previously charted by James Clark Ross and, at the very least, created a food depot that could have been used on a retreat toward the known supplies on Fury Beach—the stockpile that had in fact been used by others who had then actually been rescued. Had they gone that way, perhaps the decision to abandon the ships would have been made in 1847, after the death of Sir John, but still at a time when a sufficient number of the men remained in good enough health to manage the trek to Fury Beach. And we know—though of course Franklin's crew did not—that James Clark Ross, commanding *Enterprise* and accompanied by *Investigator*, was at Port Leopold, at the northeast corner of Somerset Island, in September 1848,[14] with each ship carrying, in addition to its own supplies, six months' provisions for the Franklin crew. So, even though, at the time when Ross was there, Prince Regent Inlet was blocked with ice, the rescue of at least a remnant of the Expedition's crew might have been possible.

Did the Franklin crew try that route, but then, repulsed by the shallow water,

7. Which Side of King William Island?

head back into the pack in Victoria Strait, or did they just ignore the possibility? Were there disagreements among the officers? What was the advice of those few on board the ships who had actual experience sailing the Arctic waters—principally Crozier and the two ice masters, Blanky and Reid—and, if given, was it even listened to? Even the miraculous discovery of log books and other records from the sunken ships, if indeed such discovery ever happens, is unlikely to shed much light on these internal deliberations. Assuming that there were in fact any deliberations. Chapter 10 takes up the question of what might have happened had Crozier, immediately upon the death of Sir John, decided to abandon the ships and head for Fury Beach. For now, though, let's return to the might-have-beens that arose as the winter of 1846–47 closed in on the Expedition.

8

Winter 1846–47 and Spring 1847

What do we know about the Expedition from the time the ships were beset in the ice in September 1846 until the surviving crew abandoned *Erebus* and *Terror* in April 1848? Almost nothing. The Victory Point Record confirms that there was at least one sledging trip, in April–May 1847. We know that because the sledge crew left a written record, on the standard admiralty form, at two locations on King William Island, and because one copy of that same record was retrieved and written over with additional information a year later, creating the hallowed Victory Point Record. But that's all we know for certain. How did the ships come to be beset in the open ice, rather than finding a winter harbor? What happened onboard over the winter? During that winter, when Sir John and his crew might reasonably have hoped that the ice would break up the next summer, did they continue with the typical Arctic winter routine of scientific observations, literacy classes, amateur theatricals and the rest? Or were they already too ill, too confused or too bereft to muster the energy for those activities?

And then what happened as that winter gave way to spring? Were there other sledge journeys besides Lieutenant Gore's in the spring of 1847? Did the ice show any signs of breaking up in the summer of 1847? Did the crews try to chop and cut their way through the ice, trying either to reach a better anchorage closer to land or to attain the shallow and potentially ice-free waters near the coast of the North American mainland? How many died that first year in the ice? And from what causes? Was there anything to hunt, and, if so, were the men of the Expedition successful in getting any fresh meat to feed themselves? We do know that there was an event of great moment to the Expedition late in the Spring. Sir John Franklin died on June 11, 1847. How, of what cause, in what manner? That's all still a mystery; all that's reported in the Victory Point Record is the bare fact and date of death. No details, no mention of the burial, if indeed there was one, whether at sea or, later, on land.

It's not likely that we'll ever know much more. Here, to recapitulate, is what we do know, as recorded in the Victory Point Record:

- In the winter of 1846–47, the ships were beset in the ice in Victoria Strait, at 70°5' North latitude, 98°23' West longitude, in other words, right in the middle of the ice stream moving slowly down through McClintock Channel and Victoria Strait from the northwest. And they had been stuck there since September 12, 1846, a bit early in the season for the ice to take control and suggestive of what may have been a short sailing season since leaving Beechey

8. Winter 1846–47 and Spring 1847

Island. Did they even have time to explore the possibilities of heading east of King William Island?

- On May 28, 1847, a sledging party led by Lieutenant Graham Gore, accompanied by Mate Charles Des Voeux and six anonymous sailors, presumably from *Erebus*, since that's where the officers were from, left a paper record in a cairn on the northwest coast of King William Island. According to the 1848 addendum to the record, that cairn was the one presumably erected by James Clark Ross back in 1831, when he had crossed the northern coast of the island on a sledge journey as part of his uncle John Ross's four-year odyssey.[1] That sledge trip was just part of the Rosses' epic four-year ordeal, during which they explored Prince Regent Inlet to the east of the Boothia Peninsula, had their ship wrecked, spent three winters in the ice, left behind the stores at Fury Beach and eventually were rescued by the whaler *Isabella*, which had been John Ross's own ship on that long-ago 1818 first attempt at a Northwest Passage.
- At the time that sledge party left *Erebus*, sometime in May 1847, Sir John was still alive, and conditions aboard the ships were good enough that the note could conclude with the words "all well," rather than with a plea for rescue or a guide to searchers who might come later. Even in the laconic, grace-under-pressure style expected of Royal Navy officers, the "all well" must imply something less dire than a complete collapse of the endeavor. A few deaths perhaps, in addition to the three who had been buried on Beechey Island, but it's unlikely that the death toll at that point was anything like the 24 reported dead when the crews abandoned the ships a year later.
- In late April 1848, but before the ships were abandoned, a party commanded by Lieutenant John Irving of *Terror* retrieved the record from the cairn where Gore had deposited it and moved it a few miles south, to 69°37'42" North latitude, 98°41' West longitude, on the King William Island shore.
- Writing around the edges of the paper, Crozier and Fitzjames added the information that they had abandoned the ships in the ice on April 22, 1848, and led the remaining crew, a total of 105 men, some 17 miles (5 nautical leagues) to the island. The 1848 addendum also noted that:
- Sir John had died on June 11, 1847, perhaps even before Gore and Des Voeux returned from their sledge trip, though that's not certain;
- Less than a year after Gore's "all well," the aggregate death toll was up to 24, nine officers (including Gore himself) and 15 men, leaving a total of just 105 of the original 129-member complement still alive, though it's not clear how many, by that point, were still capable of sustained physical effort or even of minimal mental understanding; and, finally
- "And start on tomorrow 26th for Back's Fish River."

Adding to the little we know from the Victory Point Record, in 1859 Lieutenant William Hobson, second-in-command of Leopold Francis McClintock's search mission, discovered the remains of a campsite near Cape Felix, at the north end of King William Island, three miles or so from Crozier's 1848 landing site. That camp that apparently dated from the spring or summer of 1847, a full year before the ships were abandoned and perhaps dating to more or less the time of Gore's sledge trip. Or perhaps the camp was one established in the winter of 1846–47 for yet more magnetic observations, which needed to be carried out well away from the ships with their locomotives and other metal parts. There was evidence of three tents (one of them, of course, would have been reserved for the officers), fireplaces, a flag, some needles of the type typically used for trade with the Inuit, and some fragments of clothing.[2] A few of Goldner's distinctive red cans, emptied of their contents, were scattered around the site—though far fewer than the hundreds of cans that had been left on Beechey Island the year before—making it clear that the site had been occupied by Franklin crew members, but for how long, and for what purpose, one cannot say.

And that's all we know.

That "all well" in the spring of 1847 is a clue. It's safe to conclude that, whatever recriminations there may have been among the officers for having been stuck in the middle of the ice stream for the winter, rather than in a somewhat protected anchorage, winter activities in that second winter in the Arctic proceeded more or less according to plan—the plan that had guided Royal Navy Arctic missions ever since Parry's initial almost-successful journey in 1819. The classes in literacy and navigation most likely continued. The carpentry work must have proceeded, readying sledges for exploratory journeys in the spring of 1847. There must have been occasional exercise for the crews outside the ships, though the extent of walking paths may have been limited by the jumbled-up sea ice. There most probably was a Christmas dinner and perhaps even Christmas theatricals, though no record of them has survived. Perhaps there were even onboard publications, again in the style of Parry's own *New Georgia Gazette*.[3]

One can also confidently state that, in the officers' messes, there would have been debates over what to do when summer arrived, and with it a hoped-for release of the ships from the ice. Royal Navy captains were not known for the democratic nature of their discussions with their officers, and some of the officers—especially Fitzjames and his friends without polar experience—would have happily followed their commander's optimism and simply gone forward without, one suspects, too much complaint. But those were not the only officers on board. Francis Crozier and the ice masters, Thomas Blanky on *Terror* and James Reid on *Erebus*, had far more years in the ice than the rest of the crew combined, and they understood the dangers. Crozier might well have argued for retreating to a protected anchorage, somewhere to the east and south of Victory Point, before the ships were trapped again out in the open river of ice. Even if that option turned out to be impossible—perhaps it was already too late in the season to beat a successful retreat, perhaps the ice never gave up its grip on the ships, even for a day—Crozier and the ice masters would have known that going forward was not the only option. At a minimum, they would have pressed Sir John to develop contingency plans for what to do if the ships weren't released from the ice the next summer. And perhaps even contingency plans that would include the possibility of giving up and making it safely back to England. In hindsight, a decision to retreat in the spring of 1847, rather than 1848, might have saved at least a good number of the crew. So why didn't they? A part of the otherwise unknowable answer might lie in Royal Navy disciplinary procedures; officers faced a mandatory court-martial whenever a ship was lost (though the captains were often acquitted when the loss was due to causes beyond their control, so long as they and their crews had acquitted themselves well in the emergency). Another part of the answer might lie in Franklin's character and history; having been ignominiously sent home from his governorship in Van Diemen's Land only a few years earlier, he may have been inordinately averse to another disappointing, failed command. Or, having survived one near-death experience as "the man who ate his boots" in 1819–21, and being a devout Christian, prone to leaving things up to God, he may have had a more passive approach to the problem, trusting in whatever deliverance would ultimately be provided. Or both. Whatever the reason, that early retreat from the ships, at a time when more of the crew might have been strong enough to make it to safety, never happened.

Instead, officers and crew weathered a third winter in the north and a second trapped in the ice. There is little point in even speculating as to how they might have spent those dreary months. What we do know is that, by the time the crew abandoned the ships in April 1848, another 21 officers and men had died, in addition to the three buried on Beechey Island. That's roughly 20 percent of the entire Expedition crew. Yes, the high Arctic is a challenging environment, especially in the colder-than-usual days of the late 1840s, but other expeditions, before and after Franklin, had spent multiple years in the ice with little or no loss of life. At a minimum, the Expedition had been supplied with at least three years' worth of provisions, in addition to anything they could kill and eat as they went.

In his letters and conversations with whalers near Greenland, Franklin himself had suggested that the ships' stores could support the crew for as much as five years, presumably with some judicious rationing. Could so much of the canned provisions have gone bad as to leave the crew near starvation even before the end of three years? Could scurvy have made substantial inroads, despite the large quantities of lemon juice that the Expedition carried? (Spoiler alert: yes; the efficacy of lemon juice as an anti-scorbutic declines over time, of which more in Chapter 13.) Could cold and exposure, their effects magnified by the unsuitable woolen clothing that the Expedition had been equipped with, have picked off one in every five members of the crew in less than three years? Had there been some catastrophic accident on board one of the ships? Had an entire sledge party been lost? We simply have no idea.

And what about the disproportionate number of officers who died before the retreat even began? The Victory Point Record reports that nine officers, presumably including Sir John himself, had died, compared to 15 "other ranks." That's surprising, in view of the officers' generally better health, at least when starting out, and their access to better food along the way, including supplements to the ships' stores from their own private stocks of food and wine. One Scottish study hypothesizes that relatively more officers died because of accidents suffered while they were out hunting for the (non-existent) game to supplement the ships' stores,[4] although none of the recognized Franklin Expedition experts gives the theory much credit. Whatever the cause of the disproportionate death toll, with nine officers dead by the time the men left the ships in April 1848, that would have left only six commissioned officers (not including the surgeons, pursers and other "civilian" officers), stretching the chain of command possibly beyond its limits, especially as various mental disabilities, described in more detail in Chapter 13, made discipline and even ordinary logic and reasoning, ever more difficult.

The known history of the Franklin Expedition, such as it is, suggests that things went from merely bad—pretty much the norm in multi-year Arctic voyages—to considerably worse during the second winter that they were trapped in the ice, 1847–48. Perhaps it's putting too much emphasis on the laconic "all well" in the 1847 portion of the Victory Point Record, but it seems at least more likely than not that the first winter, 1846–47, was not so different from other Arctic wintering-overs, whether those of Franklin on Beechey Island or of other Royal Navy crews before and after. We don't have any record of winter theatricals or Christmas pageants, nor any copies of the hypothetical *Illustrated Erebus & Terror News*, nor of any progress that the crew may have made in improving their reading and writing skills, though surely Sir John would have been delighted had he seen some of the crew reading the many devotional books that formed a large part of the ships' libraries.

And, just as we have no real knowledge of any sledging expeditions other than Graham Gore's in the spring of 1847, we similarly don't know how many other sledging expeditions there might have been that spring, nor what the weather was like, though, given that the ships couldn't escape the ice that summer, it's likely that the spring was no warmer than average for the period, and perhaps a good deal colder. We don't know what the sledge journeys found, whether in fact they connected the dots linking the ships' position to the known navigable waters along the continental shore, thereby completing, in some sense, the Northwest Passage, nor whether they left any of the crew buried somewhere in the snowy wilderness.

The most reasonable expectation is that at least two sledging journeys were launched in the spring of 1847, each led by a senior lieutenant from one of the ships. We know that Graham Gore was in charge of a sledge from *Erebus*, because it was Gore and his sledge crew who left an Admiralty record near Victory Point and then presumably headed south along the coast of King William Island. How much further did they go? Did they cross the ice to the continental shore at some point, perhaps as far east as the Adelaide Peninsula? How fast could they have traveled? Did they

Resolute and *Intrepid* in winter quarters configuration, Melville Island, 1852–53 (sketch by George MacDougal [mate, *Resolute*]; Royal Museums Greenwich).

see any evidence of a possible route through the ice for their ships? Did they get as far as the south coast of the island, demonstrating that Victoria Strait connected to the already-mapped east-west Simpson Strait, thus proving that a Passage did indeed exist? The Victory Point Record suggests that the Gore sledge party returned to the ships with their discoveries sometime in the spring of 1847, perhaps before Sir John died. Did their discoveries provide any sort of encouragement?

And what about *Terror*? Did First Lieutenant Edward Little set out, perhaps with Frederick Hornby or another of the Mates and half a dozen crewmen? Seems likely, for there would not have been much else to do in those spring months when there was adequate daylight, but no loosening of the hold that the ice had on the ships. But where would the *Terror* sledge have gone? Assuming that Gore and the crew from *Erebus* headed south—perhaps then turning east, foreshadowing the general retreat a year later—where, then, did the sledge from *Terror* go? Due west, across the multi-year ice toward Victoria Island? At least that would be in the direction the Expedition ultimately wanted to go, but could the sledge have even made it across the jumbled, piled-up masses of years-old, 50-foot-thick ice? Due east, toward Boothia and the known resources at Fury Beach? That would have been a useful reconnaissance if there was some inkling that the entire Expedition crew would need to seek rescue, but did anyone, even the perhaps more realistic Crozier, anticipate that need? And even if Crozier had suggested a Fury Beach journey, would Franklin have allowed it, knowing the mere undertaking of such a trip would spark rumors of retreat among the crew? Not long after the Franklin crew disappeared, a sledge party from James Clark Ross's 1849 rescue expedition actually checked Parry's old supply depot on Fury Beach, on the east coast of Somerset Island, and found that the supplies there were in good order.[5] If the Franklin crew had headed in that direction in 1847, when they still might have had the strength to get that far, perhaps some would have been saved.

Or perhaps the hypothetical *Terror* sledge party headed south and then west, in tandem with Gore's *Erebus* crew for the trip down the King William Island coast and then a right turn into Queen Maud Gulf, heading for Turnagain Point, visited by Franklin himself two decades previously? Such a route would have been the closest approximation to the Admiralty's orders directing a general southwest heading, so perhaps we can take that heading as a starting point.

And who would have gone with Little? One of the mates to be sure, and perhaps the carpenter's mate as well, to cope with the inevitable breakdowns of the sledge along the way. Perhaps, and it's the merest of speculation, but Crozier might have chosen a crew that he thought would have some solidarity, possibly based on shared origins, accents, history in the Navy, whatever. In any event, it's as good an assumption as any.

So, with their logbooks and pencils at the ready, the lieutenants set out in the Spring of 1847. But, so far at least, those logbooks have not been found. Perhaps Inuit children played with the pages and left them to be scattered in the wind. Perhaps some of the logbooks were buried beneath a cairn somewhere on King William Island, never to be seen again. We know they must have gone looking, whether for an exit or for a way to complete the Passage. But's that's all we know.

9

The Second Winter Trapped in the Ice

"All" was "well" in May 1847, when Lieutenant Graham Gore and his sledge party deposited a preprinted Admiralty form with a few navigational details in a cairn near the north end of King William Island. A year later, not so much. We know, from the few details available in the Victory Point Record, that, by the time the ships were abandoned in April 1848, 24 of the 129 men who had left Greenland on the two ships had died, including Sir John and the three crewmen who had been buried at Beechey Island. But we don't know when the rest of those 24 expired. We don't even know who most of them were, with the exception of Graham Gore, whose death was referenced in the hastily written addendum to the Victory Point document. Was the declining health situation gradual, over the nearly 20 months that the ships were stuck in the ice, or did the deaths build to a crescendo late in the voyage, perhaps prompting the abandonment?

As Franklin researcher Russ Taichman points out, it's unusual for sailors to abandon their ships, even in relatively dire circumstances.[1] John Ross and James Clark Ross abandoned *Fury* and *Hecla* only when, after four years, the ships had no chance of sailing to safety.[2] And George Back had somehow sailed an ice-damaged, leaking, nearly submerged *Terror* back across the Atlantic in 1837, its hull held together by chains wrapped around the ship, saving the vessel by beaching it on the shore in Ireland rather than abandoning it.[3] So we know that, by the time the winter darkness dissipated and the sun returned in early 1848, something was amiss on the ships. But was it an early realization that release from the ice was unlikely, and that therefore they'd need to start traveling by foot, hauling sledges, before summer slush made that kind of travel even more difficult, or was it an outbreak of disease on board that gave Captain Crozier the final nudge? We can safely assume that it wasn't Dan Simmons's monster, but there must have been a cause sufficient to impel Crozier to order the men onto King William Island in April, months before there was any likelihood that the ice would break up and allow them to sail at least some of the way toward safety.

Most researchers agree that the men of the Expedition—at least those who had survived long enough to quit the ships—died pretty soon after their trek began.[4] A minority, relying principally on Inuit accounts that seem to suggest a longer timetable for the retreat, believe that some significant number of the crew, possibly including Crozier himself, survived until 1850.[5] Those two possibilities are roughly correlated with the two types of possible reasons for quitting the ships. If a major health problem had emerged, accounting for some of the deaths reported in the Victory Point Record, it's more likely than not that the remaining men would have been dying rapidly, possibly all expiring in 1848. But if the abandonment of the ships was motivated primarily by Crozier's uncertainty over whether release from the ice would be forthcoming in the summer of 1848—about the time that their provisions would be exhausted—then a longer timeline to the end seems more possible. We don't know Crozier's reasoning for ordering the abandonment of the ships. But it wouldn't have been done without compelling reasons; Royal Navy captains were never eager to leave a ship, even one stuck in the ice.

In any event, between the time that sledge journeys got under way in the spring of 1847 (and Sir John died) and the time the crews abandoned the ships, the Expedition spent close to a year still locked in the ice after the return of whatever sledging journeys had been sent out in the more hopeful days of the spring of 1847. With no records to go by, we can only speculate as to what transpired in that period, as the crew waited for a breakup that never came, but some possibilities seem overwhelmingly likely.

First, there would, as suggested in the previous chapter, have been decisions to be made based on the discoveries of the sledging journeys that had been made in the spring of 1847. We know that one of those sledges, commanded by Graham Gore, went to King William Island. Did it go further than the northwest coast, where Gore left two copies of the Admiralty form with the date, position, and the laconic "all well" message? That seems very likely. Generally, these sledge trips were outfitted with about 40 days' provisions, easily enough for a journey of 100 miles or more and then back, assuming the men hauling the sledge with its boat and supplies were still in reasonably good physical condition.

Victory Point on King William Island, where Gore left the Admiralty form, was only 100 miles or so from Cape Herschel, on the south coast of the island, just across Simpson Strait from the Adelaide Peninsula on the North American continent. In 1838, The Hudson's Bay Company's explorers Thomas Simpson and Peter Warren Dease had filled in missing portions of the coastal map, showing that there was clear water, at least in summer, virtually all the way from Chantrey Inlet, near the mouth of the Great Fish River in the east, to Point Barrow in Alaska (taking into account Sir John Franklin's own earlier mapping in the western part of that area on his famed overland voyages). And Simpson and Dease had reached King William Island, leaving a cairn at Cape Herschel on the south shore of the island.[6] While a sledge journey to Cape Herschel wouldn't have proved the navigability of a Northwest Passage, it would in some sense have "completed" the passage, by showing a marine route that was all connected, even if it couldn't be traveled by a vessel of modest size. So, let's assume that Lieutenant Gore's sledge dropped of its Admiralty messages, then proceeded along the west and south coasts of King William Island at least as far as Cape Herschel. And let's assume that the sledge crew returned to the still-beset ships, bringing news that could be spun as a triumph (as it later was by Lady Jane Franklin, when she insisted that her late husband and his crew be given credit as the true discoverers of the Passage). After a winter in the ice, pretty much any accomplishment would justify a party, or at least an extra ration of grog for the crew. Although, unless Gore's sledge returned remarkably early, the news of a completed Passage would have had to compete with the rather different news of Sir John's death. Assuming that's what Gore and his sledge party found—a maritime link from the ships' then-current position to waters that were known to extend to the Beaufort Sea—those findings may well have influenced any decision, if indeed there was a decision being made, as to whether to wait for summer and hope the ice released the ships, on the one hand, or to begin immediate preparations for escape. After all, a mere hundred miles would be easy sailing, so long as the ice permitted.

As suggested above, in addition to Gore's trip, another sledging party may well have been sent out across the ice in the spring of 1847, aiming to fill in the very vague contours of Victoria Land, as it was then known and marked on the incomplete and misleading maps that the Expedition carried. Whether that sledge crew found the coast of Victoria Island or not, it would likely have reported back that the ice between the ships' position and the land showed no signs of going away in the coming summer of 1847, or in any summer for that matter. Given the especially low temperatures in the area in the late 1840s, and the fact that the ships had been trapped in the ice before the middle of September the previous year, that party, assuming it existed and made it back to the ships, would have had little of promise to report as to possible options.

Could there have been a third sledging trip in the spring of 1847? Jonathan Schaeffer, in his recent re-imagining of the Expedition,[7] posits a trip to confirm the location of the north magnetic pole, identified more than a decade earlier by James Clark Ross on the Boothia Peninsula, but, as the magnetic scientists of the time already knew, subject to wandering—no more than 100 years after the Expedition

King William Island, site of the Expedition camp three miles south of Cape Felix (photograph by Stephen Trafton, 1989, Creative Commons License CC-BY-SA 4.0 International; Wikimedia Commons).

9. The Second Winter Trapped in the Ice

disappeared, it was already under Prince of Wales Island, well north of Ross's location. In our day, speeding up to a rate of something like 30 miles per year, its movement has carried the magnetic pole far to the north, across the plane of the North (geographical) Pole, heading down the other side of the map toward Russia.[8] Maybe such a sledge journey happened, but one fact at least argues against it; the Victory Point Record states that nine of the Expedition's officers—nearly half the officer corps—had died by April 1848. How many of them were healthy enough to have led a sledge trip in May or June 1847? And would the need to keep a reasonably strong contingent of officers on board the ships have outweighed Franklin's undoubted interest in magnetic phenomena? Sir John didn't die until early June 1847, well after any sledge trips would have departed, and he did have a formidable history of making sure the magnetic observations were undertaken, whether in his previous travels or when settled in as Governor of Van Diemen's Land. But, after two winters in the north, would he have still insisted on doing his perceived scientific duty, or would he, by this time, have been single-mindedly focused on searching for a way out of his Arctic trap?

For now, in the absence of any proof to the contrary, we'll guess that there were just two sledge trips in the spring of 1847: the one led by Graham Gore, which may well have mapped the link between the ships' position and Simpson and Dease's cairn at Cape Herschel, thus in some sense completing the Passage; and the other, as imagined in the previous chapter, that headed west across the ice, looking to fill in the blank space on the Admiralty's map. And we can assume that both sledges returned safely to the ships sometime in June, before anyone could reasonably have expected the ice to break up and release *Erebus* and *Terror*.

What next? With the sun fully above the horizon all day (and night), everyone must have been focused on looking for signs of the ice breaking up and, in particular, looking for leads—cracks in the ice that widened enough to permit a ship to sail, to be propelled by its locomotive, or at least to be hauled along by ropes pulled by sailors walking on the ice. But, as we've seen, the summer of 1847 was probably not warm, even by Arctic standards. In addition, we know that not only did the ships fail to escape from the ice; they barely moved from the time they were beset in September 1846 until they were abandoned in April 1848, drifting just a few miles in all that time in a virtually unmoving sea of ice. So whatever leads in the ice may have been seen by the ice masters perched high up on the vessels' masts, those leads were apparently neither close enough for the crew to cut their way through the ice to reach them nor big enough to permit the ships to move through them, or both. By late August or early September 1847 it would have become apparent to everyone on the ships that they would be spending another winter in the ice. And by mid–September, the temperature would have dropped sufficiently, the sun would have become merely a part-time visitor, and the ice would have shown signs of growing again, after whatever small retreat it had made during the summer. Time for some contingency planning, or at least for some serious depression, perhaps accompanied, as imagined in Dan Simmons's novel, by serious drinking. In any event, not a happy winter with a crew no longer reasonably looking forward to sailing out of their frigid prison.

With Francis Crozier—never known as someone full of good cheer—now in command of the Expedition after Sir John's death, the optimism that Sir John had done so much to promote would have faded. Undoubtedly some of the remaining officers would have still harbored the hope of the ships' escaping the ice and reaching some sort of safety the next summer. But that optimism would need to have been tempered by the evidence of the summer just past, which had permitted no such escape, by the recognition of their dwindling food supplies, and by the deterioration in the health of the crew. Clearly, summer 1848 would be their last hope for finding a way out of the ice, but what direction would that escape take, and how soon were decisions made and actions begun to facilitate it?

So, what was going on that second winter locked in the ice off King William Island—the third winter in Arctic darkness? With an ever-sicker group of officers and men, we can be pretty sure that there was no repetition of the Christmas theatricals and no new editions of the Expedition's occasional broadsheet. Perhaps not even enough enthusiasm, either, for the regular Royal Navy winter routine of literacy and navigation classes. The basic work of insulating and protecting the ships would have been done, if there were enough healthy men to do it, once again creating a tent-like structure on deck and building up a protective barrier of snow around each vessel. Beyond that, perhaps not much, as the remaining officers and crew waited for Crozier to decide on a course of action.

Sometime between September 1847, when it became obvious there would be no sailing that summer, and April 1848, when the Victory Point Record was updated, someone must have done some planning. Did Crozier involve the (surviving) officers in the process? Or did he retreat to his cabin through the winter darkness and try to imagine, or drink, his way out of the situation? At the very least, there would have been instructions to the carpenters to fashion sledges for some kind of journey in the spring of 1848, or to repair the sledges that had been used on whatever trips had been made the previous spring. There would have been some attempt to inventory the remaining stores and estimate how long they could sustain the crew. One must assume wardroom debates about which way to go once travel again became possible, but we don't know that such debates occurred. One can equally imagine Francis Crozier sitting alone, perhaps reading the Rosses' accounts of previous expeditions gone wrong and deciding on a plan of retreat. But which way to retreat?

That final winter in the ice, then, is the blankest of blank slates. Time to move on.

10

Where Did They Go?

Francis Crozier and his crew were not without options when they abandoned *Erebus* and *Terror* in April 1848. They had at least four possible routes, some with a variety of sub-options:

1. Head due north for a rescue. They could have used some of the ships' boats, hauled on sledges until they reached open water, to travel back north toward Barrow Strait and Lancaster Sound, retracing the route they had followed in the summer of 1846. If successful, that would have put them in the most likely place to find a rescue party or even, as John Ross and his crew had many years earlier, to encounter a whaling ship. In fact, James Clark Ross, in one of the first Franklin rescue missions, with the ships *Enterprise* and *Investigator*, arrived at Port Leopold, on the northern tip of Somerset Island, in September 1848, when some of the Franklin crew may still have been alive.[1]

2. Head northeast, across King William Island and the Boothia Peninsula, toward Fury Beach, which Crozier knew first-hand from his time as an officer on Parry's third attempt at the Northwest Passage in *Hecla* in the winter of 1824–25, where *Hecla*'s companion ship *Fury* had been abandoned there, though both crews made it safely back to England in *Hecla*. Crozier might have assumed (correctly) that there might still be food cached at Fury Beach. James Clark Ross in fact sent a search party to Fury Beach in the spring of 1849, hoping that at least some of the Franklin crew had had the good sense to head toward the one known source of food in the area. Apparently they didn't. Ross found no sign of the Franklin crew at Fury Beach, and no one since has discovered the slightest hint that anyone from the Franklin Expedition might have headed there.

A variation on that route for the retreat from the ships would have been for those who left *Erebus* and *Terror* to have first headed east and then north, aiming for the food and other supplies still thought to be at Fury Beach.

As it happens, there is a well-documented traditional Inuit trail from the west that passes just south of King William Island and connects up, nowadays, with the Inuit settlement at Pelly Bay, to the east of the Boothia Peninsula.[2] Had the Expedition's crew been able to follow that trail, they might have ended up not far from Fury Beach, just a bit to the north of the latitude where their ships were abandoned and within reach of rescue and food. But did they know how to get there? Were they on the trail at the right time in the season to encounter Inuit who could give them advice or guide them? Would they have been able to communicate with the Inuit sufficiently to find out about the existence of the trail? Inuit "trails" weren't like those marked on the neat, printed maps of the Appalachian Mountain Club, or stored in today's ubiquitous GPS devices. Unlike trails in other, less harsh, geographic settings, Inuit trails are not even permanent features of the landscape that endure independent of the knowledge of those who follow the trails. There are no signposts to guide travelers or tourists, no arrows carved into the bark of trees—because there are no trees— no guardrails or yellow lines painted on the ice. Nor are there well-worn paths in the dirt. All the trails disappear whenever snow covers the landscape, and trails across the sea ice disappear annually when the ice melts during the Arctic summer. But "the spatial memory, however, remains in people's memory and materializes again when the next trailbreaker makes the trip."[3] The routes that the trails take are well known in Inuit collective memory, even though the trails themselves are ephemeral, vanishing in the next blizzard or when the sea ice melts in the summer.[4]

The Expedition's maps would have been of little or no use, other than perhaps in providing the latitude and longitude of the hoped-for destination. The only way crew members could have followed the trail would have been to link up with Inuit who happened to be going in that direction and who knew how to get there. But, as the Inuit oral history recounted by David Woodman suggests, such communication as there was between the two groups was mostly a matter of pantomime or dumb-show, despite Crozier's and others' purported knowledge of Inuktitut. And, even if they had found Inuit who could (a) understand what the kabloonas were saying and (b) happened to know the specific trail that led to Fury Beach, were any of the crew still healthy enough in 1848 to have made it that far, considering that none of them, so far as we know, made it to their stated objective of the mouth of the Great Fish River.

    3. Head southeast toward Repulse Bay, in the hope of encountering ships that had come through Hudson's Bay, or at least in the hope of making contact with the Hudson's Bay Company agents in that area who might be able to provide some food and, just as important, send news of the survivors' whereabouts to those who might have been able to send potential rescuers.

    4. Head for Back's Fish River, at the base of Chantrey Inlet, and from there up the river to the Hudson's Bay Company post on Great Slave Lake, more than 600 miles away.[5] This seems to be the way that at least a good part of the surviving crew actually went. Crozier's handwritten addendum to the Victory Point Record said: "and start on tomorrow 26th for Back's Fish River," and enough bones and relics of the crew have been recovered along the route from Victory Point to Chantrey Inlet to suggest that it was indeed the intended destination. But to recognize that only raises more questions. If, as we'll see, Back's Fish River was the worst possible choice for a retreat—at least if the object of the retreat was to save the lives of any of the crew—then why did Crozier choose it? Was that choice the key mistake that doomed every man on the Expedition? Could some have been saved if they had gone in a different direction from the start?

How do we know that the Back's Fish River route was the worst possible choice? First, we have the testimony of those who had traveled the river and written about the journey. When John Rae and his party descended the river in 1853, in a small boat considerably lighter and more maneuverable than those hauled from the ships by Franklin's crew, they found the going very difficult indeed. And Rae and his party were going downstream, with the current, and not upstream against it, as the

Franklin crew would have had to do if they had made it that far. Rae reported that most of the river was shallow, with strong currents and many rapids, as well as frequent waterfalls, as much as 25 feet high, that required arduous portaging.[6] A subsequent Hudson's Bay Company expedition in 1855 counted 83 rapids along the river before giving up for the season.[7] The number of 19th-century expeditions that succeeded in descending the river—all consisting of just a handful of voyagers using light canoes—can be counted on the fingers of one hand. The number that succeeded in ascending the river from Chantrey Inlet to Great Slave Lake can be counted without using any of those fingers.

By the time the crews left *Erebus* and *Terror* in April 1848, a full year later than what would have been, in hindsight, the best time to go with some hope of being healthy enough to reach a place from which they might have been rescued, their chances for survival might have been slim no matter what direction they went in. But why did they choose the least survivable of all routes? Even at the time, and throughout the 19th century, informed opinion in England, including the views of George Back, who had traveled down the Great Fish River a decade earlier, and of James Clark Ross, was strongly that the most likely places for Franklin's crew to have gone, if forced, as they indeed were, to abandon the ships, were either east to Fury Beach and then north to Barrow Strait or, alternatively, toward the mouths of the Coppermine River to the west, in territory that had previously been surveyed by Franklin himself.[8]

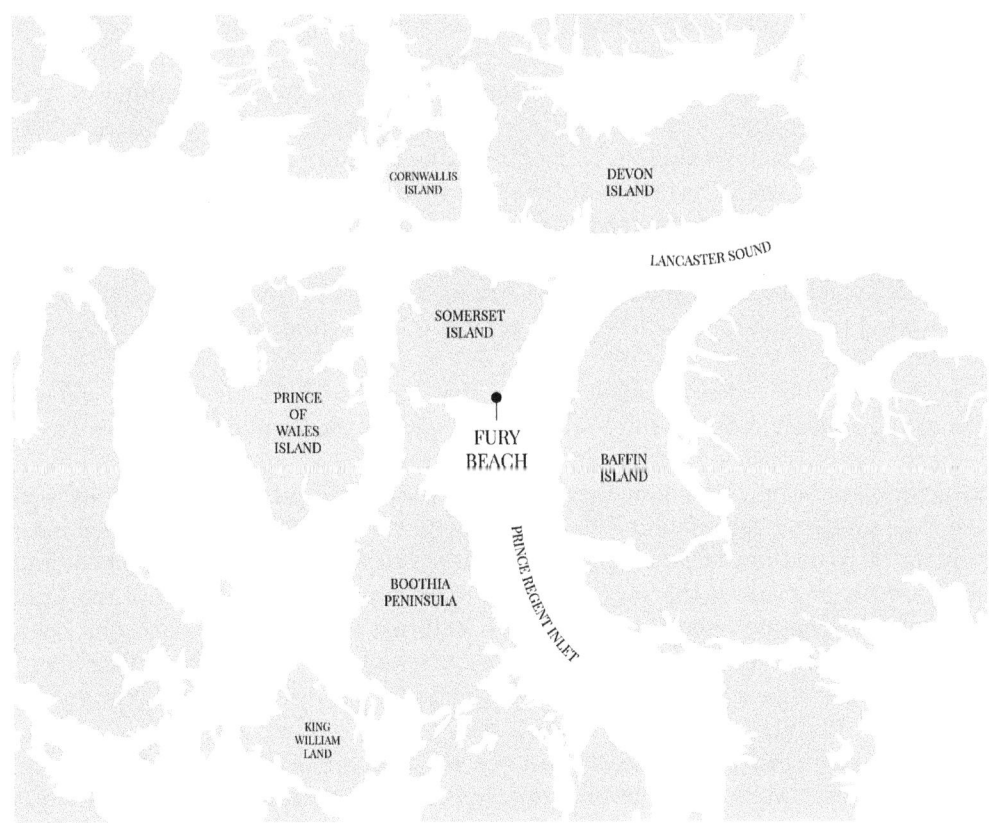

Possible retreat destinations, including Fury Beach (map by David Veller).

One plausible theory for the choice of the Chantrey Inlet/Great Fish River destination is that the chances of finding food were better in that direction than elsewhere, better even than at Fury Beach. George Back's journey a decade earlier had shown that the river was indeed full of fish, and it would have been sensible to assume that the further south they went, the more likely they would be to encounter four-footed game as well. Perhaps the thought was: if we can make it that far (the mouth of the river), we can live off the land for as long as it takes us to reach a safe haven.

Choosing Back's River as a destination would have served two purposes. Crozier could reasonably have assumed that rescue missions would have been sent out sooner than they actually were, and that those rescue efforts would have discovered the Victory Point Record long before McClintock's team actually found it in 1859. it might have focused the attention of the rescue missions and sent them in the right direction. But that would have been true of any intended destination that was put on the Record. And any use of a message to indicate location would have depended on the rescuers' finding that message; as it turned out, almost all the rescue missions concentrated their efforts well to the north, in Lancaster Sound and Barrow Strait, with the result that Crozier's and Fitzjames's message wasn't found for 11 years after it was deposited, far too late to be of any use. Once again, from hindsight, it seems obvious that the rescuers, if they came, would come the same way the Expedition itself had come, via Lancaster Sound and Barrow Strait. So why not go in that direction?

A second reason for the choice of Back's Fish River is darker. Perhaps Crozier had, by April 1848, made up his mind that there was no way out, that one way or the other they were all doomed. If that had been the case, the actual destination wouldn't have mattered, as, in Crozier's mind, they'd all be dead before they got there, wherever *there* was. The only reason for picking a destination at all, the only reason for leaving the ships, would have been to give the remaining officers and crew something to do, a project to work on, instead of just sitting on board the ships or in their initial campsite in the northwest of King William Island. The Royal Navy had a strong aversion to leaving hands idle, and the crew's having a project to work on—even if in Crozier's mind the project was pure folly—provided something, anything, to do while waiting to die.

The record of bones and relics shows that they did, to a considerable extent, follow the route laid out in the Victory Point Record. Some of the crew, we know, went to the "boat place" at the southern end of Erebus Bay, on the western shore of King William Island. We know that because Hobson and then McClintock in 1859 reported finding a 28-foot ship's boat there, with the partial skeletons of two individuals.[9] Two decades later, Frederick Schwatka's expedition also found a boat and some bones, perhaps at the same place, but certainly quite close.[10] And the Inuk In-nook-poo-zhe-jook told Charles Francis Hall that there had been a second boat, abandoned somewhere quite near where Hobson and McClintock found the first.[11] Perhaps there was even a third boat abandoned or at least sojourning at the southern end of Erebus Bay. How many more of the crew passed that way, and whether the boat found there by Hobson and McClintock, its bow pointed back toward where the ships were beset in the ice rather than toward the southeast and intended salvation at Back's River, had simply been left behind by the crew heading southeast, or whether it represented at efforts of some of the crew to return to the ships, cannot be known.

What we do know is that a number of the Expedition's crew died in this area, and that they left behind a lot of artifacts.

The archaeologists Doug Stenton and Robert Park, who surveyed the area repeatedly in the past decades, concluded that they had definitively identified Hobson and McClintock's "boat place," but that the location of the other[12] (or possibly two) boat sites reported by the Inuit remained uncertain.[13] They, and others, never found evidence of the tent foundations and cooking sites mentioned by In-nook-poo-zhe-jook, but that doesn't necessarily mean there wasn't some sort of large campsite there nearly two centuries ago. Wind, water, animals, Inuit hunters and other agents may well have obliterated whatever might have remained of such a site. All we know for sure is that some of the crew passed that way, some died there, and the bones of some of the dead showed evidence of cannibalism.

Despite an earlier historical consensus that the relics found near Victory Point and at the boat place near the southern end of Erebus Bay showed that the crew abandoning *Erebus* and *Terror* had taken with them a mad jumble of mostly useless objects, closer study suggests that they actually knew what they were doing, and, in particular, were modifying shipboard material so that it might be of use on the march to a hoped-for salvation.[14] As Robin Rondeau argues, much of what was found along the route, at the boat place and elsewhere, was not "junk" but rather survival gear like knives, bags of musket shot, sails and paddles for the boats and material for repairing the boats as they progressed.[15] That would be consistent with an organized retreat, with sailors being assigned tasks and, for a while at least, taking their responsibilities seriously.

If, indeed, the destination was the mouth of Back's River, then the shortest way there would have been to cross from King William Island's south shore across the narrowest point of Simpson Strait, some miles west of Tulloch Point and even further west of the Todd Islets. This was also the most likely place to find caribou crossing from the mainland to the island and back again, as evidenced by the abundance of caribou bones left behind by Inuit hunters that Hudson's Bay Company's William Gibson found at the site in 1931. When Gibson was there, he also found human bones that appeared to be those of the Franklin Expedition crew, some as far east as the Todd Islets.[16] Why would the crew have gone that far, away from their stated objective of the river mouth?

And then there was the backgammon board found on Montreal Island, reported in 1855 by James Anderson of the Hudson's Bay Company.[17] Despite the somewhat plausible explanations for much of the retreat route, there remain questions about the route that the main party of for-the-moment survivors seem to have taken that simply can't be answered, and that never will be answered, even if *Terror*'s captain's log eventually emerges from the wreck in Terror Bay.

Had any of the Expedition members made it as far as Back's River, what would they have found? When Hudson's Bay Company Chief Factor James Anderson descended the river in 1855, he reported encountering ice in the water as late as June near Great Slave Lake and well into August at Montreal Island, in the river's estuary.[18] The notion of proceeding upriver through ice and rapids and hauling the ships' boats over multiple portages is, to say the least, daunting.

But, back in 1837, George Back had reached the mouth of the Great Fish River, traveling downstream, and reported that there was an abundance of game at the estuary.[19] The Expedition certainly had Back's narrative in the libraries on both ships, and perhaps the promise of food, rather than any realistic escape route, was the reason for Crozier's decision to head southeast. Montreal Island, not far from the mouth of the river, was a little closer to the abandoned Expedition ships as was Fury Beach—where James Clark Ross actually looked in 1849.[20] But Crozier, who might have heard rumors that whalers had plundered the Fury Beach cache, didn't know that.

There is, however, at least some evidence that one group of the Expedition's crew headed north-northeast, which would eventually have led them to Fury Beach and its cache of stores.[21] Whether this party headed off by design or, perhaps more likely, simply got lost, is a matter of conjecture. But it wouldn't have been the dumbest decision they could have made, whoever they were.

In 1949, Inspector H.A. Larsen of the Royal Canadian Mounted Police searched much of King William Island and, among other discoveries, found a Caucasian skull, probably from the Franklin Expedition's crew, near Cape Felix at the north end of the island.[22] But that only raises more possibilities; did the sailor whose skull was found there die on board the ships and was later buried ashore? Did he die at some point while the ships were beset in the ice and sledge parties were sent out to the island? Was he resident at a scientific observation post established by the crew at Cape Felix, and, if so, for how long was that post manned? And, finally, was that skull from one of the retreating crew who died after April 1848, suggesting that at least some of the Expedition members had headed north instead of southeast? Or was it from an earlier death, perhaps someone assigned to make magnetic observations during the nearly 20 months the ships were stuck in the ice? In the next few chapters we address some of the many remaining questions about the retreat.

11

Return to the Ships? Mutiny?

Nearly a century ago, the Canadian engineer and Franklin searcher L.T. Burwash hypothesized that as many as half the crew that initially left *Erebus* and *Terror* in April 1848 had returned to the ships.[1] The recent (2016) discovery of the wreck of *Terror*, essentially in one piece and upright on the seafloor of Terror Bay, just off the south coast of King William Island, strongly suggests that the ship was sailed there—or even reached there under what remained of its steam power. According to the divers who examined the wreck in the summer of 2019, the ship's propeller was deployed as if there had been an attempt, or at least an intention, to use its locomotive to help move the ship through the ice.[2] Also, the ship's anchor had been set, again suggesting that human agency, not merely the movements of the tides, currents and ice, had been at work.[3] No unmanned ship trailing a heavy anchor could have drifted into Terror Bay on its own; the *Terror*'s anchor must have been catted and fished (i.e., raised and stowed on the ship) en route to its final destination and then set in the presumably at least somewhat open water of the bay once the ship had reached its intended destination.

The late Louie Kamookak, the leading Inuit historian of the Franklin Expedition, believed that at least a part of the Expedition crew had re-manned *Terror* and sailed her around King William Island, perhaps in the summer of 1848 or perhaps as late as 1850, to her final resting place in Terror Bay.[4] Or maybe both ships were re-manned, and *Erebus* also sailed on from wherever she was when the ice finally released her, continuing on to the south and ending up in her final resting place off the west side of the Adelaide Peninsula. Less likely, perhaps, than *Terror*'s hypothetical odyssey, because *Erebus* could easily have ended up where she did simply by being carried along as the ice drifted, and a somewhat uncontrolled drift might better account for the substantial damage that was observed when she was found in 2014. But, like so much in our story, alternatives are not impossible; perhaps some of the crew returned to *Erebus* as well. Both *Erebus* and *Terror* were found in relatively enclosed locations, suggesting purposeful sailing rather than mere drift, and both were in a winter-quarters configuration,[5] with topmasts removed and, in the case of *Terror* at least, fore-and-aft spars deployed in a way that could have served as ridgepoles for the canvas covering above the deck that was usually used when a ship was occupied during an Arctic winter.

Another argument for a return to the ships is the finding of what is believed to be Lieutenant John Irving's body—if, indeed that body, now buried in Edinburgh, is that of John Irving himself and not that of a more senior officer, perhaps even Sir John, that was misidentified because Irving's maths medal that he had earned at the

Royal Naval College was found at the site—very near the place from which the crew had departed after abandoning the ships. The grave also contained a telescope, gilt buttons from an officer's uniform, and a silk handkerchief, all suggesting that it was definitely the final (until Lieutenant Schwatka disinterred the remains and returned them to Scotland) resting place of a commissioned officer, though not necessarily that of Lieutenant Irving. One reason to doubt the identification is that, if Irving was in good enough health to be sent off to locate the Victory Point Record in 1848, as is noted in the record itself, would it have been likely that he died there, even before the retreat toward Back's River could fairly begin? Or more likely, perhaps, that he was part of a group that later returned to or toward the ships and died at some later date?[6] Of course, if the body is not that of Lieutenant Irving at all,[7] any thoughts of his having returned to the ships after April 1848 are mere speculation.

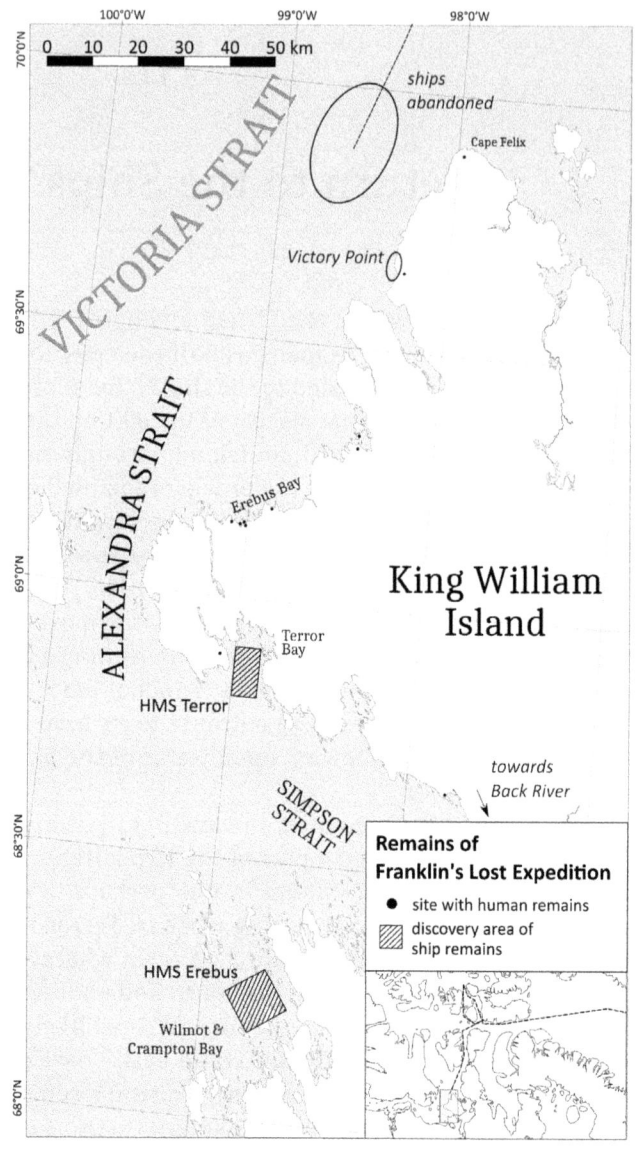

Location of *Erebus* and *Terror* wrecks (wreck areas per Parks Canada, Creative Commons Attribution-Share Alike 4.0 International License; Wikimedia Commons).

If Crozier had ordered everyone to leave the ships, and if some of the crew then returned, they were probably disobeying their commander's orders in doing so. And that suggests a mutiny. The destruction of the chain of command lends credence to the theory that there may have been one or more mutinies soon after the ships were abandoned. When Franklin sailed from England in 1845, there were 110 men (of whom five were sent home from Greenland) and 24 officers, a ratio of 4.6 crew to each officer, a figure in line with the staffing of most smaller Royal Navy ships (as *Erebus* and *Terror* were) at the time. But, by the time the crew left the ships in April 1848, only 90 men and 15 officers remained, a ratio of six to one.[8] Did the reduction in officer ranks, along with all the Expedition's other tribulations, so undermine the chain of command that a mutiny became more likely and more feasible? Were the officers who remained even in a fit state to command, addled as they might have been by lead poisoning, scurvy and possible mental illness?[9] As the ever-diminishing number of Franklinites split into smaller sub-groups—perhaps some staying behind on the west coast of King William Island, perhaps some back on the ships, almost certainly a number remaining near Terror Bay, and then smaller groups whose bones were found scattered across the southern coast of King William Island and on the mainland's Adelaide Peninsula—the senior officers, assuming Crozier and Fitzjames survived for at least a while, would have inevitably lost their ability to command all the men, and leadership of at least some of the smaller groups would have devolved on those perhaps more endowed with charisma, strength or force of personality than with a commissioned officer's epaulets. Even at the start of the retreat, there would have likely been only one officer per boat as the boats were being hauled along. Undoubtedly, some of those remaining officers either died early in the retreat or became too weak to carry on, leaving their sledges in the hands of warrant officers or even of exceptionally persuasive or brutal sailors (picture the fictional version of Cornelius Hickey in the television adaptation of Dan Simmons's *The Terror*).

Casual discussion of the splintering of the Expedition's crew after abandoning the ships often refers to a mutiny on the part of some of the crew.[10] But, whatever happened leading to the fragmenting of the crew on its death march, it's unlikely to have been a mutiny in the sense that the term was understood at the time.

Unlike more traditional mutinies, which might involve putting the officers overboard in a small boat or even killing them, a revolt among the *Erebus* and *Terror* survivors would not necessarily have been an occasion for violence. As the officer in charge of a sledge or of a temporary encampment became too feeble to exert command, or even died, what remained of a chain of command, if not already frayed to the breaking point by the disasters of the past three years, would have simply faded away, leaving it up to those still left to take action to decide for themselves how to proceed. If a group became isolated, away from officers with navigation skills, the only route they knew might have been the one that took them back to the ships, retracing their steps and hoping for a miracle. While that would have constituted disobedience to orders, if there were still any orders to be disobeyed, it was not the sort of active revolt against authority that the word mutiny usually conveys.

In fact, the Articles of War, the harsh code that governed the Royal Navy at the time, doesn't even define a "mutiny" per se. Article 19 referred instead to a "mutinous assembly," suggesting that the Royal Navy was not about to admit, even in an accident of language, that an attempt at mutiny could ever succeed:

> If any person in or belonging to the fleet shall make or endeavor to make any mutinous assembly upon any pretence whatsoever, every person offending herein, and being convicted thereof by the sentence of the court martial, shall suffer death: and if any person in or belonging to the fleet shall utter any words of sedition or mutiny, he shall suffer death, or such other punishment as a court martial shall deem him to deserve: and if any officer, mariner, or soldier on or belonging to the fleet, shall behave himself with contempt to his superior officer, being in the execution of his office, he shall be punished according to the nature of his offence by the judgement of a court martial.

The Articles of War, first adopted in the mid–18th century, also proscribed concealing a "traitorous or mutinous practice or design," striking or disobeying a superior officer, "stirring up a disturbance" and numerous actions thought injurious to good shipboard discipline. But their wording suggests a continuing problem: how to define what a mutiny actually is. Several studies of mutinies confirm that it's not an easy task. First, can mutiny be the action of a single rebel? The reference in the Articles of War to a "mutinous assembly" suggests that some kind of collective action is necessary. Further, what kind of intent must the mutineers have? Is the aim of overthrowing authority and seizing control (of the ship, the expedition or mission) a required element, or is mere refusal to obey orders, even to the extent of a sit-down strike, enough?[11] Can there be a non-violent mutiny?

The Royal Navy was not unfamiliar with mutinies and other forms of rebellion against hierarchy and authority. In the century or so before Franklin's voyage, there had been more than a dozen well-known instances of ships' crews refusing to follow orders and, in some cases, putting their officers overboard, as with Captain William Bligh of the *Bounty* in 1789[12] or even killing them. Bligh himself was the target of at least three shipboard mutinies, not to mention the "Rum Rebellion" that ended his tenure as Lieutenant-Governor in New South Wales.[13]

But there were many more precedents for those who knew their naval history. More than a century Franklin sailed before, the crew of HMS *Wager* had rebelled after their ship was wrecked on a desolate island in the south of Chile, abandoning Captain David Cheap and his loyalists on the island. Like Franklin's crew, many of the *Wager* sailors were suffering from scurvy, although, unlike Franklin, a few survivors of the mutineers and the loyalists, including Cheap himself, a future Vice-Admiral, eventually returned to England.[14]

The Royal Navy's devotion to keeping the crew hard at work, even on meaningless activities, had its origins in the fear of mutiny. As one Admiral reportedly commented "the only way to keep large bodies of men in order is by dividing and subdividing them" and keeping them constantly employed, with minimal time for rest, meals, washing and mending their clothes.[15] Leave the crew idle for any amount of time and, according to British Admiral Richard Klempenfelt, "the people, left to themselves, become sottish, slovenly and lazy, form cabals, and spirit each other to insolence and mutiny."[16] By April 1848 there would have been plenty of idle time on the Franklin Expedition.

In 1797, in the midst of the Napoleonic Wars, the crews of dozens of Royal Navy ships had mutinied at Spithead, a major Navy anchorage near Portsmouth, refusing to sail until their demands for a pay raise, better food, and the removal of officers the men deemed too harsh and unjust were met. After the Navy agreed to those demands, other ships' crews tried the same tactic somewhat less successfully at the Nore, a naval assembly point in the Thames estuary. These revolts were primarily about pay and work conditions—seamen's pay had remained unchanged for 150 years, despite considerable inflation in the late 18th century, and what little there was often arrived months, if not years, late. So it is not surprising that these revolts were primarily about pay and working conditions, though the unsuccessful mutiny at the Nore also involved some political demands, including an immediate end to the war with France, which were firmly resisted by the authorities.

The rebellions at Spithead and the Nore, occurring when John Franklin had just joined the Royal Navy as a boy, were part of an explosion of discontent that cascaded across the British, French and Dutch navies in the 1790s.[17] Conservative estimates are that the decade saw at least 150 single-ship mutinies and half a dozen fleetwide rebellions among what were then the world's three leading navies and that by the end of the decade somewhere between one-third and one-half of all the sailors in those navies had been involved in at least one mutiny.[18] While the Royal Navy perhaps was less affected by the spirit of the 1789 French Revolution than its counterparts across the Channel, it was certainly not immune to lower-deck protest, as Spithead and the Nore demonstrated. In fact, some historians point to an unofficial tradition permitting a certain amount of resistance, typically expressed in illegal shipboard assemblies or riots to give voice to discontent and met with officers' responses that tended to quietly deal with whatever specific issues might have triggered the mutiny, along with the not-so-quiet punishment of a few of the ringleaders as a way of re-establishing the appearance of proper discipline.[19] But that tradition of quiet, perhaps under-reported mutinies certainly changed during the Napoleonic Wars, when the Royal Navy was forced to augment its numbers with many sailors who were conscripted (often press-ganged off the streets) or foreigners who had signed on not for love of country but perhaps as the only jobs they could get.[20] Mutinies became more frequent, and retribution became ever more unforgiving. In perhaps the most violent episode, the crew of HMS *Hermione*, on the Royal Navy's West Indies station, killed 10 of the ship's officers, delivered the ship to the Spanish authorities in Caracas and then, for the most part, melted away into the Americas.[21]

But that peak of mutinous activity in the 1790s was a half-century removed from the Franklin Expedition. After the end of the Napoleonic Wars in 1815, and the resulting drastic reduction in the size of the Royal Navy, the ranks of sailors once again were made up primarily of volunteers. There were only a few reported shipboard revolts, none of any great significance. In fact, no mutinies of note occurred again until the world wars of the 20th century, when the Royal Navy was once again staffed with large numbers of less than fully willing participants. Moreover, Franklin's crew, unlike the mutinous crews of the Napoleonic period, was made up of volunteers—and volunteers for double pay, at that—and almost entirely of native English, Scots and Irish.[22] In the 1797 mutinies at Spithead and the Nore, by way of contrast, only 20 percent of the Royal Navy crews were volunteers.[23] So perhaps whatever led to the fragmenting of the Franklin Expedition crew once they reached land was less about resistance to authority than about the absence of sufficient authority—too few officers to maintain effective command—leaving the men to muddle through as best they could.

Franklin had a reputation throughout the Royal Navy as a very light disciplinarian. And Crozier, though not the subject of nearly so much hagiography as his expedition commander, was also not known for frequent use of the "cat" or other harsh discipline. In good times, these commanders might well have fostered happy ships, with the crews showing more than adequate trust in their leadership. But, by April 1848, the good times were long gone. And, as discussed in some detail below in Chapter 13, everyone's ability to make rational decisions may have been seriously compromised by illness and starvation at the time of the retreat. What little discipline remained might have been enforceable only at the point of a gun, a conclusion suggested by Lt. Hobson's report that the likely final survivor at the boat place was found with two loaded shotguns in his skeletal arms.

Both the Inuit stories told to John Rae, Charles Francis Hall and others, and recounted in David Woodman's books,[24] and the often-noted story of the Erebus Bay boat place, with its ship's boat turned around and pointing back in the direction of the abandoned ships,[25] strongly suggest a return to the ships, although the three or more crewmen who made it back as far as the "boat place"—one of whom may have provided some last nourishment for the other two[26]—probably never did reach the temporary refuge of the ships. In fact, there were at least two, and possibly three, boat places near Erebus Bay[27] and maybe, as Stenton and Park suggest, not all of them reflected attempts to reject the plan of retreat and head for the ships. But maybe they did.

A breakdown in order and discipline? Certainly, with the result that the Expedition crew almost certainly splintered into ever-smaller groups struggling to survive as best they could. Mutiny? Not so likely, at least in the accepted meaning of the term. More likely that the structures of authority, rather than being overthrown, just gradually melted away.

12

Off the Beaten Path— But Where?

Filled with optimism, Franklin searchers pursued the Expedition not only where it was ordered to go—through Lancaster Sound and into the unmapped Canadian Arctic, but also where it was expected to emerge, somewhere past the Bering Strait, or perhaps even into the South Seas. We can be pretty certain that none of the expedition's crew made it that far, but mere facts were never enough to fully quell the imaginations of searchers.

The Franklin Expedition may have been the subject of more searches than any other ever was or probably will be. Beginning in 1847 when the Expedition had been gone only two years, the Royal Navy began alerting ships going in that direction to keep an eye out for Franklin, and whaler William Penny did in fact sail into Lancaster Sound that year in *Saint Andrew* in an unsuccessful attempt to locate the missing ships.[1] The next year, James Clark Ross in *Enterprise*, accompanied by *Investigator*, made the first of what might be called official search voyages, while John Richardson and John Rae set off overland from Hudson's Bay. Altogether, W. Gillies Ross counts some 39 search expeditions from 1847 through McClintock's 1859 trip. These 39 searches—32 by sea and seven by land—involved a total of 47 ships and 40 ship-winterings in the Arctic. In addition, Ross counts another four "dual-purpose" expeditions, including that first voyage of William Penny. While many of the searches bore the imprimatur of the Admiralty, its support ceased after 1854, when the crew were officially declared dead. But, even after the Admiralty's declaration, and after John Rae had returned from the Arctic with evidence of the fate of the Expedition, Lady Jane Franklin kept up an unending campaign for ever more searches, petitioning the Admiralty, the Prime Minister, and even the United States government, raising money from friends and public subscriptions as well as from her own resources, and sponsoring further search efforts until McClintock returned in 1859 with incontrovertible proof of the Expedition's fate.

None of the search teams found a living survivor. Nor did they find the ships. They did, however, find a number of dead bodies and many traces of gear from the Expedition. The most telling evidence that any members of the Franklin crew had passed by, and when, came not from the search parties, but, much later, from Charles Francis Hall and U.S. Army Lieutenant Frederick Schwatka, who interviewed Inuit, found human remains and relics, and whose collections of Inuit narratives provided the basis for later reconstructions of what might have happened.[2]

North to Alaska

Several of the Franklin search expeditions headed through the Bering Strait to the north and east coasts of Alaska, hoping to find evidence of the Expedition somewhere along the coast west of the area that Franklin himself had surveyed on his two overland missions some decades earlier. The most notable of these was the 1850 voyage of Captain Richard Collinson's *Enterprise* and Commander Robert McClure's *Investigator* to Alaska—a trip that included being towed through the Strait of Magellan by a paddle-steamer, followed by a long sail up the coast of South America and to Hawaii (then called the Sandwich Islands). Despite orders to stay together, the ships were never in company after leaving the southern tip of South America, in part because McClure was at least as concerned with making the first transit of the Northwest Passage himself, albeit in a west-to-east direction, as he was with finding Franklin.

In the event, McClure did complete a passage of a sort. After intentionally missing several planned rendezvous with Collinson, he sailed *Investigator* eastward along the continental coast of Alaska and Canada, reaching the Mackenzie Delta in late summer 1851, then headed north before he was finally trapped for the first winter in the ice in the Prince of Wales Strait, between Banks and Victoria Islands. The next spring, McClure took a sledge party to the northwest tip of Banks Island, from which he could look out toward the point on Melville Island that had marked Parry's furthest west some 30 years earlier, thus supporting a claim of sorts for McClure's having "discovered" a Passage. Trapped in the ice for two more winters, and with his crew suffering badly from scurvy, McClure and his men were finally rescued in March 1853 by a sledge party sent out from one of the Lancaster Sound–based Franklin search expeditions. Eventually, McClure and his crew sledged across the ice to Dealy Island, joining Captain Henry Kellett's ship *Resolute* there—Kellett was part of the five-ship official search expedition sent out by the Admiralty in 1852–54—and returning to England in 1854, giving McClure yet another claim to have discovered, or at least traversed, the Passage, a claim that was eventually recognized by Parliament, much to the dismay of Lady Jane Franklin.[3]

Another would-be Bering Strait rescue mission was entrusted by Lady Franklin to William Kennedy, a former Hudson's Bay Company employee who had most recently commanded the *Prince Albert* in an 1851–52 search expedition near Somerset and Prince of Wales Islands, accompanied by French enseigne-de-vaisseau Joseph-René Bellot. For the 1853 expedition,[4] which was largely financed by contributions from Franklin's former colonial subjects in Van Diemen's Land, Kennedy was given charge of the 149-ton steam sloop *Isabel* and sent on the long trip around the southern tip of South America, through the Straits of Magellan and then, in theory, to proceed up the Pacific coast of South, Central and North America, through the Bering Strait and onward into the unmapped area where, Lady Jane must have hoped, Franklin and his men still might remain. But Kennedy never made it. In fact, he reached only as far as Valparaiso, toward the southern end of the Pacific coast of Chile. There, in their first landing since braving the Southern Ocean, his crew, perhaps enticed by tales of gold rushes in Australia and California, perhaps simply tired of Kennedy's teetotaling style of command, deserted the ship, leaving Kennedy to carry on for a few years of coastal trading in *Isabel* before returning to England in 1855.

12. Off the Beaten Path—But Where?

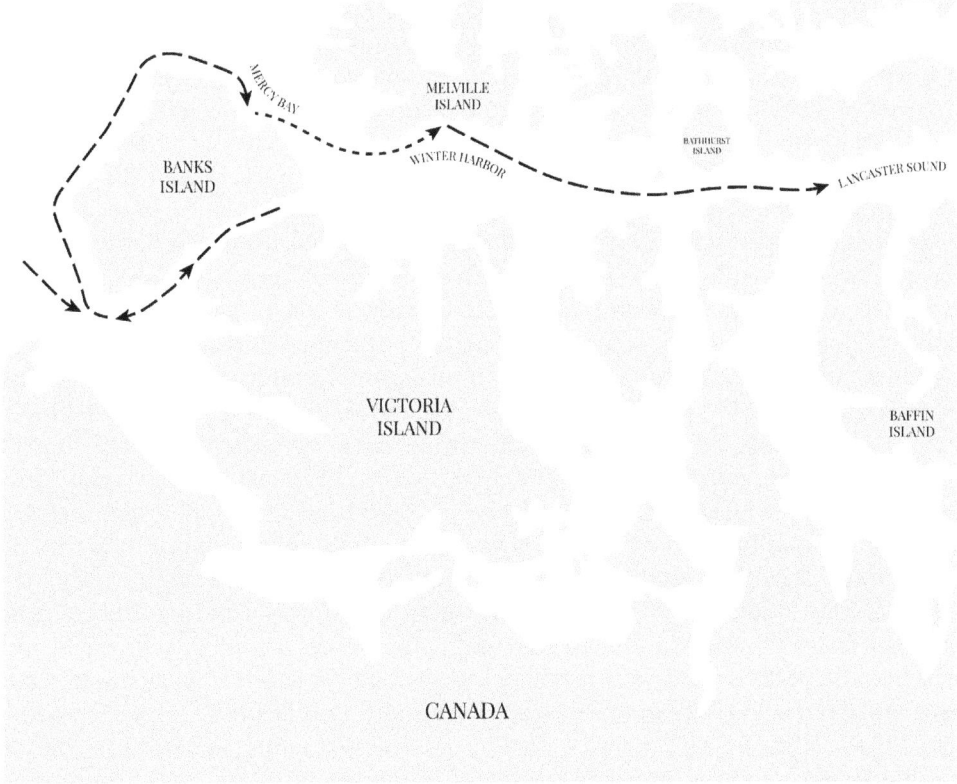

Map of McClure's northwest passage from west to east (map by David Veller).

Alive and Well in Russia?

From the beginning of the Expedition, there was a Russian connection. Fitzjames had expressed the hope that, when the ships reached the Bering Strait, he might be put ashore in the Kamchatka Peninsula, either to dash home to London through Siberia before the onset of winter or, more likely, to spend the winter months making magnetic observations in Okhotsk, Tomsk, or Yakutsk and then heading west across Eurasia to bring the news.[5] Even if he had to over-winter in Siberia, Fitzjames would have been in London before the ships, which would almost certainly have stopped for a lengthy refitting in the Sandwich Islands had they ever made it through to the Pacific.

Fitzjames, as always the optimist, never made it that far, but one well-traveled rumor, originating with the Pond Inlet Inuit and traveling, via a series of whaling ships and other unreliable informants, including the British consul at Mazatlán in Mexico, started with Franklin's ships supposedly beset in Prince Regent Inlet and ended with the Expedition somehow making its way through the Arctic to Kamchatka.[6] Never happened, of course, but this was not the only suggestion that perhaps the Expedition, or at least a part of it, had made its way to Russia, thus completing the Passage. A young Lieutenant Bedford Clapperton Trevelyan Pim, who had been on board the search vessel *Plover* when she wintered over near then-Russian Alaska in 1849–50, had heard rumors of mysterious white men—who could they be but Franklin survivors?—somewhere in the interior. So Pim set out to find them, taking a sledge across the Seward Peninsula and over the frozen ice of Norton Sound.[7] No luck finding survivors there, but coincidentally, he would rescue a different but related group of survivors, just four years later, when another of his sledging expeditions came across McClure and the ragged remnants of his crew from *Investigator*. Searching for Franklin, Pim rescued another party of Franklin searchers instead.

Still, in the face of endless ice and death, leavened only by the occasional lucky rescue, the dream of the survival of at least some of the Franklin crew refused to die. What if some of Franklin's crew had made it not just as far as Alaska, but then across the Bering (or, as it then was, Behring's) Strait to Russian Siberia? Alternatively, as Pim seemed to believe, what if Franklin's crew had sailed through the open Polar Sea and reached Siberia that way? After all, the putative existence of that open sea wasn't definitively disproved until Fridtjof Nansen's *Fram* drifted across the Arctic, all the while firmly locked in the ice, in 1893–96.[8] But, back in England in 1851, Pim still believed in the myth, and proposed to the Admiralty that he travel overland, mostly by sledge, from St. Petersburg, at the far eastern end of the Baltic Sea, to the mouth of the Kolyma River, the easternmost of Siberia's four great rivers, and from there search the coast eastward, along the coast of the Chukchi Sea as far as the Bering Strait if necessary, to find the Franklinites. The Admiralty was decidedly lukewarm to the whole idea, but Pim massaged his connections to pull together funding for a minimal expedition and got as far as an audience with the Tsar in St. Petersburg. There, however, his project ended, perhaps because of Russian fears of letting an English officer conduct what amounted to a survey of the country's northern coast. Eventually, Pim headed back to the Arctic across the Atlantic in 1852 and, as we've seen, ended up rescuing not Franklin, but the starving, scurvy-suffering crew of McClure's *Investigator*.

But what if some of Franklin's crew had actually made it to Siberia? Perhaps some of their descendants would even have crossed paths with the perhaps imaginary descendants of the crew of the ill-fated *Karluk*, flagship of a Canadian Arctic expedition that was caught in the ice and crushed in 1913. Her captain Bob Bartlett, the survivors of his crew and his Chukchi Inuit companions endured extreme cold and hunger—11 of them died—before they were rescued.[9]

As late as 1870, Lady Jane Franklin still apparently believed that the expedition might have headed toward Russia, or at least made it some of the way into the western part of the North American Arctic archipelago. That year she and Sophia Cracroft traveled to Alaska, newly purchased from Russia by the U.S. government, in the hope of finding artifacts of the Franklin Expedition that had somehow made their way into Russian hands.[10] Once again, they found no evidence that any of the Expedition crew had made it that far, as, indeed, there was no evidence to be found.

Melville Peninsula, Pond Inlet or Hudson's Bay?

David Woodman, in *Strangers Among Us*,[11] his companion book to *Unravelling the Franklin Mystery*,[12] compiles the Inuit narratives recorded principally by Charles Francis Hall, but also by John Rae, Frederick Schwatka and others, that point to the presence of white men in the Melville Peninsula, some hundreds of miles east of where *Erebus and Terror* were abandoned, in the years immediately after the crew left the ships, and possibly as late as 1854. While many Franklin Expedition historians have attributed

12. Off the Beaten Path—But Where?

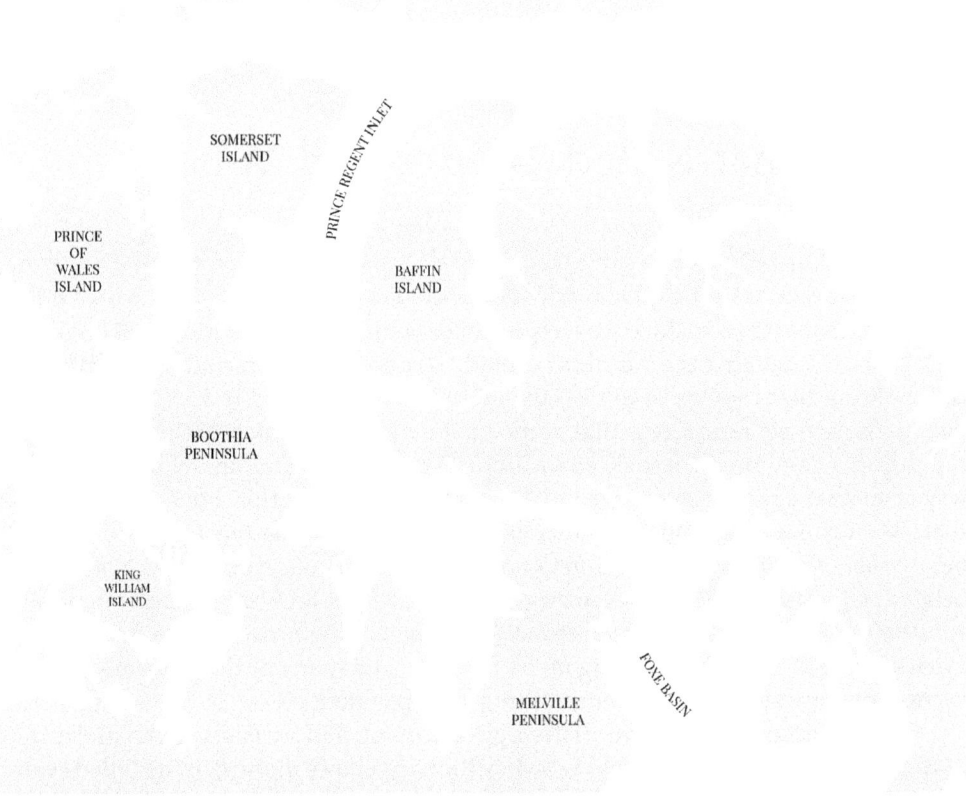

Map of King William Island to Melville Peninsula (map by David Veller).

these sightings either to memories of John Rae, who did indeed travel to parts of the peninsula in 1847 and again in 1854, or to memories of Captain Edward Parry, who had been in the area back in the 1820s, Woodman marshals substantial evidence, from close analysis of the Inuit narratives, to argue that it is at least possible that there were two groups of Franklin survivors—or perhaps a single group that later split in two—that made it at least as far as the northwest corner of that peninsula, near the area now known as Parry Bay.[13] These sightings by the Inuit, if they are in fact sightings of Franklin's men and not of other European explorers or even non–Inuit Indians, are in addition to the various stories of white hunting parties seen at locations much closer to the areas where the Expedition's ships were abandoned. In the most optimistic of these Melville Peninsula stories, most of which were recorded in the 1860s by Charles Francis Hall, a group of more than a dozen white men headed off to the east, aiming, perhaps, for the large Inuit settlements near Repulse Bay or perhaps to the north, toward Pond Inlet on Baffin Island, a known location that was known to be occasionally visited by English whalers; and a smaller, separate or breakaway group was said to have been sheltered by Inuit and then headed off to the south, across Chesterfield Inlet, toward the known Hudson's Bay Company depots far down the coast of Hudson's Bay, at Churchill and York Factory. For now, all we can say is that there was no physical evidence to show that the Franklins had been there, just the Inuit oral history.

13

What Killed Them—and When?

If we were to take the official archives of the Royal Navy at their word, we'd have to believe that virtually the entire Franklin expedition died on March 31, 1854.[1] At least that's the day that the Admiralty finally struck the officers and men off the payroll, holding that they were to be "considered as having died in Her Majesty's Service."[2] Though we now know that some of the men died much earlier, even before the sailors abandoned their ships in Victoria Strait, and that many died along the way after that abandonment, we don't even know for sure that none survived after that. We surmise, based on the scanty evidence available, that most of them died well before that official date, but we don't know that to an absolute certainty. Not that any belated requests for back pay were ever received by John Barrow's successors (though some £80,000 in wages already accrued at the time of the formal announcement was eventually paid out to the missing men's families, and many of the presumed widows eventually were awarded pensions or lump-sum payments[3]).

But by January 1854, the British government had already spent more than £610,000 (about £73 million or U.S. $94 million in today's money), in its failed search attempts, and that sum doesn't even include the costs of Jane Franklin's and others' privately financed expeditions and of the American search efforts.[4] With the Crimean War heating up, those in the British government with fiscal authority evidently decided that it was time to stop spending the taxpayers' money on an expedition that was almost certainly gone forever. So, no more regular paychecks for the crewmembers' families, just one final lump-sum payment of accrued back wages and pensions for the officers' wives. And, by the end of 1855, the Admiralty had officially killed off the ships as well, assigning the names *Erebus* and *Terror* to new vessels.[5]

According to the Victory Point Record, that some 24 members of the expedition, or nearly 19 percent of those who set out across Davis Strait after their last contact with other Europeans in 1845, had died even before the ships were abandoned. Compare that to the overall death rate of just 3 percent on Royal Navy discovery expeditions between 1819 and 1836, or of just 4 percent on the four-year Antarctic mission of James Clark Ross and Francis Crozier in 1839–1843.[6] And the eventual death rate of 100 percent represents a major accomplishment, if of a somewhat unwanted kind. For the 19th century as a whole, the death rate on government-funded Arctic expeditions like Franklin's was under 6 percent (barely more than 2 percent if we exclude Franklin's total loss, which skews the overall statistics), and, on privately funded expeditions, less than 1 percent.[7]

But we don't know what happened to the 21 Expedition crew members who reportedly died between the time the ships left Beechey Island in summer 1846 and the time when *Erebus* and *Terror* were abandoned in April 1848. Nor do we know the precise fate of each of the 105 Expedition crew members who left the ships in April 1848. A comprehensive survey by archaeologist Doug Stenton[8] in 2018 could conclude only that the remains of somewhere between 45 and 85 of the crew had been discovered and confirmed as likely being the bones of members of the Expedition. So at least 20, and perhaps many more, of the crew have simply vanished into the Arctic fog and mist. Did some of those die along with others whose remains have been recovered, but simply left no trace? Did some end up far off the well-beaten searchers' path? We do know that a lot of them must have died in a relatively short period of time, and the logical question, one asked many times by historians of the Expedition, is: what killed them?

That's a question that has concentrated the minds of many scholars of Arctic expeditions, not just those who have looked at what happened to the Franklin crew. Even as simple a case as the deaths of the three balloonists of the Swedish Andrée North Pole expedition in 1897—a case where all the bones and most of their campsite supplies were ultimately recovered—remains unresolved, with nearly a dozen plausible, if not provable, causes of death.[9]

Despite heroic work by an army of scientific sleuths, amateur and professional, we don't really know. Here are some possibilities: hunger, exposure to the cold, tuberculosis, lead poisoning, botulism, trichinosis, zinc deficiency, beriberi, scurvy, simple exhaustion, and, oh yes, cannibalism. Let's start with the easiest cases. As we'll see, even those are not completely free from doubt and controversy.

The Beechey Island Graves

Three members of the Expedition were buried on Beechey Island, the site of the first over-wintering in 1845–46. Thanks to the work of Owen Beattie and his team of forensic anthropologists who visited the island in 1984 and 1986 and conducted autopsies, we pretty much know what killed them, though some doubts remain.[10] The team's first autopsy, in 1984, was of *Terror*'s leading stoker, 19-year-old John Torrington, who had died on January 1, 1846, and who, the scientists agreed, had died from pneumonia brought on by an underlying case of tuberculosis, though, like the remains of many of the Expedition members, Torrington's body also exhibited high lead levels.[11] On their second visit to the island in 1986, Beattie and his colleagues autopsied the other two frozen bodies, those of Royal Marine William Braine and Able Seaman John Hartnell, both of *Erebus*, and reached a similar conclusion—pneumonia and tuberculosis. At least one of these early dead, Private Braine, appears to have died while on a sledging trip, so perhaps exhaustion or the cold may have been a factor as well.[12] All three bodies, Beattie found, showed very high levels of lead in bones and tissues. In Beattie's view, the lead poisoning may have contributed to the early deaths and may have been even more important later in the expedition as lead levels—which Beattie attributed to solder in the tins containing the canned food—increased over time. Tuberculosis and pneumonia? Lead poisoning? All of the above? Even in what seem to be the easiest cases, where the bodies were frozen for over a century and were available for autopsy using modern scientific techniques, there's an inescapable residue of uncertainty. For the rest of the crew, where the only evidence that we have is a scattering of bones, how much less certain will we be?

The balance of this chapter examines the various causes of death that have been proposed over the years by one or another Franklin historian. For most of these putative causes, the verdict must remain: not proven.

Was There Enough Food?

> "As I hurtled through space, one thought kept crossing my mind—
> every part of this rocket was supplied by the lowest bidder."
> —U.S. astronaut John Glenn[13]

Like American rockets, British 19th-century naval discovery expeditions were outfitted, most of the time, by the lowest bidder. And, like those rockets, once the ships left home, there wasn't much that could be done to resupply them or to fix problems. As a later polar explorer noted, "The little ship which bears the hope of a polar expedition must contain in its restricted space everything to supply all the needs of its people for two or three years in a region where nothing can be obtained but meat, and even that only by those who possess the 'know-how.' Even when the needs are reduced to an almost primeval simplicity, the multiplicity of essential things is great."[14]

Despite the earlier examples of John Rae and other Hudson's Bay Company stalwarts, it wasn't until Lieutenant Frederick Schwatka's expedition in 1878–1880 that any group of more than a handful of non–Inuit were able to live off the land in the Arctic; Schwatka's men fed themselves on the trail by killing more than 500 caribou, plus assorted musk-oxen, seals and bears.[15] But, even so, Schwatka had only 17 men with him, compared to the 105 who left *Erebus* and *Terror* in April 1848, and he appears to have had better weaponry than Franklin's expedition possessed. Also, Schwatka generally was for the most part in a different area; only a small portion of his travel time was actually spent on King William Island, a place the Inuit knew to be a terrible hunting ground in the mid–1840s, just when the Franklin crew would have been searching for food there. So, after three years on the ships, and despite Franklin's boast to the whalers off Greenland that he had enough supplies for five years, it's likely that the ships' stores were running low by that final April in the ice. No fresh meat or vegetables, the rest of the stores suffering from the inevitable effects of three years' storage. And the problem would have been even more acute if some of the canned food, perhaps in the extra-large and likely under-sterilized tins that supplier Stephen Goldner had rushed to the dock at Greenhithe at the very last minute, turned out to be inedible. So, the men probably left the ships already hungry, and without many ways to remedy that situation.

Accidental Death

Perhaps none of the various causes of death hypothesized over the years satisfies; in that case, one can always fall back on, well, bad luck. Keith Millar's and his colleagues' study of the health of nine other Royal Navy crews operating in the Arctic[16] suggests that all the standard explanations for the high death rates—nine officers and 15 sailors by the time the ships were locked in the ice—didn't match the experience of the other mid-century voyages, and so one must be open to the possibility of a catastrophic accident. Only one problem: there's no evidence for such an accident. But, of course, absence of evidence is not evidence of absence; we can't know for certain that nothing catastrophic happened before the Victory Point Record was first deposited in 1847 (when the situation was described as "all well") or even by the time the ships were deserted and the handwritten note left around the margin in 1848, updating the situation as something decidedly less than "all well." But, if such a thing had occurred, one might reasonably think it would have at least been mentioned in the one piece of paper the Expedition is known to have left behind. So was it possible that there had been a catastrophe? Of course. Probable? Not really. A few of the 21 men who died between Beechey Island and Victory Point may have succumbed to accidents—falling from a mast or slipping into a crevice in the ice during a sledging trip—but not all of them. And, after the men left the ships in 1848, there would have been fewer opportunities for accidental death on a mass scale.

Beriberi

Beriberi is a disease caused by Vitamin B-1 or thiamine deficiency. It occurs in two variants, one generally leading to heart failure and the other to nervous system degeneration and paralysis. Anyone with a decent, well-rounded diet won't get it, but a decent well-rounded diet hardly describes the food that was available to the Franklin Expedition, especially by 1848.

And beriberi wasn't unknown in the Navy. Several cases had been reported in sailors off Labrador and Newfoundland around the turn of the 19th century, and the disease's rise was associated with a switch from dark bread to white, the latter made with milled flour that effectively eliminated the B-1-containing roughage needed to prevent the disease.[17]

Like scurvy and zinc deficiency, beriberi is a slow-acting, cumulative killer. By the time it struck, its victims would likely have been suffering from a variety of other problems as well. Was it a problem for the Franklin crews? Perhaps; the more they used refined flour for their biscuits and bread, rather than whole wheat, the less Vitamin B-1 it would contain. The ships' manifests[18] simply list "biscuit" and "flour," without specifying the type, so yet another area of apparently irreducible uncertainty.

Symptoms of beriberi often mimicked, and were confused with, the symptoms of scurvy.[19] Add it to the list of ailments the crew may have been suffering from, but perhaps not as the primary cause of death.

Botulism[20]

An early candidate for the primary cause of death was botulism, supposed to have been carried in the improperly canned, so it was said, food provided to the Expedition by Stephen Goldner's cannery in Houndsditch, East London. When

Captain Erasmus Ommanney, on one of the early search missions, reported on the cache of tins that his Franklin search found on Beechey Island in 1850, he claimed that many of them contained rotten meat—though by that time any meat would have been exposed to the air for half a dozen years—and that the tins had been filled with "putrid abominations ... thus fatally diminishing the three years' provisions which were supposed to be on board."[21] More recently, one of the books that revived interest in the Franklin Expedition late in the 20th century, Scott Cookman's *Ice Blink*,[22] tried to lay the entire blame on Goldner, portraying him, not without a hint of anti–Semitism, as an evil genius, intent on enriching himself by ignoring the requirements of public health and deliberately sending the Expedition off into the unknown with ships full of poison. It is true that some years later, after Goldner had supplied many Royal Navy vessels with perfectly edible and untainted canned food, a number of the tins he had supplied under later contracts were examined and found to be dangerous and inedible.[23] But the Franklin Expedition was Goldner's first significant Royal Navy contract, and he supplied the Navy for a number of years before any complaints arose. It's likely that the reason he was the low bidder on the Franklin contract was not, as Cookman claims, that he never planned to deliver what he had promised, but rather that he had an innovative supply chain, reaching back into Moldavia, from which he could obtain meat far more cheaply than English canners relying on local sources. According to one estimate, Goldner, who had established his Moldavian raw-materials supplier in 1844, was able to obtain beef for five pence per pound, while his competitors had to pay 7½ pence in England.[24] Some of his canning methods may have been less than perfect, though not necessarily to the gruesome extent that Cookman suggests, but the evidence of when and where the Franklin crew members died doesn't appear to support a major role for botulism.

And the attacks on Goldner, both at the time of the Franklin search in the 1850s and later, undoubtedly drew on the then-still-rampant anti–Semitism in England. The English statute *De Judaismo* of 1275, which barred Jews from most professions, limited where they could live, required them to wear yellow badges, and relieved debtors of any obligations they might owe to Jewish lenders, was officially repealed only in 1846, and the 1852 Parliamentary inquiry into Goldner's Royal Navy contracts was riddled with overt references to his very non–British religion.[25]

Goldner and his associate John Wertheimer had each obtained British patents for their canning method, which involved heating the cans in a chemical bath at temperatures as high as 280° Fahrenheit to drive off any air in the can as it was being sealed.[26] Goldner's method was commented upon favorably by at least two professors of chemistry in London, and the Admiralty didn't complain about Goldner's quality until 1849. Could many of the tins sent out with *Erebus* and *Terror* have gone bad? Of course, And Goldner's use of tins larger than the two- to six-pound units originally contracted for may have been associated with inadequate cooking and sterilization in the center of the larger tins, but Cookman's accusation, at best, rates a verdict of "not proven." A comprehensive survey of medical issues in the Royal Navy concludes that there is no evidence that Goldner's supplies to Expedition were of such poor quality that botulism from the canned goods can be blamed for any significant number of deaths.[27]

In any event, there are a number of reasons why it's highly unlikely that Goldner's canned goods killed off the men of the Expedition: (1) the Franklin Expedition contract was Goldner's first major deal with the Royal Navy; it's unlikely he would have handled it as badly as his critics suggest. (2) No complaints about Goldner's provisions arose until seven years after he had supplied Franklin, and after changes in the Navy's specifications, calling for larger tins and tighter filling of them, had made the canning process more problematic. (3) Several Expedition members, including Fitzjames and Engineer James Thompson, sent back letters from Greenland attesting to the good quality of the tinned food.[28]

Moreover, the Navy had been using tinned food, albeit from other suppliers, since at least 1814, when a supply was sent to St. Helena, and some cans were supplied to John Ross's *Isabella* and *Alexander* on their 1818 voyage to Baffin Bay.[29] These canned supplies, prepared by the pioneering British firm of Donkin & Gamble, were routinely used on Arctic voyages through the first half of the 19th century. And when Parry's *Fury* was wrecked in 1825, some of those tines were offloaded and later found by John Ross on his 1829–1833 expedition; Ross reported that the food was in perfect condition.[30] Supplies of canned food from other British firms were also used by Royal Navy Arctic expeditions in the pre–Franklin period without any apparent complaints. Goldner did indeed deliver a substantial portion of the canned provisions quite late in the process of preparing the ships for sea, and that late delivery may well have raised questions that wouldn't arise in the case of the Navy's older, and perhaps more punctual, suppliers. But the encomiums sent back in letters from Greenland attesting to the provisions' quality make one a bit hesitant to put all, or maybe even any, of the blame on Goldner.

A number of authors cite the large number of empty tin cans found on Beechey Island as evidence that something must have been wrong with the tinned food for that many of the cans to have been jettisoned that early in the expedition. But May Fluhmann, in her biography of Francis Crozier, calculates that, if the Expedition relied primarily on tinned rations over the winter of 1845–46, it would have used up well over 1,000 of Goldner's cans. So, the tins found on the island may have just been a large garbage pile, with no significance as to the health or safety of the food.[31] The expedition started with some 64,000 pounds of tinned food[32]; if those 1,000 tins left behind at Beechey Island averaged 4 pounds apiece, that would still have been only a small fraction of the Expedition's supplies. Based on that quantitative evidence, there's no reason to suppose that the canned food supply as a whole was contaminated.

It's true that botulism spores were detected when Marine William Braine's body, buried on Beechey Island, was analyzed. But there is at least an argument that Braine's botulism was caused by eating seal or walrus meat, rather than the Expedition's canned food.[33] And botulism may have been involved in some of the 24 deaths that the Expedition suffered before the ships were abandoned in April 1848. But whether those botulism attacks, if they occurred, were from the canned food or from fresh seal or walrus meat is something that is unknowable. There is ample evidence that "fresh" meat, if subjected to repeated freeze-and-thaw cycles before being eaten, can lead to botulism,[34] though none of the search expeditions that went looking for Franklin reported botulism symptoms among their crews, and many of those expeditions also relied to some extent on fresh seal and walrus meat.[35] Notwithstanding the luck of the Franklin search expeditions, there is ample evidence that the available Arctic meat and seafood diet provides substantial opportunity for botulism to occur.[36]

Exposure and the Cold

The Arctic is cold. The fleets of Franklin searchers in the 1850s recorded winter temperatures of -50° Fahrenheit (roughly -45° Celsius) or colder. And the Royal Navy's clothing for the Discovery Service was hardly adapted to the region. No Gore-Tex parkas. No fur clothing of the kind developed by the Inuit over thousands of years in the region, although some of the officers, at least, were apparently issued sealskin outer coats and caps.[37] Lots of wool, though, which absorbed water and then froze, rendering the wearer uncomfortable and nearly immobile. And the crew's thin leather boots, worn outside blue cloth leggings and perhaps stuffed with a bit of straw or rag as insulation, were hardly a match for the sealskin *kamik* of their Inuit counterparts.

The Arctic in 1845–48 was a lot colder than it is today, or even than it was before the onset of global warming in the last half of the 20th century. A careful study of ice cores from the region has led the researchers to suggest that the Franklin Expedition (as well as much of the subsequent search effort) took place during one of the coldest and least navigable periods in the region within the past 1,000 years.[38] At best, the ice-core evidence suggests, only one year in five would have allowed any sort of sailing through open water in the summer, while perhaps two of the other five years would have allowed some progress to have arduously been made by hauling the ships through leads in the ice that temporarily appeared and then, just as quickly, froze over. If the summer of 1846, when the ships made it from Beechey Island to the waters off King William Island, was that one open-water year, it's not surprising that sailing conditions, and, by implication, the cold, were worse the next two years.

And, despite the best efforts of evolution, humans actually don't have very good tolerance for extreme cold. As one leading text notes, "humans for the most part show surprisingly little physiological adaptation to low temperature.... Clothing and shelter remain the first line of defense against the cold."[39] And, as Peter Marchand notes, while "we have exploited the coldest places on earth, [w]e remain essentially tropical creatures."[40] Even the Inuit, perhaps the most cold-weather adapted of all humans, rely not so much on body fat, in the way that layers of blubber protect whales, seals and other Arctic creatures, but more on subtle evolutionary mechanisms: quicker dilation of blood vessels to permit more blood flow to the extremities, lower normal heart rates and lower blood pressure[41]—that make life in the Arctic just barely possible. For the Franklin Expedition's crew, without benefit of centuries of such evolution, the cold would have been a serious threat to human life.

In fact, the mid–1840s don't seem to have been a good time for travel anywhere in North America. Perhaps it's just a coincidence, but the Donner Party, which also resorted to cannibalism, was trapped in the snow in the Sierra Nevada in the winter of 1846–47, the result, similar to Franklin's, of a late start and an early winter.[42] Although some of the Donner Party eventually made it out of the Sierra Nevada to safety, many died in the cold, and a few, at least, resorted to cannibalism.

Mental Illness

The Arctic isn't a place conducive to mental health. On the American Greeley expedition to Ellesmere Island in 1881–84, for example, the crew "was driven to mutiny, madness, suicide and cannibalism, leaving six survivors of a crew of 25 men."[43] The causes of mental stress, if not outright mental illness, aren't difficult to determine. First, there are the physiological issues: daylight-related disruption of circadian rhythms, resulting from 24 hours of daylight in summer and of darkness in winter. Then there are temperature-related changes in circulation in the extremities, hypothermia and frostbite, as well as suppression of the immune system and hormonal changes.[44] Many of the personality changes are largely attributable to changes in thyroid hormones caused by the harsh physical environment. In Antarctica today, that's easily treatable with thyroid supplements.[45] In Victoria Strait in the winters of 1846–47 and 1847–48, not so much. No vitamin supplements, no tanning lamps, no therapists available for Zoom calls.

Even if it were not for all those physical stresses, the artificial conditions of life in a closed-in frozen world leads, almost inevitably, to mental strain as well. Twentieth-century studies of Arctic and Antarctic expeditions reveal a wide range of symptoms that occur with considerable regularity. Among the more common are headaches, boredom, fatigue, inattention to personal hygiene, "intellectual inertia," and increased appetite and weight gain. In turn, these issues often lead to sleep disruption, cognitive impairment, negative affect (often coupled with anxiety and irritability) and increased interpersonal tension.[46] Some expeditions that followed Franklin by a half-century or more still suffered from widespread depression, lethargy and even psychosis.[47] And if that's what happens in a well-supplied permanent polar station late in the 20th century, with no food shortages, plenty of artificial light, modern medicine and contact with the outside world, how much worse would the situation have been for the crews of *Erebus* and *Terror*, condemned to spend a third year—and who knew how many more?—completely alone, with their food spoiled or running out, and in darkness?

Even after the intense psychological screening that's part of the selection process for modern Arctic and Antarctic assignments, some five percent of over-wintering Antarctic crew members and military personnel on Arctic winter duty show diagnostic signs of seasonal affective disorder or some other defined psychiatric problem.[48] While modern psychiatric diagnostic tools weren't available to the ships' surgeons on the Franklin expedition, we do have reports from other expeditions, ranging from the literally insane Greeley crew to other lesser examples of antisocial behavior. Even if we accept the modern five percent prevalence of seasonal affective disorder among selected scientists and service members or the six percent reported among members of modern native Arctic peoples, who presumably have centuries of adaptational evolution working for them,[49] that would indicate that at least some six or seven of the Franklin expedition's 125 crew were certifiably crazy. In fact, given the lack of psychological selection and the additional stresses of the voyage, the number was probably far higher. Unlike modern Antarctic and space voyages, there was virtually no screening of the Franklin Expedition's crew for psychological factors, other than Fitzjames's choice of some officers because he had gotten on well with them in the Mediterranean, Iraq or China. Recent research has identified certain clusters of psychological traits as even more important than technical skill or training factors in the success or otherwise of long-term missions in isolated circumstances.[50] Just being in the Arctic for three long dark winters could have affected the mental health of the officers and crew, making decisions less rational or just harder to make, increasing the likelihood of senseless violence on the one hand or lethargy and torpor on the other. Extreme mental illness, leading to murder or suicide, may not have been common, or perhaps not even present at all, among the Franklin crew, but it's almost certain that there was a significant amount of mental impairment just from the environment. Now let's add a bit of lead poisoning to the mix, in an already difficult and apparently hopeless situation, and we get a recipe for very bad outcomes.

Lead Poisoning[51]

A considerable body of writing on the Franklin Expedition treats lead poisoning as a major contributor to the deaths that occurred, either directly or as a result of the lead poisoning's impairment of the crew's mental faculties. As early as the 1980s, Franklin researchers knew that the remains of the Expedition's crew evidenced very high levels of lead.[52] In fact, "every member of the expedition whose bone has been analyzed shows very high levels of lead contamination."[53] That lead was there, in the crew's bodies, is not at issue. What is still a matter of scientific debate, however, is (1) what the source of that lead was, and (2) whether lead poisoning killed them or contributed materially to their deaths.

Lead poisoning certainly doesn't help people to perform at their best. Its mental and physical effects include memory lapses, headaches, irritability, dizziness and general physical weakness. At high exposure levels, similar to those found in many of the crew members' bones, there is significant impairment of brain function as well as physical symptoms that include vomiting, convulsions, coma and, eventually, death.[54]

Not everyone agrees that lead was a major factor.[55] For one thing, the three men buried on Beechey Island, who all died very early in the voyage, also had very high lead levels, but died from other causes, as did Assistant Surgeon Harry Goodsir, whose subsequently-identified body was found on the south coast of King William Island, where he may have died during a sledging trip before the ships were abandoned.[56] Goodsir, in fact, may have died from sepsis that originated in a toothache![57] One recent study suggests that, while Goodsir may have evidenced some impairment in both cognitive and motor functions as a result of lifelong lead exposure, it's unlikely that lead poisoning either caused his death or substantially interfered with his performance of his duties.[58]

While lead levels found in the remains of the Franklin crew were high—astonishingly high by modern standards—they may not have been all that extreme by then-current standards among the English working class.[59] And there is a great variance between individuals in the extent to which they may be affected by lead poisoning.[60] And, as Keith Millar and his colleagues point out, the bad effects of lead poisoning could well have been exacerbated by other factors that were also plaguing the Franklin Expedition, among them, stress, vitamin C deficiency and starvation.

As for the source of the lead, Beattie and those drawing on his research argue that it came from the solder in Goldner's cans; the solder, it is thought, must have leached out into the food contents of the cans. But the Navy had been using tinned food for several decades before Franklin sailed, with no more lead poisoning than the average for everyone at the time, and no reports of crews gone amok or dying from its effects. Moreover, *Erebus* and *Terror* themselves had spent four years in Antarctica, relying to a considerable degree on tinned food supplies, without obvious evidence of widespread lead poisoning. Even if the solder used to seal the tins was rather casually applied, it's more likely than not that electrolytic action would have caused the lead in the solder to migrate to the tins, and not into their contents.[61] Nor did the various rescue missions sent after Franklin seem to have a problem with lead exposure; the symptoms of lead poisoning were well known, even at that time, and none of the search expeditions' medical officers reported it as an issue.[62]

And several studies question whether it would even have been possible for the sailors' bodies to have absorbed as much lead in just three years as was found when their bones were analyzed.[63] An alternative hypothesis is that the extra lead came from the ships' fresh water systems, which had been modified for the 1845 Expedition and which were different from those used elsewhere in the Royal Navy.[64] And, in fact, this last theory might even explain why officers died disproportionately before the ships were abandoned; they would likely have been given the first draw of water from the melting tank, which might have carried the highest lead levels.[65] If, in fact, lead poisoning was what killed them or contributed to their deaths.

And the various rescue missions sent after Franklin didn't seem to have a problem with lead exposure; as stated above the symptoms of lead poisoning were well known, and none of the search expeditions' medical officers reported it as an issue.[66] So, yes, the officers and men of the Franklin Expedition did indeed have high levels of lead in their bodies, at least by modern standards. But, no, it's highly unlikely that lead poisoning was what killed them. But did exposure to and absorption of lead so interfere with their ability to plan and reason, so as to lead to decisions that hastened or even caused their deaths? Especially when, after April 1848, the Royal Navy's vaunted command system must have broken down, leaving at least some of the groups of men without recognized decision-makers to give them orders. Most of the subsequent rescue missions didn't last as long as Franklin's voyage had; perhaps it just takes that long, and a large enough crew, for there to be sufficient symptoms of lead poisoning to disrupt things. So, what can we conclude? Maybe lead didn't kill a lot of the crew directly, but perhaps it made it easier for other causes to do the job.

Scurvy[67]

By the time Franklin sailed in 1845, scurvy was well known as a primary threat to sailors on extended voyages. From the time of Columbus through the mid–19th century, in fact, scurvy is estimated to have killed some two million sailors, more than storms, shipwrecks and wars combined.[68] The first reports of scurvy began with the advent of long, ocean-crossing voyages at the end of the 15th century; Vasco da Gama's crew reported many cases and deaths on his journey round Africa to Asia in 1497, and a century later, the first Dutch East Indies Company fleet sailed with 249 men and returned with only 88—most of the deaths due to scurvy.[69] The company's second fleet took along lemon juice and grew "scurvy grass," a form of cress, on board and lost only 15. Spain's navy also had extensive experience with scurvy on its voyages to Central and South America and learned early on that lemons and oranges were an effective preventative and cure. In fact, most of the world's actual sailors—if not those who gave them orders from desks in European capitals—were well aware of the value of lemon juice in particular, and citrus fruits and fresh vegetables more generally, as an effective remedy for scurvy.[70] The lesson should have been brought home to the Admiralty when, during the 1740–1744 circumnavigation of the globe by Commodore George Anson's squadron, nearly 1,000 of the original complement of 1,854 died of scurvy; on the advice of the College of Physicians, the Admiralty had stocked the ships with "elixir of vitriol" (a combination of sulfuric acid, alcohol, sugar and spices) instead of lemon juice. After that, individual British commanders made sure their ships carried lemon juice, though it took the Admiralty another 50 years, until 1794, to make the treatment mandatory.[71]

Although vitamins weren't discovered until much later—and the actual cause of scurvy is Vitamin C (ascorbic acid) deficiency—the British doctor James Lind had, back in the mid–18th century, conducted a kind of clinical trial that showed that lemon juice could prevent and even cure scurvy,[72] though Lind himself never actively advocated for the use of lemon juice as a general cure. While Lind's primacy as a discoverer of the cure for and prevention of scurvy is contested, he generally gets the credit. By the time Franklin sailed, the Royal Navy knew that the disease would be an issue on long voyages without adequate fresh food, and, as described below, the Navy tried to equip the ships as well as the science of the time allowed.[73]

Among scurvy's symptoms are hemorrhaging, especially in the gums and mouth tissues, loosening teeth, joint stiffness and pain, swelling of the extremities, re-emergence of old wounds, and eventually a general debility leading to death. A normally healthy person typically has sufficient Vitamin C in their system to ward off the disease for about 90 days,[74] but after that the body's supplies of ascorbic acid needs to be replaced. On Vasco da Gama's voyage around Africa, beginning in 1497, the dreaded symptoms began to appear just after more than 100 days at sea without fresh food. Similar symptoms appeared on Magellan's voyage around the globe in 1520–21.[75]

Franklin's ships did carry some 9,300 pounds of lemon juice,[76] and sailors were given a daily ration of an ounce a day, which should have been enough for the entire crew until some months after the date on which the ships were first abandoned. But the Navy didn't know then that Vitamin C in lemon juice deteriorates over time, and so, by the end of the first winter in the ice, it's likely that the anti-scorbutic effects of the lemon juice were diminishing, although the Royal Navy generally tried to prevent deterioration by storing lemon juice in containers with olive oil poured on top to keep air out, a precaution that closed the barn door long after the horse, in the form of effective Vitamin C, had escaped.[77] And, just to ensure its ineffectiveness, it's likely the lemon juice would also have been boiled at some point before being given to Franklin's sailors; heating lemon juice to 90° Celsius (194° Fahrenheit) for just three minutes would be sufficient to reduce its potency by half.[78] With similar precautions, or lack thereof. John Ross's 1829 expedition had spent four winters in the Arctic with only a single death from scurvy, and then only in the final year.[79] But it's likely that potent lemon juice was not what had saved Ross's crew; that expedition had access to fresh meat, which would have been a reasonably effective substitute, as would fresh herbs and vegetables grown on board, as some expeditions did. As explorer Robert Peary points out above, securing fresh meat depends on some skill in hunting, which it's not clear that the Franklin Expedition had in abundance. A few other expeditions did ward off scurvy without the help of lemon juice, but those were generally (1) much smaller than Franklin's and (2) made up of travelers who, like John Rae and his colleagues, were skilled at living off the land, good hunters, and well acclimated to the Arctic, and eschewed alcohol.[80]

Other Arctic expeditions certainly suffered from scurvy. When McClure's *Investigator* crew was finally rescued in 1853, all but four of the men aboard were found to have significant symptoms.[81] In fact, scurvy affected every one of the Franklin search ships that spent two or more winters in the ice, and even appeared in two ships that wintered over for only a single year. In all those cases, as with Franklin, the crews had access to ample supplies of lemon juice, at least for the first two years.[82]

So, did the crews of *Erebus* and *Terror* suffer from scurvy, and did it get worse as the expedition went on into its third and fourth year? In the voluminous Franklin literature, scurvy has been seen as an important, perhaps leading, cause of death, beginning with McClintock's observations in 1859 and continuing through Cyriax's once-definitive history of the Expedition, their conclusions repeated by a host of other writers. Cyriax concluded, from an examination of the contents of a medicine chest that had been found by Lieutenant Hobson of the McClintock search party in 1859, that the medicines normally carried on board to deal with scurvy had been used up and that, therefore, scurvy was at least one of the maladies affecting those leaving the ships.[83] Scurvy was almost certainly present among the crew of the Expedition, but how important was it in killing the men or in accelerating their demise?

The most exhaustive study of the issue is in Simon Mays's and colleagues' report in the *International Journal of Osteoarchaeology* in 2013.[84] They studied the bones of at least 15 individuals—possibly more—that had been found along the Expedition's line of retreat on the coast of King William Island and concluded, based primarily on the absence of the bone lesions typically associated with scurvy, that "the skeletal remains provide little support for the idea that scurvy was a serious problem for the expedition prior to the desertion of the ships ... contrary to claims in the literature, the remains do not provide support for the scurvy hypothesis."[85] And in the same vein, the "all well" message in the Victory Point Record as of spring 1847 suggests that the disease had not, at least by then—two years into the mission—become a major health problem. But that 2013 study was of only a small sample of the crews—some 15 individuals or thereabouts—so it does not necessarily prove that scurvy was not, by 1848, affecting a significant portion of the men. And Inuit testimony about encountering men with black teeth and gums—clear symptoms of scurvy—supports a conclusion that some of the men may well have had the disease. Sooner or later, it would have killed them, but did some other cause get there faster?

Starvation

It seems the men died with at least some supplies still available. When Lieutenant Hobson, and subsequently Captain McClintock, found the "boat place" at Erebus Bay on King William Island in May 1859, among the supplies in the boat with the two bodies was some 40 pounds of "chocolate." If the Franklin Expedition crew had actually been dying of hunger, wouldn't they have at least tried to eat the chocolate?

It might not have been a complete food group—though my children would have disagreed—but surely it could have sustained the sailors a bit longer. After all, Franklin himself had, once upon a time, eaten his boots.

Chocolate, though, wasn't the same thing in mid–19th-century naval stores as it is in a modern supermarket.[86] The "chocolate" found at the "Boat Place" was almost certainly hard-pressed cocoa cake—roasted cocoa beans that had been ground into a paste, then cooled into a storable cake form. That cake was hard as a rock, and just about as tasty. A bit could be broken off, then mixed with hot water and sugar to make a palatable hot drink. If the exhausted Franklin sailors were too weak, or lacked the means, to melt and heat water and mix it with the chocolate, it would have been of little use, other than as something that would have broken what few remaining teeth they had as they tried to chew it.

More generally, by the spring of 1848, supplies were almost certainly running short. By way of comparison, here's the standard Royal Navy weekly ration at the time the Expedition left Greenhithe, a ration that would have supplied something like 5,000 calories per day, not at all excessive for men working in a cold, difficult environment:

7 pounds of biscuits or bread
7 gallons of weak (2–3 percent alcohol) beer
4 pounds of beef
2 pounds of pork
2 pints of peas or other vegetables
3 pints of oatmeal
6 ounces of butter and
12 ounces of cheese.[87]

Even if supplies remained on board in April 1848 to provide anything like the recommended amounts, how would the crew have hauled them on their retreat? The list adds up, even leaving behind the beer, to pretty close to a ton per week in the aggregate for each of the 100 or so who left the ships. And if they anticipated a journey of two or three months, at a minimum, that would be some 12 tons, added to the weight of the boats they were hauling and the camping and cooking equipment. Even spread over eight or 10 sledges—likely more than they actually dragged with them—the required food alone would have resulted in some 250 to 300 pounds of haul-weight per man, right at the limit of what a healthy sailor could pull. Add the weight of the boats and the other equipment, and consider their debilitated state of health, and it's immediately apparent that they couldn't have been carrying as much food as they needed, even if that food had somehow survived until they left the ships.

What about eating what you kill? To some, the death of the Expedition crew was a simple problem of ignorance and preconception leading to starvation. Vilhjalmur Steffansson, for example, claimed that Franklin's men starved to death in an area where game was abundant, because of their fixed belief in the barrenness of the Arctic.[88] But Steffansson was wrong. The problem was, as John Rae had reported as early as 1854, the Expedition was trekking across an area where game was notoriously scarce. Not far from the south coast of King William Island, Rae's own party, "all practiced sport-men, picked men, and in full strength and training," had shot "one deer [caribou] only and a few partridges" in nearly two months.[89] The area on the west and south coasts of King William Island was generally one that Inuit hunters avoided, because of the lack of easily killable game.

Charles Francis Hall reported an Inuit story of a small group of white men who had met an Inuit family group and had appealed to them, unsuccessfully, for help in obtaining food.[90] Hall faulted the Inuit for abandoning the white men—notwithstanding that the Inuit themselves probably had rather less food than they needed for their own survival—but, if Hall is to be believed, that the whites were a "large company of starving men,"[91] then it's unlikely that the Inuit could have been of much help, in an area sparse in game to begin with, in satisfying the crew's nutritional needs. Certainly, the members of the Expedition who reached King William Island in the spring of 1848 were short of food, and especially short of protein. Not eating enough will definitely kill you. And, as they progressed down the coast and past Terror Bay, the lack of food would have incapacitated more and more of the remaining crewmen, leaving ever fewer to haul the heavy sledges and to hunt for what little food might be found within their reach. Not until they had reached the area near Starvation Cove, on the eastern side of the Adelaide Peninsula near Chantrey Inlet, would they have been in areas where game might have been more plentiful, and even then only at the right time of year. Only during the short migration periods at the beginning and end of summer would the caribou have been present in sufficient numbers. And by then most of the men would have been too weak to hunt.

Trichinosis via Polar Bear

Author Ken McGoogan has suggested that a number of the deaths, especially those of the nine officers and 15 sailors reported in the Victory Point record as having died before the ships were abandoned, might have been caused by trichinosis, contracted by eating raw or undercooked polar bear meat.[92] Drawing on the history of the Danish Jens Munk Arctic expedition of 1619–20, in which 62 of the 65 expedition members died, and in which there was documented evidence that they had eaten at least some polar bear meat, McGoogan hypothesizes not only that the onboard deaths reported in the Victory Point Record could have been the result of polar-bear-borne trichinosis, but also that the disease might have been a factor leading to the abandonment of *Erebus* and *Terror* in 1848. True, the Royal Navy knew that one had to cook polar bear meat extremely well to avoid sickness, but, in the exigent circumstances of the expedition in 1847–48, perhaps best practices were not always able to be followed. Unless the meat was cooked thoroughly, trichinosis could still infect those who ate it. An earlier study suggests that it would have taken many hours to heat polar beat meat to a temperature (137° Fahrenheit) that would be effective to neutralize the trichinella larvae.[93] It seems highly unlikely that the hungry Expedition crew, desperate for fresh meat, would have waited around the stove that long on the rare occasions when they did have polar beat meat available.

The trichinosis theory assumes, though, that the Franklin Expedition crew had polar bear meat available, an unlikely hypothesis. How many polar bears were there in the vicinity of King William Island in 1848? Highly doubtful that there were enough to elevate *nanuk*-borne trichinosis into a major cause of death. When Jens Munk's crew were struck down, arguably by the disease, they were at what is now Churchill, Manitoba, in the heart of polar bear country. Even today, polar bears still rummage through the garbage cans on the outskirts of Churchill. But the ice-covered water off the west coast of King William Island is a very different place, and with many fewer reports of large game. Enough polar bears to kill off the whole crew, or even the nine officers and 15 sailors? Perhaps, but by no means a certainty. And, if there had been regular polar bear feasts, one would have expected to find some of those polar bear bones around the known campsites of the Expedition's retreat, such as Crozier's Landing and Terror Bay. No such finds have been reported, despite the large number of human bones found over the years.

Tuberculosis[94]

Tuberculosis was endemic in the 19th century. And all three of the bodies the Franklin crew buried at Beechey Island showed evidence of small tissue masses in the lungs that are typical of a variety of lung infections, including but not limited to tuberculosis.[95] But a subsequent analysis of the bones, looking for the genetic markers of the tuberculosis organism, failed to find any.[96] And tuberculosis doesn't kill in a hurry. If, indeed, more than 100 of the Expedition's crew died in a few months after leaving the ship in April 1848, one needs to look for a cause that operates with more ferocity than TB. Most likely, a substantial proportion of the Expedition's men were infected with tuberculosis, as were many working-class Englishmen at the time, but that doesn't mean it was their primary cause of death.

Zinc Deficiency

In 2016, a team of scientists published a rigorous analysis of the toenail and thumbnail of John Hartnell, one of the three Expedition crew members buried on Beechey Island.[97] The analysis found that Hartnell, after only six months away from England—he died in early January 1846—suffered from chronic and severe zinc deficiency, which could have led to, among other ailments, depression, diarrhea, immune-suppression and increased risk of infection. The last of these could be an explanation for the pneumonia that was found to be Hartnell's proximate cause of death.

But, really, zinc deficiency as a cause of any significant number of the Expedition's deaths? While the analysis of Hartnell's toenail may cast doubt on the lead-poisoning theory,[98] and also may suggest that, if others on the Expedition were also zinc-deficient, a likely result of a diet lacking in fresh meat and fish, and therefore susceptible to a variety of infections, including pneumonia, it's still a leap from that suggestion to a conclusion that zinc played a crucial role in the Expedition's mortality rate, especially since most of the deaths occurred two years and more after Hartnell had died. Hunger, exposure to the cold and scurvy all provide more likely paths to death.

Summary

So, what *did* they die from? Cold (exposure) and starvation, most likely, though we could certainly use a good team of forensic pathologists—perhaps like those on the television shows *Silent Witness* (from the UK) or *Balthazar* (from France)—to work their magic and tell us for sure. Perhaps just from the bad luck of being under the command of Sir John Franklin, the most lethal commanding officer in the history of the Royal Navy's Arctic service. Were they short of edible food? Were they weakened by a variety of debilitating ailments—scurvy, tuberculosis, lead poisoning, etc.? Of course. Did they make bad decisions, their reasoning impaired by physical and mental weaknesses? Very likely. But no, there was no single magic bullet that killed them all. The reductionism that seeks to lay all the blame on a single cause, or on a single individual, whether John Franklin or Stephen Goldner, is simple, and therefore satisfying to some, but, in the end, just not true.

14

Cannibalism

> Instead of being an aberration, practiced by only a few prehistoric Donner Parties, killing people for food may have been a standard human behavior.[1]
>
> It is difficult to write a popular narrative of polar exploration that is not a disaster narrative with details as formulaic as any Gothic romance: madness, mutiny, murder and cannibalism.[2]

Charles Dickens to the contrary,[3] they ate each other.[4] In most retellings, the Franklin expedition becomes a story of heroic failure—the brave sailors discovering, or at least passing across, the missing portion of the Northwest Passage before they nobly died. In this literary myth, Franklin's men display the supposed British values of duty, discipline, faith and camaraderie.[5] Eating each other? Englishmen simply wouldn't. But they did. Or at least that's what all but a few remaining skeptics agree.[6]

That Victorian faith in the civilized nature of English gentlemen came up against horrible reality when John Rae returned to London in 1854, carrying relics of the expedition that he had purchased from the Inuit and reporting the latter's' stories of a ragged band of Franklin survivors last seen alive struggling toward the southeast from King William Island and then later seen dead by other Inuit. As Rae reported, "from the mutilated state of the corpses and the contents of the kettles, it is evident that our wretched countrymen had been driven to the last resource—cannibalism—as a means of prolonging existence."[7]

Not that cannibalism was unknown to those mid–19th-century Victorians. Across the Atlantic, Edgar Allan Poe's *The Narrative of Arthur Gordon Pym*[8] and Herman Melville's *Moby-Dick*[9] had at least raised the possibility, but it was generally a possibility that was reserved for "savages," not for Englishmen.[10] The pervasive racism of Victorian England, a perhaps unintended product of the Enlightenment,[11] reserved cannibalism, along with such other "self-exterminating" habits as infanticide and widow-strangling or burning, for the lesser humans.[12]

Dickens, famous in England both as a novelist and as a prolific journalist, whose own periodical, *Household Words,* had championed Arctic exploration, leapt into the argument soon after John Rae brought the grim news to England. In two articles published on May 2 and 9, 1854, he argued that either the Inuit did the eating or, in the alternative, inveterate liars that they in Dickens's view must have been, invented the stories they told to Rae. And, as a last line of defense, Dickens mused, if there was cannibalism among the crews of *Erebus* and *Terror,* it could only have been among the lower ranks; gentlemen would not eat gentlemen. (Leaving aside the question of whether gentlemen would eat the lower orders.)[13] This view was reinforced by at least one of the Inuit narratives that was reported by Charles Francis Hall, a story that told of Inuit meeting a group of four Expedition survivors, three of whom were fat from eating their dead comrades, but one, who Hall thought was perhaps Francis Crozier, was thin because he refused to partake of human flesh.[14] On the other hand, some studies of cannibalism suggest that nautical tradition would have cannibalism start from the top of the chain of command: "[i]n perilous situations, the captain, officers and crew had the knowledge necessary to survive. The more expendable survivors (such as slaves, young boys, and passengers) were often the food supply for the knowledgeable."[15] Dickens, like many "gentlemen" of the time, was happy to believe anything negative about those who were not Englishmen, and especially so if those informants could be relegated by race to a lower category of humanity. Because his views were so colored by his prejudices, one shouldn't put much faith in anything he said.[16] Others also expressed some doubt in the cannibalism stories, pointing out that most of the stories originated with an informant, In-nook-poo-zhe-jook, whom even Hall suspected of going beyond the facts.[17]

But eat each other they did. Perhaps the officers abstained; after all, there were only half a dozen officers still alive when they abandoned the ships in April 1848, and Lt. John Irving, or at any rate some officer initially identified as Irving, appears to have died not long after they retreat began, while a body identified as that of Lieutenant Henry Le Vesconte was discovered on the south shore of King William Island by the explorer Charles Francis Hall in 1869.[18] But there is now conclusive evidence of cannibalism at Erebus Bay, Terror Bay and Starvation Cove. The knife marks on the bones and the severed limbs examined by forensic archaeologists Owen Beattie, Anne Keenleyside and others strongly suggest that *homo sapiens* made up a significant part of the dwindling food supply.[19] Specifically, in a study of the bones found near Erebus Bay, Keenleyside and her colleagues found 92 separate bones with cut marks that were distinguishable from animal tooth marks and that were consistent with "defleshing," or removal of muscle tissue from the bones.[20] While archaeologists have varying criteria for classifying bones as showing evidence of cannibalism,[21] with some giving more weight to similar butchering techniques for human and animal bones found at the same site—a situation that could not occur where the Franklinites' bones were found, because the only bones were human—all agree that the presence of purposeful straight cut marks, like those studied by Keenleyside and her colleagues, is a key indicator.

Cannibalism comes in various flavors.[22] At least 150 different species of non-human animals have been identified as cannibals.[23] Our close evolutionary relatives, chimpanzees and gorillas, show no lack of appetite for their own kind, though, like shipwrecked mariners who seem to have a propensity for cabin boys, these great apes tend to kill and eat youngsters.[24] And cannibalism is certainly not a modern invention even among humans and our direct ancestors; archaeologists have found convincing evidence of cannibalism in several pre-modern hominids, as far back as *homo antecessor* nearly a million years ago.[25] More recently, there is conclusive evidence of humans eating other humans at Anasazi sites linked to the Chaco Canyon culture in the southwestern United States between 900 and 1300 CE.[26] Closer to home, for our purposes, the Saunatuk site in the Northwest Territories of the Canadian Arctic yielded the bones of at least 37 individuals showing signs of violent death and cannibalism.[27] Other examples in relatively recent times come from the islands of the Pacific: Fiji, the Cook Islands, Papua, New Guinea, and from the Maori culture of New Zealand[28] as well as from the Amazon Basin.[29]

The Franklin survivors practiced what some would call "survival cannibalism," much like that carried out by the ever-diminishing number of survivors of the Donner party in the Sierra[30] at roughly the same time that the Franklin crew were attempting their retreat, or by the Uruguayan rugby team's plane crash survivors in the Andes in 1972.[31] But there are other varieties as well. The Fore people of Papua New Guinea ate the brains of deceased family members, as a way of honoring their memory and, by doing so, unwittingly spread the neurological disease kuru, a form of transmissible spongiform encephalopathy.[32] Similar mortuary ritual cannibalism, sometimes described as a way of keeping the corpse from being eaten by animals or rotting, has been found elsewhere as well.[33] And certainly some societies, notably in Fiji and other Melanesian islands in the southwest Pacific, show evidence of a warlike cannibalism where killing and eating one's enemies was part of the victory celebration.[34]

Broadly speaking, we can characterize cannibalism as being either (1) for survival, in desperate circumstances; (2) for ritualistic purposes, including honoring the dead, appeasing the gods through human sacrifices, or celebrating victories by eating the vanquished; or (3) for individual or (usually) psychotic reasons, as in the case of American murderer Jeffrey Dahmer.[35] Alternatively, one could classify cannibalism with a sort of decision tree, asking, in each case, (1) was it necessary or by choice; (2) was it culturally driven or idiosyncratic; (3) did the cannibals eat the entire body or just selected parts; and (4) was the eating for nutritional purposes or symbolic?[36] In any case, the Franklin Expedition falls into the category of survival or necessity.

As a general rule eating modern humans is not a particularly efficient means of gaining the calories necessary for daily activity,[37] although a recent study[38] suggests that cannibalism may have been fairly common among pre-modern hominid populations, and that Neanderthals, in particular, provided food value not all that different from the large animals that those populations may have hunted. In hominid species, large muscles, like thighs, calves, and upper arms, are particularly nutritious, and some organs, especially the heart and liver, have relatively high calorie values.[39]

Heavily muscled Neanderthals may have been a more attractive food source, vis-à-vis other animals, than would modern, comparatively scrawny *homo sapiens*, but *in extremis*, in the absence of any other food source, one may well eat what's there. It was easy for Dickens to insist that eating another Englishman was just not on. After all, he wasn't at the tether end of three long years in the Arctic. And even Dickens didn't say it never happened. But when white people ate other humans, they were, in the civilized English view, "going native." In Dickens's own *Great Expectations*, it's the escaped convict Abel Magwitch who threatens to tear out plucky Pip's heart and liver and roast and eat them.[40] Not a gentleman, our Abel. And "going native" was not a purely fictional device. As Sir Harry Johnston, the administrator of British Central Africa in the latter 19th century, remarked: "I have been increasingly struck with the rapidity with which such members of the white race as are not of the best class, can throw over the restraints of civilization and develop into savages of unbridled lust and abominable cruelty."[41]

"As are not of the best class." One didn't even have to be non-white to be non-human from this English upper-class perspective. Recall that the Franklin

Expedition was happening at the same time as the great famine in Ireland, a disaster caused as much by conscious English policy to protect aristocratic landowners as it was by the potato blight.[42] And there is substantial evidence that, at least in the poorest Irish counties of Cork, Galway, Kerry and Mayo, starving Irish tenant-farmers did in fact turn *in extremis* to human flesh.[43]

Of course, one could go native without embracing cannibalism. New recruits to the colonial service were warned that failure to dress for dinner or to use a butter knife were the first steps on that downward slide. But, for some colonials, cannibalism did seem a part of going native, as in the case of the Pakeha Maori, white convicts who fled Australia to join Maori societies in New Zealand.[44] (Of course, convicts transported to the Antipodes, like Irish peasants and even like the lower ranks on an average Royal Navy ship, could easily be "othered" by those, like Dickens, who clung to a class-bound perspective.) But surely, Dickens believed, Franklin's crew, the flower of English civilization, an expedition that had been the talk of the entire nation, would not stoop so low.

Modern research has proved conclusively that people do eat other people, especially when the eaters are *in extremis*, as, for example, during famine times in ancient Egypt and Greece. One researcher cites 11 separate documented instances of cannibalism in Europe, all caused by severe food shortages, from the 8th to the 15th centuries,[45] while other reports document similar episodes in the 16th and 17th centuries.[46] In the 20th century, there are many examples of extreme widespread privation resulting in cannibalism: famines in Russia and Ukraine in the 1920s and 1930s (the latter the infamous *holodomor* caused by Josef Stalin's command that Ukrainian wheat be shipped to Russia rather than used at home)[47]; during the World War II siege of Leningrad[48]; in the concentration camps of Treblinka and Bergen-Belsen at the end of the war[49]; and by Japanese soldiers, cut off in remote locations in New Guinea by the Allied advance, who ate Asian prisoners of war, New Guinean natives and, eventually, each other.[50] Those Japanese soldiers, in fact, occasionally removed flesh from still-living victims, perhaps to guarantee a fresh food supply in a wet jungle environment where decomposition would rapidly occur.[51] That, at least, would not have been a concern for the Franklin survivors. There was also well-documented cannibalism in rural areas during the Chinese famine caused by the Great Leap Forward in 1958–62, when human meat was sometimes offered for sale in open markets.[52]

It's hard to imagine a situation more extreme than that of the Franklin Expedition survivors straggling down the coast of King William Island in the spring of 1848, their food supplies dwindling or gone, no wildlife in sight, and suffering from both the physical and mental effects of disease and hunger. It would have been astonishing if they hadn't eaten each other. And cannibalism was by no means unknown at sea. Shipwrecks and lifeboat survival stories, in fact, represent the largest single category of cannibalism incidents in modern times. At least until the end of the 19th century, well after the Franklin crew had been written off as dead, the "proper tradition of the sea" was that, *in extremis*, survivors could eat those who had died before them and even, if there were no more bodies left, could draw lots to determine who would be sacrificed for the benefit of those remaining.[53] Only in cases where there was not a fair lottery—as, for example, if the victims were chosen only from the ranks of slaves, foreigners or cabin boys—were questions raised about the culpability of the survivors.[54]

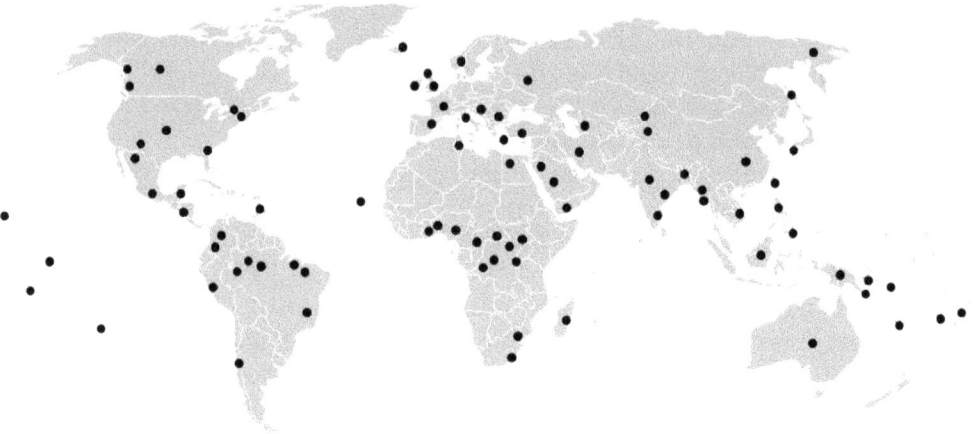

Documented human cannibalism sites around the world (Wikimedia Commons, based on Edwin Loeb, *The Blood Sacrifice Complex* [American Anthropological Association, 1923]).

It was not until 1884, in fact, in the famous English legal case of *The Queen v. Dudley and Stephens*,[55] that the courts rejected the defense of necessity and held that sailors who killed and ate one of their shipmates while lost at sea after a wreck could be prosecuted for homicide. In that case, involving the wreck of the yacht *Mignonette* on a voyage from Falmouth to Sydney, the captain stabbed cabin boy Richard Parker, perhaps believing Parker near death; the survivors eventually returned to Falmouth and were convicted, but only because they failed to use a lottery. Earlier English cases, though not setting any precedent due to the technical complexities of English law, had in fact resulted in the freeing of sailors who had killed and eaten their comrades in extreme circumstances.[56]

Even before Franklin sailed, there were well-known cases of shipwreck and cannibalism. The most famous is the wreck of the French warship *Médusa* which ran aground off the coast of West Africa in 1816, leaving more than 100 survivors adrift on a raft where not only cannibalism but fighting and anarchy reigned, inspiring (if that's the right word) the famous painting by Théodore Géricault, *Le Radeau de la Méduse* (now in the Louvre), as well as a host of films and novels.[57] The *Médusa* was far from the only well-known shipwreck that led to cannibalism among the survivors; others included the *Nottingham Galley*, which ran aground off the coast of Maine in 1710; the *Peggy*, an American ship disabled at sea in 1761 (where the victim, supposedly chosen by lot, just happened to be an Ethiopian slave); the whaler *Essex*—inspiration for Herman Melville's *Moby-Dick*—sunk by a whale in 1820; the *Frances Mary*, disabled at sea in 1826—an unusual case because the survivors included two women; usually, women were thrown overboard as bad luck—and the *Francis Spaight*, sunk in 1844, where there was a lottery, but only among the cabin boys, two of whom were killed and eaten.[58]

And years after the Franklin Expedition, another ill-planned Arctic voyage apparently also resulted in the survivors eating not only their boots, as Franklin himself had famously done on his first overland journey in 1819–21, but also each other. When relief ships intended to supply the crew of the Greeley Expedition of 1881–84 failed to arrive in the summer of 1882 and again the next summer, the survivors apparently began to consume those who died, if not, at least not certainly, some of those still living.[59]

Apart from shipwrecks, other kinds of journeys gone awry have also led to cannibalism. The Donner Party, trapped in the Sierra Nevada in 1884, is perhaps the most famous, but another Western overland trip, by Alfred Packer and five companions, who headed from Utah to southwest Colorado in the winter of 1873–74, is almost equally celebrated. Packer, the only one to emerge alive, admitted the cannibalism but denied killing the others—except for one, whom Packer said he killed in self-defense. He eventually served 18 years in prison for manslaughter. Subsequent forensics showed that all five of his companions had been violently killed.[60]

And, bringing the story into a more modern travel era, the Uruguayan national rugby team's plane crashed in the Andes in 1972. Only 16 of the 45 on board survived, after some 72 days in the mountains. The survivors admitted to eating the bodies of those already dead, though they refrained from eating any of the three women who had been killed in the crash.[61] Because many of those who were eaten had died in the crash, rather than wasting away through hunger, like the victims in many of the shipwreck stories, their bodies were more nutritious, with higher levels of body fat, benefiting those who survived.[62]

People, especially when in distress, will resort to eating other people. Thus, if some of the Franklinites ate those who had already died—or followed the "proper tradition of the sea" by using a lottery to choose a living crew member to sacrifice—they would not by any means be alone in history. Ironically, human behavior now seems to be driving other species toward cannibalism. In the Arctic, for example, human-induced climate change has shrunk the extent of sea ice, leading polar bears to eat their young more frequently as traditional food supplies become unavailable.[63] While examples of cannibalism in the animal kingdom are by no means rare, it's probably not something that English gentlemen, comfortably ensconced in their London clubs and fond of their dogs and horses, would have encouraged.

When, where, and how the Franklin Expedition sailors resorted to cannibalism is still uncertain. Bones from the crewmen have been discovered all the way from Victory Point in the north of King William Island, along the western and southern coasts of the island, and beyond, at Starvation Cove and elsewhere on the Adelaide Peninsula, and on the Todd Islets near the southeast point of King William Island. The knife-marked bones described by Beattie, Keenleyside and their colleagues were found in at least three locations, representing different stages of the retreat from the ships. At Terror Bay, the bones most likely belonged to those left behind in a camp for crew members who were physically unable to travel further. In the Starvation Cove area, the bones were from those who had gone as far as they could before collapsing. At the Erebus Bay site, the bones may have been from those who died relatively early in the retreat, or they may have been from members of a group that had turned away from the line of march and headed back toward the supposed safety of the ships or Crozier's landing site, with its abundance of abandoned material. Keenleyside's analysis, the most thorough-going of those that have examined the remains for evidence of cannibalism, looked at an assemblage of bones on a small island in Erebus Bay, amounting to some 400 human bones, mixed in with a small number of animal bones.[64] Of those 400 bones, 92, or nearly one quarter, showed evidence of cut marks.[65] The bones in Erebus Bay did not have all the indices of cannibalism that some researchers had reported in other locations. For example, in their work on the Chaco Canyon Anasazi sites, Christy and Jacqueline Turner used a five-factor test: cut marks, perimortem breakage, anvil or hammerstone abrasions, burning, and missing vertebrae. The bones from the Erebus bay site that Keenleyside analyzed met only two of these tests: cut marks and perimortem breakage.[66] The cuts were clearly made with metal knives, rather than bone tools that might have indicated Inuit origin or animal teeth; even with only two of the tests satisfied, this was strong evidence of cannibalism by members of the Expedition crew. The cut marks were concentrated near joint surfaces on the arms and legs, suggesting that the fleshier body parts may have been cut off for use as a portable food supply, and at least three long bones were fractured, possibly indicating an attempt to extract the marrow.[67] But Keenleyside study revealed few signs of violent death, such as blunt-force trauma to skulls. All this suggests that those who were consumed most probably died of natural causes (if scurvy, exposure and starvation can be considered natural) before being eaten, rather than being killed by others with the aim of increasing the killers' protein intake.

Keenleyside's conclusions are consistent with earlier work by archaeologist Owen Beattie and his colleagues on the bones of more than 35 individuals recovered from other sites on King William Island. Those bones, like the ones analyzed by Keenleyside, showed cut marks and breakage; they also showed signs of polishing and having been boiled in water, further supporting the Inuit stories of "the contents of the kettles" reported by John Rae.[68] Based on the work by these two teams of professional archaeologists, there's little doubt that Rae was right; the survivors of the Franklin Expedition were indeed reduced, like so many others in similar, desperate circumstances, to eating their dead comrades.

Dickens's and others' mid–Victorian views notwithstanding, modern audiences are probably likely to regard the evil Cornelius Hickey, as portrayed in Dan Simmons's novel *The Terror*[69] and in its hugely popular television adaptation,[70] as the authoritative source of information on the Franklin crew's anthropophagia. And, alas, that source pretty closely mirrors those old prejudices of Dickens and his London clubmates. In *The Terror*, it's the lower-class Hickey who stokes the cannibal fires, while the upper-class assistant surgeon Harry Goodsir, forced to perform some defleshing under Hickey's threats, embodies the gentlemanly virtues and, of course, refuses to partake of the available calories.

While there is now little doubt that some of the survivors—while they were still survivors and not themselves yet dead—did eat parts of those who had died before them, questions still remain. Who ate whom? And were all of those who were eaten dead from "natural" causes, or, despite the absence of skull fractures, were some of them, at least, killed for their protein by stronger survivors? Did the extra food make any difference? Were there any regrets? The answers to those questions will not come from newly discovered documents, or even from yet more scientific analysis of the bones in the Arctic. Just another instance of the uncertainties that will forever remain.

15

Survivors?

The Franklin literature, both fiction and non-, has no shortage of imagined survivors. Even in such self-styled havens of rational debate as the British Parliament, speakers strongly championed the cause of Franklin Expedition members who had supposedly lived on long after the searches were mostly abandoned and after all the crew members had officially been declared dead.

Often, the fictional survivor died a peaceful, almost serene, death, firm in his imagined Englishness. Thus, the Irish poet Edmund Falconer's "Last of the Crew" spent his last few minutes at peace, thinking of "England's dewy skies," of "a murmuring rill, / The humming of a busy bee, / The clack perchance of some old mill, / 'A ploughboy whistling o'er the lea."[1] Less likely, perhaps, than thoughts of "I'm cold, hungry and in pain," but more poetic.

Quite apart from Mordecai Richler's mythical survivor, Ephraim Gursky,[2] and William Vollmann's lingering-on spirit of Sir John Franklin,[3] among many others in the literature, there have been a number of more-or-less credible reports of Franklin Expedition survivors who lived for some years after the ships were abandoned.

A number of 19th-century searchers came across Inuit reports of survivors. One of Charles Francis Hall's informants, Too-shoo-art-thariu, reported seeing Crozier somewhere near Pelly Bay, far to the east of the Expedition's generally accepted route of retreat, in the winter of 1853–54,[4] though it's not entirely clear that the "Aglooka" of this story was in fact Crozier—"Aglooka" was an Inuktitut label applied, over the years, to a number of different Europeans. And numerous stories collected by Hall refer to a group of three or four white men seen no earlier than 1850 in the area of Ootjoolik, not far from where the wreck of *Erebus* was eventually discovered in 2014.[5] One of the narratives recounted by Hall, in fact, reports that Crozier and one last companion had not died until 1864, when they perished at Southampton Island, at the northern end of Hudson's Bay, while they were trying to reach whaling stations.[6] One should perhaps note, though, Hall's "obsessive belief that a few hardy survivors of the lost expedition remained alive."[7] Wishing, however, does not necessarily make it so.

But, to this day, writers continue to repeat and credit 19th-century reports. Farley Mowat, perhaps not the most believable of reporters, recounts what he says are Inuit stories of Crozier and another white man having been seen in the Baker Lake area, several hundred kilometers south of the mouth of Back's River, sometime between 1852 and 1858.[8] And David Woodman collected and analyzed a number of Inuit narratives that suggested the possibility that survivors had reached the Melville Peninsula, far to the east of the location where *Erebus* and *Terror* were abandoned.[9]

The most comprehensive attempt to identify the number, if not the identity, of Expedition survivors in the years since Woodman's 1995 work has been undertaken by retired Canadian geologist John Roobol. In his 2019 paper, "Status of the History of the Lost Franklin Expedition of 1845,"[10] Roobol suggests the following potential groups of post–1848 survivors:

- Some of the group of then-surviving crew members who returned to *Erebus* (and perhaps to *Terror* as well) in the late spring or summer of 1848;
- A senior officer (perhaps Crozier), who died in 1849 or 1850 and was buried with the ships' papers in a cemented vault on King William Island that was mentioned in Inuit accounts but has yet to be found;
- The 40 or so men encountered by an Inuit party in the summer of 1850, according to the narratives told to John Rae and others, near Washington Bay on the south coast of King William Island. Whether it was 1848 or 1850, the crewmen were showing visible signs of scurvy;
- Four survivors found by Inook-poo-zhe-juk, two of whom reportedly died in the winter of 1850–51, but the other two left in the spring of 1851, heading for Fort Churchill in what is now Manitoba, never to be heard from again;
- A dozen or so men who sailed or drifted in *Erebus* toward the continental shore arriving near what is now known to be the ship's final resting place sometime in 1850, and whence some of the survivors may have headed west in a boat, toward the Mackenzie River delta; and
- Four crewmen and the ship's dog Neptune, who remained on board *Erebus* until 1851 or 1852, leaving at times for hunting expeditions and then setting out overland for Iwillik on Repulse Bay, some hundreds of miles to the east, once again never to be heard from again.

Apart from the last two categories, it's pretty clear that most of the "survivors" posited by Roobol and by David Woodman's careful reconstruction of Inuit narratives[11] didn't last long. Giving maximum credence to the Inuit narratives, and accepting the dates reported from those narratives by Rae, Hall and others as accurate, we still end up with everyone pretty much dead by 1852 at the latest—long before Hobson's and McClintock's discovery of the actual fate of the expedition in 1859. Perhaps, had John Ross's ill-equipped and unlucky search expedition of 1850 been able to press on as far as King William Island or even to the Adelaide Peninsula, Ross might have found a few still-barely-alive remnants of the Franklin Expedition. But Ross never got that far, and the flotilla of search ships that followed in the next few years steadfastly avoided the area where Franklin's crew had actually gone.[12]

David Woodman suggests yet another group of survivors, based on Inuit narratives from the Melville Peninsula area, far to the east of King William Island.[13] Those stories rely principally on Inuit narratives collected by Hall, and there is considerable doubt as to whether particular stories refer to Crozier, to Edward Parry or to John Rae.[14] But, assuming for the moment that the stories are about the Franklin Expedition, the most compelling version, as summarized by Woodman, is that "the survivors of the disaster at King William Island arrive at Pelly Bay in their effort to reach Iwillik/Repulse Bay. Presumably their aim is to get to either Pond Inlet or Churchill. Aglooka [Crozier?] and two or three men fall behind and are cared for by 'Tooshooarthariu' for one winter. The next they are sent southward."[15] All the men, those who forged ahead as well as those who stayed behind for the winter, are last seen, according to these narratives, somewhere on the Melville Peninsula, but are never reported as having been seen to die. Beyond Woodman's reconstruction of the Hall material, there's nothing to support this survivors' tale, but that doesn't mean it's not true.

All we know for sure about when the Expedition crew members died is the dates of just four of those deaths: the three who died on Beechey Island that first winter of 1845–46, and the date of Sir John's death, recorded in the Victory Point Record. Without additional written records, preferably of the type scrawled by Captain Robert Falcon Scott shortly before he died on his ill-fated South Pole expedition,[16] we will never know exactly who died where and when, even if we might conclude that some of the crew members lasted beyond the summer and autumn of 1848.

The bones of the expedition crew have been found pretty much where one might expect to find them, given the paucity of information that we have from the Victory Point Record. Some made it across Simpson Strait to the Canadian mainland, though generally not very far; a skull found some 10 miles south of the strait, on the Adelaide Peninsula, has been identified as a European, though not necessarily as a Franklin crew member.[17]

In the end, they were all dead. And, as Alexa Price suggests, "survival of the body was overwritten by survival of the British spirit."[18] Even before McClintock returned in 1859 with the Victory Point Record, the process of creating the heroic narrative—conveniently unanchored by facts, of which there were so few—had begun.

But what if there had been one or more who had learned from the Inuit how to survive in that harsh landscape and who were never rescued, but simply faded into a new—for them—society. Dan Simmons's *The Terror* ends with Crozier more or less reincarnated as an Inuit shaman. One does not have to be that fanciful to entertain the possibility that someone from the Expedition, perhaps Crozier, perhaps a young sailor lucky enough not to be eaten, someone, survived.

16

Sir John's Grave

With such little evidence left—a bundle of bones, a few relics recovered from Inuit, a single piece of paper—it's no wonder that graves, both real and imagined, have become almost magical talismans in the Franklin literature. Captain Edward Inglefield, part of the Franklin search efforts in 1852, reported from the known gravesite on Beechey Island that he had seen a "huge bear ... continually sitting on one of the graves, keeping a silent vigil over the dead."[1] Whether Inglefield's bear existed is beside the point; from the searchers' point of view, he should have. A gravesite, or for that matter, a tomb in Westminster Abbey, with or without a body inside, would be one of the few visible and lasting remnants of the Expedition. While that tomb at Westminster was never built, there is a memorial there, a white marble bust, commissioned by Lady Jane Franklin and bearing Sir John's relation by marriage Alfred Lord Tennyson's famous poem, tucked away in one of the Abbey's chapels. Not quite the same thing as a grave with real remains in it, though.

Of all the presumed but as yet undiscovered relics of the Franklin Expedition, none has so engaged the attention of researchers as that grave, wherever it may be. If only it could be found, if only there was a grave, surely it would reveal the answers to oh so many of the questions that remain. Thinking, no doubt, of future search expeditions, those who buried Sir John would have made sure that the grave was filled with records carefully detailing the Expedition's progress up to June 1847 and filling in all those blank places on the Admiralty maps. Just because they neglected to do it at Victory Point doesn't mean they would have been so careless of the interests of future historians when it came to Sir John's resting place. Or perhaps not.

As far back as 1867, whaling captain Peter Bayne recounted a story of a cemented vault containing Franklin's body, located on King William Island somewhere near Victory Point and adjoined by three other graves and a cairn containing written records. Around the same time, Charles Francis Hall's Inuit informant Su-pung-er told a similar story, describing an underground stone vault in a similar location (somewhere on the west coast of King William Island between Cape Felix and Erebus Bay).[2] While many later searches turned up no evidence of such a vault, the story, in many variations, persists.[3] And in many of those stories, not only does the gravesite contain what's left of Franklin's body, but also those almost-mythical records that would, it is thought, reveal how Franklin and at least some of his men—those who died before Sir John himself did—came to such an unhappy end.

Not all the experts agree as to the likely location of Sir John's grave, or even if such a thing exists. As early as 1969, Franklin scholar Richard Cyriax addressed the problem and concluded that there was no particular reason why the Navy would have buried the Expedition's records—whether or not they buried Sir John—and that whatever records might have come ashore with the men who deserted the ships in 1848 had probably long since been scattered to the wind.[4] Cyriax also noted that Franklin's "omission to mark his route by means of records left on land was quite contrary to what was expected by the Arctic authorities," specifically mentioning the expectations of Parry, Sabine and Richardson that the Franklin searchers would find written records left behind at regular intervals in well-marked cairns.[5] Other explorers had left multiple cairns with position notes and records. Why not Franklin?

Back in 1967, the Canadian government spent well over $100,000 searching for Franklin's grave. In August that year, some 50 or so soldiers of the Canadian Light Infantry decamped in the general direction of where the Franklin expedition had last been reported.[6] After a few weeks of exploring in awful weather conditions, the Army concluded that "time has erased any signs on land left by Franklin's expedition."[7] From then until the present, intrepid researchers have combed the coast of King William Island, with pretty much nothing to show for the searches. In the last two decades, a renewed search effort focused on minute smudge marks that have been observed on satellite images has been underway, but again without any success.

In addition to the marble bust in Westminster Abbey, there is another monument to Franklin, in Waterloo Place, London. It shows a burial on the ice, with Francis Crozier reading the service over Sir John's coffin.[8] What happened to the coffin after the service is unclear. Was it lowered into the ocean through a hole cut in the ice? Was it taken back to *Erebus* in anticipation of some later interment? Was it put on a sledge and sent off to King William Island for a land burial? That last alternative, which is the one Franklin devotees most want, seems the least likely. We certainly can't look to the Waterloo Place monument for guidance. The frieze there depicting the burial is based, so far as we know, on exactly zero evidence. Nonetheless, with its depiction of a formal burial service on the ice, it tells, or at least begins, a story that some find satisfying, no matter how much the scene conflicts with what would have been typical Royal Navy practice.

In the normal course of events, sailors, including officers, who died at sea were buried there: "Sailors developed a highly ritualized funerary service, distinct from burials on land and reflective of the unique context of shipboard life. The dead sailor would be shrouded, weighted, carried in a brief procession and then slid overboard after a brief service."[9] Burial at sea, complete with weighting down the shroud and sewing the final stitch in that shroud through the dead sailor's nose also served the valuable purpose of making sure that the sailor didn't rise from the deep and return to the ship as a ghost.[10] So, the weight of tradition and custom suggests that a sea, or rather ice, burial for Sir John would have been most likely, assuming that such a burial would have been possible in whatever the circumstances of the Expedition were in June 1847.

On the other hand, we have the evidence of the three Expedition sailors buried during the winter of 1845–46 on Beechey Island. The unwritten rules regarding burial at sea may not have applied when the ship was resting at anchor, even if at a temporary winter harbor. And the ever-religious Sir John might have thought a land burial, even in the relatively shallow graves dug into the permafrost, was more suitable. A ship beset in the ice of Victoria Strait would not, though, have had the same burial options available as one at rest in winter quarters within easy walking distance of a minimally suitable burial ground. On the other hand, we know that the crew did in fact walk from the ships to King William Island when they abandoned the ships the following year, so some sort of a pilgrimage across the ice to the island in 1847 would have been possible.

So, what could have happened to Sir John's body? Is it still safe in its as yet undiscovered tomb on King William Island? Was it left behind on *Erebus*? Perhaps his was the very large body with long teeth that features in a number of the Inuit stories about finding the ship and removing useful bits of it for repurposing in Inuit life.[11] Was it buried at sea, or on the ice, as depicted on the Waterloo Place monument, perhaps in the summer of 1847, when the ice might have thinned out enough for a burial, if not enough for navigating the Northwest Passage? Or was it loaded onto one of the ship's boats and carted off on the retreat, perhaps with the hope of getting the body back to England for a proper burial, in the same way that, after Admiral Horatio Nelson had been killed at Trafalgar, his body was pickled in a cask of rum or brandy and brought back home for a hero's funeral?[12] As he lay dying, Nelson reportedly urged his surviving officers to make sure his body was returned to England and not left behind in the Mediterranean.[13] Nelson's body, however, required only two months—and three changes of brandy[14]—to make it home safely from Gibraltar. By the time the Expedition crew abandoned the ships in April 1848 Sir John had already been some 10 months dead, and would have been facing the prospect of at least another year on the road before making it home for the Westminster Abbey burial that Lady Jane would surely have insisted upon.[15] In any case, sailors knew it was bad luck to carry a dead body on board for any length of time, despite the precedent of the Nelson case.[16] Moreover, the Expedition may have been running out of brandy, and Sir John wasn't the thinnest of men; adding his body to a sledge already weighted down with a ship's boat and weeks' worth of provisions doesn't seem like the best sort of planning for the retreat.

Pickled or not, if Franklin's body was part of the jumble of supplies and artifacts that set off from somewhere near Victory Point in April 1848, we can be pretty sure that it ended up in much the same state as the bodies of those who were living when they set forth—its bones buried or scattered somewhere along the way. One would expect that Sir John would have been buried, or transported, in full naval regalia, and so it would have been likely that something more than his Royal Hanoverian medal and some pieces of his silverware would have been recovered. Thus far, those are the only relics that have been definitely identified with Franklin, and that's not enough to decide with any degree of certainty what happened to his body. The medal would have been among the most likely memorabilia objects for someone to have carried, in the hope of returning it to Lady Jane. It was small, but of great significance, a perfect substitute for the whole body. As for the forks and spoons—Goodsir tells us that each officer was required to supply his own, at what the young, probably impecunious and definitely penny-wise doctor thought was an inordinate cost.[17] The fact that those items were recovered from the Inuit by Rae, Hall and others tells us very little about the provenance of any particular item, or how many hands that item might have passed through before being recovered. With most of the officers dead, their silver forks and spoons could have been the hands of quite a few of the crew members before coming to rest alongside one or another of the dying sailors.

On the other hand, there's a suggestion that a body recovered from a carefully dug grave on King William Island and tentatively identified as that of Lieutenant John Irving, since Irving's maths medal from the Royal Naval College was found near the gravesite, is actually that of Sir John himself.[18] It must be Franklin, so the reasoning goes, because who else would have merited the backbreaking work that went into constructing the (for the Arctic) elaborate gravesite? Interesting theory, but, if it had been Franklin, wouldn't there have been a bit more by way of identification for the first searchers to come upon the scene—U.S. Army Lieutenant Frederick Schwatka's party some two decades later—to find? When Schwatka discovered the grave, he reported that it had been carefully dug and was surrounded by a ring of stones, suggesting that it had, perhaps, originally been covered over as well—quite an undertaking if done in 1848, when the crew was undoubtedly already sick, exhausted and underfed. But perhaps when Sir John died, in June 1847, there would have been enough men in good enough physical condition to complete the task. As Russell Potter comments regarding the identification of that gravesite as Irving's:

> And yet supposing this was Irving's skeleton, one wonders: how could this be his grave? Its site, quite close to that of Ross's cairn at Victory Point, poses a difficult puzzle; since Irving is mentioned in the Victory Point note as having *found* Ross's cairn, we can safely presume that he was fit enough to be sent on such a mission [in 1848]. And yet here, a stone's throw distant, lies his grave? Did Irving meet with some sudden end, so soon after the Victory Point Record that the main body of the expedition had not yet moved on?[19]

One might also question why Irving, an officer, yes, but only the third lieutenant on *Terror* (though, with the multiple officers' deaths before the ships were abandoned, one might assume that he'd moved up a bit in seniority by April 1848), would have received such an elaborate burial. He was identified, remember, on the basis of his naval college maths medal that was found at the gravesite, and the body was then shipped back to Edinburgh, where it was buried with great ceremony in a funeral presided over by Irving's brother, a major general in the Army. Unlike other bodies from the Expedition, there was no identification on the basis

The grave of Lt. John Irving—or is it really Sir John Franklin's? (photograph by Ronnie Leask, Creative Commons License BY-SA 2.0; Wikimedia Commons).

of, say, dental records (Henry Le Vesconte) or, more recently, DNA matches (John Gregory). Maybe it's time to dig up that body in Edinburgh, see if one can still extract DNA samples, and compare them to Sir John's currently living relatives? And then, maybe, we could call off the search on King William Island.

Or perhaps what was left of Sir John is still aboard the decaying wreck of *Erebus*, and maybe the Parks Canada archaeologist-divers will find it before the wreck becomes too unsafe to examine further. Or maybe there really was that burial on the ice as depicted on the monument in London, and whatever is left of Sir John is on the bottom of Victoria Strait. Just another mystery.

17

Franklin's Legacy

> The Franklin expedition demonstrated that Europeans struggled to survive in the Far North, had difficulty navigating the waters and failed, tragically, to adapt when their expedition ran into extreme difficulties.[1]
> Reserved, good-natured, dutiful but unlucky.[2]
> Whatever admirable qualities Franklin possessed; he was no genius.[3]

Some historical figures are vivid, their personalities and actions speaking to us across centuries. Sir John Franklin, not so much. As the perhaps overly critical Andrew Lambert comments, "While Franklin had many biographers, he remains an elusive subject."[4] For a century after he disappeared, Franklin was generally regarded as a hero-saint, a noble figure who carried out his duty to Queen and country.[5] For example, Henry Traill's biography of Sir John justifies its appearance, after half a century of narratives describing in excruciating detail the search for Franklin, as follows: "What Franklin *did* may be sufficiently well known to his countrymen already. What he *was*—how kindly and affectionate, how modest and magnanimous, how faithful in his allegiance to duty, how deeply and unaffectedly religious—has never been and could never be known to any but his intimates.... The character of such men as Franklin is, in truth, as much a national possession as their fame and work."[6]

But, more recently after that hagiographic portrayal and a number of 19th- and early 20th-century tales of the noble hero(es), bravely confronting the unknown in the service of Queen and Country, complemented by epic poems praising Franklin's legacy,[7] the story of the Franklin Expedition turned more critical. First, in the last half of the 20th century, a revisionist perspective portrayed Franklin and his crew as blundering imperialists, doomed by their ignorance of local conditions and their unwillingness to ask those who knew the condition—the Inuit—for help.[8] Author Ken McGoogan, a champion of John Rae's claim to be the discoverer of the Northwest Passage, called Franklin "a plodding man, gloomy, bumbling and bovine."[9]

The 21st century has seen the emergence of what might be called a Hegelian synthesis of these two disparate viewpoints. In this newest perspective, Franklin appears as neither heroic explorer nor dim-witted fool, but as someone dutifully, if unimaginatively, doing his duty, despite having been sent on a fool's errand that could only end in death and defeat. For example, Martyn Beardsley, in his recent biography of Franklin, says that Sir John was sent on an "impossible task, and even his critics do not attempt to blame him personally for the deaths of the officers and men."[10] Well, maybe not all the critics; Sir John did have a distressing tendency to lose his men wherever he went.

Back to that mid–20th century critical perspective. Beginning as early as 1951,[11] and increasingly after the publication of Pierre Berton's *The Arctic Grail*[12] in 1988, a revisionist view largely took hold, in which Franklin is seen either as a rule-bound bureaucrat, unable to adjust to the changing demands of an Arctic expedition and incapable, as a good Englishman, of seeing the virtue in Inuit ways of being in the North, or as simply a bumbling incompetent. Even more recently, Franklin and the hundreds of other British sailors who swarmed the Arctic archipelago in the mid–19th century have been seen as part of the European colonialist impulse that subjected so many of the world's peoples to foreign rule and exploitation. The 21st century, with the Canadian government's push for mining development in the far north, is a different story; the assertion of Canadian sovereignty, especially by then–Prime Minister Stephen Harper's government in the 2010s, demanded Canadian heroes, and Sir John, even though not Canadian and regardless of whether he had become, as Richard King warned at the time, "the nucleus of an iceberg"[13] or not, was one of those necessary heroes.

Let's start with the hagiography. To some, Sir John was an incarnation of an Arthurian knight, giving his life in the pursuit of a noble, if completely useless quest—in Sir John's case, the non-navigable Northwest Passage.[14] A later avatar of this heroic concept, Navy Captain Robert Falcon Scott, seemed to express much the same notion as he man-hauled his way to death in the Antarctic in 1912.[15]

Even while the Expedition was still en route, though, Sir John was less than universally respected. In his last letter home, addressed to James Clark Ross from Greenland, Francis Crozier commented that "I cannot bear going on board *Erebus*," followed by a thoroughly scratched-out passage in which he presumably made critical comments about Franklin that he then thought better of before posting the letter.[16] And more recently, as Sherrill Grace points out, criticism of Sir John serves the modern purpose of celebrating a Canadian identity that's different from Canada's English roots: "At times, I think, Canadians have delighted in the image of this

Statue of Rear Admiral (as he became posthumously) Sir John Franklin in Hobart, Tasmania (photograph by Debra Widdicombe, Creative Commons License BY-SA 4.0; Wikimedia Commons).

arrogant, foolish old Brit trapped in *our* Arctic ice; the telling of Franklin can sometimes be a Canadian revenge tragedy."[17]

As for the criticism that Franklin and his Expedition would have done better had they only been willing to "go native"—if, that is, they had been able to hunt, fish and shelter themselves in Inuit fashion on the 1848 retreat—it's not at all clear that they could have, nor that this supposed prejudice ran all that deep. Franklin himself had depended on Indian efforts to rescue and feed him in his first expedition to the shores of the Arctic, in 1819–22, and Janice Cavell convincingly demonstrates that mid-century British opinion was not nearly so resolutely anti-native as is sometimes supposed.[18] In fact, much of the discussion of the Expedition's fate in the British press in the early 1850s revolved around the likelihood of the survivors' having adapted to Inuit ways of hunting and of clothing and sheltering themselves.[19] And Franklin was conveniently dead by the time the Expedition survivors most needed to adapt Inuit ways of being in and travelling over the land on their retreat in 1848. That Crozier and Fitzjames equally failed to "go native," if in fact they did, was perhaps more a consequence of the by-then appalling physical and mental condition of the survivors, as well as of the simple fact that there were far too many men to have any reasonable hope of surviving by killing what they needed to eat and by travelling light, as the small Inuit hunting parties did. By that point, there was likely no escape, no matter what tactics the Expedition's leaders adopted.

As the later Franklin search expeditions demonstrated, sending a large ship into, say, Lancaster Sound and then using it as a base—and a refuge in winter—from which small sledging parties could explore and map the unknown was not a bad strategy for dealing with the polar climate. Trying to force a large ship through the shallow waters to the east of King William Island might have been a mistake, but we don't even know for sure that Franklin attempted that.

Most of our sources agree that Franklin was deeply religious. His letters to his family throughout his naval career are filled with religious sentiment and conventional piety.[20] (The letters also show a profoundly insecure man, uncertain of his ability to write the required narratives of his journeys and doubting his place in the naval and social hierarchies of the day.) While nominally a member of the Church of England, he had shown a strong susceptibility early in life to the more emotional, evangelical habits of the Methodists and, early in his naval career, came under the influence of Lady Lucy Barry, a distant cousin of Methodism's founder John Wesley and a relentless proselytizer for evangelical views.[21] Franklin's evangelical zeal apparently faded after his marriage to his first wife, the distinctly mainstream, Church of England, Eleanor Porden in 1823, but the theme of enduring suffering with a stoic Christian resignation, or, as Janice Cavell puts it, dying "with a prayer on his lips and a religious book in his hand,"[22] persisted well into Victorian times.

Franklin, unlike some other Royal Navy commanders, preached to his men every Sunday at Divine Service, and he made sure that the ships' libraries contained a very large number of religious tracts and devotional manuals, likely adapted, after the fashion of children's readers, to the varying literacy levels of the crew. By the middle of the 18th century, the religious battles that had occupied English society since the 1500s had largely subsided, leaving behind a comfortable mainstream set of beliefs that didn't ask too much of the Church of England's adherents. Whatever remained of his earlier evangelical fervor, one suspects, had been channeled by Sir John into a public-facing religion that didn't ask too much more of his parishioners—the Expedition crew—than to trust in God and obey their superiors. Appropriate for the times, perhaps, but not much of an inspiration in our current cynical days.

So, what are we, in the end, to make of Sir John's, and the Expedition's, legacy? The Franklin Expedition remains a prodigious source of tales, some well-told, some not. And perhaps, by the sheer size of the disaster—whether by the numbers (129) or percentage (100 percent!) of men killed, it ranks as the most fatal exploration attempt in history—it provides an antidote to the stirring tales of noble heroes that are, to this day, relied on to convince young men to join up to kill for their country, even though Sir John's own mission had relatively little of the aggressive or colonialist about it. And so we are left not far from where we started: some facts known, more unknown, and some that, despite the efforts of a worldwide network of Franklin sleuths, are destined to remain unknown. Perhaps it's time, truly, finally, to sit back and let the mystery be.

18

What Do the Recent Discoveries Mean?

Before it became a factor in Canada's attempts to assert jurisdiction over the Arctic, the Northwest Passage was, as archaeologist Lisa Hodgetts notes, "a holy grail for European explorers."[1] No matter that the search for the passage was often more reminiscent of *Monty Python and the Holy Grail* than of the valiant individual heroics of Malory's epic; in the end, Britain, and the Industrial Revolution that gave birth to industrial capitalism, created the Anthropocene, warmed the planet, melted the ice and made the Passage navigable for cruise ships full of faux-heroic self-styled adventurers. Perhaps not exactly what Parliament had in mind when it created the reward for discovery of the Passage, but history sometimes works in mysterious ways.

For white Canadians, as Ian MacLaren has pointed out, the Northwest Passage was, and remains, a mystery: "remote, serene, and inaccessible to most Canadians, it lies above us in latitude, in virtue, in fortitude."[2] Although the Expedition was a product of English, not Canadian, ambition, and although few, if any, of the Expedition crew had any ties to what would become the nation of Canada, the voyage has become an historic touchstone for Canadians, most of whom, safe in their cities within a few miles of the U.S. border, have never ventured anywhere near the northern continental coast where the Passage lies.

The now decades-long involvement of Parks Canada in the Franklin search and especially the discoveries of the sunken *Erebus* (in 2014) and *Terror* (in 2016) have served to intensify that national identification with the Expedition. Perhaps impelled by a desire to assert the Canadian-ness of these waters in international law,[3] former Prime Minister Stephen Harper closely followed the Parks Canada and private expeditions—some financed in part by oil companies eager for new drilling sites[4]—and closely managed the public release of their discoveries.[5] The English, only too happy to divest themselves of what many might regard as an embarrassing failure, have happily transferred all claims and responsibilities to Canada.[6] But what story can white Canadians, who now effectively own the Franklin Expedition, with all its history and uncertainties, tell?

From the beginning, white folks distrusted Inuit and Indian stories of the Franklin Expedition. Franklin himself, in his original journals of his 1819–22 expedition, much sanitized for publication, first by Franklin himself and then presumably by the Admiralty's John Barrow and publisher John Murray, started out with a number of comments regarding the untrustworthiness of his native hunters.[7] While Franklin himself may, upon reflection, have softened his view of the natives,[8] and while others who actually explored the Arctic came to rely on Inuit technology, their experience didn't changes attitudes back in England, where the general tendency to discount Inuit testimony persisted throughout most of the century and a half after the Expedition disappeared.

But, if we ignore Inuit stories, what are we left with? A century and a half of on-site investigation by (mostly) white men, and even the indefatigable researches of the 3000-plus-strong band of Franklinites on Facebook, have given us, at best, only a small and fractured part of the story. We know a few of the relevant dates: when Franklin left England, when the ships were beset in the ice of Victoria Strait, when the crew first abandoned the ships in 1848. And we've identified a few of the dead—the three buried on Beechey Island and a few more whose remains were found with sufficient evidence to allow reasonably certain identification. We've found the sunken ships, thanks to both modern technology and Inuit guidance, and we've collected innumerable artifacts and bones along the route(s) that some or all of the Franklin crew may have taken after leaving the ships. And yes, it's possible that further discoveries—especially the not very likely retrieval of the now-mythical captain's log from *Terror* or the finding a cache of papers at some as yet unexplored site on King William Island—will shed further light on the Franklin history and fill in some of the many gaps in our current knowledge. The summer 2022 diving season, the most recent at this writing, yielded a trove of some 275 artifacts from *Erebus*'s steward's pantry, mostly plates and serving dishes but also including a leather portfolio with a few sheets of paper inside.[9] That paper was taken to Parks Canada's lab in Ottawa, where it may reveal secrets of the Expedition, or may turn out to be the steward's laundry list. Needless to say, enthusiasm on Facebook was instantaneous.[10] As for that Holy Grail on board *Terror*, the Crozier logbook, we still aren't sure that it even exists; the divers didn't return to that wreck in 2022, because the wreck of *Terror* is in much better condition than that of *Erebus* and presumably can wait a few years longer to unlock its secrets, if such there be.

But, even with the addition of Inuit narratives, many of them collected during Charles Francis Hall's years of living with and talking to the people who actually lived in the area and had first-hand knowledge or had heard the stories from those who did, and with David Woodman's painstaking reconstruction of those Inuit stories,[11] as well as Dorothy Eber's work collecting tales from those currently living in the Franklin area,[12] there's still an irreducible minimum of unknowns. We know, or at least we're pretty sure, that the Expedition set out from Greenland with 129 men aboard and that 105 men left *Erebus* and *Terror* in April 1848. We know that a large number of them made it as far as a camp near Terror Bay, in the southwest corner of King William Island. We know some of that number made it much farther to the southeast, in the direction of their stated objective, the mouth of Back's River. We're pretty sure that at least some of the crew broke off from the main party and returned to either or both of the ships, perhaps sailing them toward where the vessels were eventually found, perhaps drifting with them when the ice eased its grip. We're almost certain that some of the crew encountered Inuit hunting parties, either along the west and south coasts of King William Island or south, toward "Ootjoolik," in the area where *Erebus* eventually came to rest. And we're pretty sure that they all died not very long after leaving the ships, whether that time was a few weeks, months or (a very few) years.

18. What Do the Recent Discoveries Mean?

Poster announcing reward for relief of the crews or information as to their fate (Library and Archives Canada/Library and Archives Canada Miscellaneous Poster Collection/e010754422).

Beyond that, as the previous chapters demonstrate, much is conjecture, some of it better informed and better reasoned than other parts, but, nonetheless, to some degree unknowable. Post-mortems on centuries-old bones may tell us something about lead levels or identify other toxic clues to the men's demise and may persuade us that there probably was not a single catastrophe or a single cause of all the deaths, but are unlikely to identify the particular cause of each death. With each year, Arctic weather further degrades whatever cairns, bones and artifacts are still out there, waiting to be discovered. With former Canadian Prime Minister Stephen Harper, the champion of national sovereignty in the North, gone from the political scene and with budget constraints increasingly impinging on research institutions' ability to do fieldwork in the Arctic, the chances for any extraordinary breakthrough in Franklin Expedition research may be dimming. And the two-year hiatus in on-site research caused by the COVID-19 pandemic in 2020–21 only further diminished the chances of finding meaningful records. With every year, the already fragile *Erebus* continues to slowly fall apart. So, we know what we know and, increasingly, we know what we don't know and may never know. Time, or perhaps past time, for a reckoning.

As the previous chapter suggests, Franklin's legacy is at best mixed. In his defense, the Northwest Passage, at least in the mid–19th century, could never have been conquered by the relatively large ships that the Admiralty sent to traverse it. And no amount of consultation with the Inuit would have found enough food for a crew that was 10 times as large as an average Inuit hunting band. Nor would the crew that eventually left the ships, weakened by scurvy and increasingly hungry, have ever completed the long march to Back's River and the nearly impossible ascension of the river as far as the nearest Hudson's Bay Company outpost. Sir John Barrow and his Admiralty colleagues simply didn't know what they didn't know, and so they sent the Expedition's crew off to what would inevitably be either failure—if Franklin had turned back earlier in the journey—or death if, as eventually happened, they failed to turn back in time. Nothing in the remaining discoveries will change those basic truths. Beyond that, the Franklin Expedition will always, to some considerable extent, remain a mystery.

Despite the heroic efforts of Franklin searchers and researchers—official, unofficial and armchair—there is still much information that's missing and that we're unlikely ever to know with certainty. What killed Sir John Franklin, and, for that matter, what killed the rest of the crew? Where did all of the men go after leaving the ships in 1848? Did any survive the first few months of the retreat and, if so, for how long and where? Why did they delay leaving the ships until it was too late for survival? And, even were there to be tangible evidence like logbooks or other records recovered from the wrecks, there are still a host of intangibles that we'll most likely never know. What were they all thinking as the Expedition and then the retreat progressed? What kind of leaders were Sir John and Francis Crozier? What kind of men were they? But even as we accept that no amount of searching will ever enable us to fully reconstruct the past, especially its psychological elements, that doesn't mean we will or should give up the search for more knowledge. Just like everyone else, I'll be looking for every new bit of evidence that's uncovered, watching for new finds by Parks Canada in the hope that more information, more useful clues, will come to light. That's what we humans do; we search for information, we try to make sense of the world and of history. In that spirit, this book has set out what we know and what remains uncertain; it's up to each of us to fill in the gaps with our imaginations.

Appendix I

The Victory Point Record

28 of May 1847 H.M.S.hips Erebus and Terror Wintered in the Ice in Lat. 70°5'N Long. 98°.23'W Having wintered in 1846–7 [sic] at Beechey Island in Lat 74°43'28"N Long 91°39'15"W After having ascended Wellington Channel to Lat 77° and returned by the West side of Cornwallis Island. Sir John Franklin commanding the Expedition. *All well* Party consisting of 2 Officers and 6 Men left the ships on Monday May 24 1847.—Gm. Gore, Lieut., Chas. F. Des Voeux, Mate

April 25 1848 HMShips Terror and Erebus were deserted on the April 22 5 leagues NNW of this having been beset since 12th Sept 1846. The officers and crew consisting of 105 souls under the command of Captain F.R.M. Crozier landed here—in Lat. 69°37'42" Long. 98°41' This paper was found by Lt. Irving under the cairn supposed to have been built by Sir James Ross in 1831–4 miles to the Northward—where it had been deposited by the late Commander Gore in May 1847. Sir James Ross' pillar has not however been found and the paper has been transferred to this position which is that in which Sir J. Ross' pillar was erected—Sir John Franklin died on the 11th of June 1847 and the total loss by deaths in the Expedition has been to this date 9 officers and 15 men.—James Fitzjames Captain HMS Erebus F.R.M. Crozier Captain and Senior Offr And start on tomorrow 26th for Back's Fish River.

Appendix II

Erebus and *Terror* Muster Rolls

(From Richard J. Cyriax, *Sir John Franklin's Last Arctic Expedition* and from the originals at the Public Record Office, London: files ADM 38/672 and 38/1962)

HMS Erebus

Officers
Sir John Franklin, Captain, Commanding the Expedition
James Fitzjames, Commander
Graham Gore, Lieutenant
H.T.D. Le Vesconte, Lieutenant
James Walter Fairholme, Lieutenant
Robert Orme Sergeant, Mate
Charles Frederick Des Voeux, Mate
Edward Couch, Mate
Henry Foster Collins, Second Master
James Reid, Ice Master
Stephen Samuel Stanley, Surgeon
Harry D.S. Goodsir, Assistant Surgeon
Charles Hamilton Osmer, Purser

Warrant Officers
John Gregory, Engineer
Thomas Terry, Boatswain
John Weekes, Carpenter

Petty Officers
John Murray, Sailmaker, age 43
William Smith, Blacksmith, age 28
Thomas Burt, Armorer, age 22
James W. Brown, Caulker, age 28
Francis Dunn, Caulker's Mate, age 25
Thomas Watson, Carpenter's Mate, age 40
Samuel Brown, Boatswain's Mate, age 27
Richard Wall, Ship's Cook, age 45
James Rigden, Captain's Coxwain, age 32
William Bell, Quartermaster, age 36
Daniel Arthur, Quartermaster, age 35
John Downing, Quartermaster
Robert Sinclair, Captain of the Foretop, age 25
John Sullivan, Captain of the Maintop, age 28
Phillip Reddington, Captain of the Forecastle, age 28
Joseph Andrews, Captain of the Hold, age 35
Edmund Hoar, Captain's Steward, age 23
John Bridgens, Subordinate Officers' Steward, age 26
Richard Aylmore, Gunroom Steward, age 24
William Fowler, Purser's Steward, age 26
John Cowie, Stoker
Thomas Plater, Stoker

Able Seamen
George Thompson, age 27
John Hartnell, age 25
John Stickland, age 24

Thomas Hartnell, age 23
William Orren, age 34
William Closson, age 25
Charles Coombs, age 28
John Morfin, age 25
Charles Best, age 23
Thomas McConvey, age 24
Henry Lloyd, age 26
Thomas Work, age 41
Robert Ferrier, age 29
Josephus Geater, age 32
Thomas Tadman, age 28
Abraham Seeley, age 34
Francis Pocock, age 24
Robert Johns, age 24
William Mark, age 24

Royal Marines

David Bryant, Sergeant, age 31
Alexander Pearson, Corporal, age 30
Robert Hopcraft, Private, age 38
William Pilkington, Private, age 28
William Braine, Private, age 31
Joseph Healey, Private, age 29
William Reed, Private, age 28

Boys

George Chambers, age 18
David Young, age 18

HMS Terror

Officers

Francis Rawden Moira Crozier, Captain
Edward Little, Lieutenant
George Henry Hodgson, Lieutenant
John Irving, Lieutenant
Frederick John Hornby, Mate
Robert Thomas, Mate
Giles Alexander McBean, Second Master
Thomas Blanky, Ice Master
John Smart Peddie, Surgeon
Alexander McDonald, Assistant Surgeon
E.J. Helpman, Clerk in Charge

Warrant Officers

James Thompson, Engineer
John Lane, Boatswain
Thomas Honey, Carpenter

Petty Officers

Thomas Johnson, Boatswain's Mate, age 28
Alexander Wilson, Carpenter's Mate, age 27
Reuben Male, Captain of the Forecastle, age 27
David McDonald, Quartermaster, age 45
John Kenley, Quartermaster
William Rhodes, Quartermaster, age 31
Thomas Darlington, Caulker, age 29
Samuel Honey, Blacksmith, age 22
John Torrington, Leading Stoker, age 19
John Diggle, Cook, age 36
John Wilson, Captain's Coxswain, age 33
Thomas R. Farr, Captain of the Maintop, age 32
Harry Peglar, Captain of the Foretop, age 37
William Goddard, Captain of the Hold, age 39
Cornelius Hickey, Caulker's Mate, age 24
Thomas Jopson, Captain's Steward, age 27
Thomas Armitage, Gun-room Steward, age 40
William Gibson, Subordinate Officers' Steward, age 22
Edward Genge, Subordinate Officers' Steward, age 21
Luke Smith, Stoker, age 27
William Johnson, Stoker, age 45

Able Seamen

George J. Cann, age 23
William Strong, age 22
David Sims, age 24
John Bailey, age 21

William Jerry, age 29
Henry Sait, age 23
Alexander Berry, age 32
John Handford, age 28
John Bates, age 24
Samuel Crispe, age 24
Charles Johnson, age 28
William Shanks, age 29
David Leys, age 37
William Sinclair, age 30
George Kinnaird, age 23
Edwin Lawrence, age 30
Magnus Manson, age 28

James Walker, age 29
William Wentzall, age 33

Royal Marines

Solomon Tozer, Sergeant, age 34
William Hedges, Corporal, age 30
William Heather, Private, age 37
Henry Wilkes, Private, age 28
John Hammond, Private, age 32
James Daly, Private, age 30

Boys

Robert Golding, age 19

Appendix III

Sir John Franklin's Sailing Orders

(Available at https://www.canadianmysteries.ca/
sites/franklin/archive/text/InstructionsToFranklin_en.htm)

By the Commissioners for executing the office of Lord High Admiral of the United Kingdom of Great Britain and Ireland.

1. Her Majesty's Government having deemed it expedient that further attempt should be made for the accomplishment of a north-west passage by sea from the Atlantic to the Pacific Ocean, of which passage a small portion only remains to be completed, we have thought proper to appoint you to the command of the expedition to be fitted out for that service, consisting of Her Majesty's Ships "Erebus," under your command, taking with you Her Majesty's ship "Terror," her Captain (Crozier), having been placed by us under your orders, taking also with you the "Barretto Junior" transport, which has been directed to be put at your disposal for the purpose of carrying out portions of your provisions, clothing and other stores.

2. On putting to sea, you are to proceed, in the first place, by such a route as from the wind and weather, you may deem to be the most suitable for despatch, to Davis' Strait, taking the transport with you to such a distance up the Strait as you may be able to proceed without impediment from ice, being careful not to risk the vessel by allowing her to be set in the ice, or exposed to any violent contact with it ; you will then avail yourself of the earliest opportunity of clearing the transport of the provisions and stores with which she is charged for the use of the expedition, and you are then to send her back to England, giving to the agent or master such directions for his guidance as may appear to you most proper, and reporting by that opportunity your proceedings to our secretary, for our information.

3. You will then proceed in the execution of your orders into Baffin's Bay, and get as soon as possible to the western side of the Strait, provided it should appear to you that the ice chiefly prevails on the eastern side, or near the middle ; but as no specific directions can be given, owning to the position of the ice varying from year to year, you will, of course, be guided by your own observations as to the course most eligible to be taken, in order to ensure a speedy arrival in the Sound above mentioned.

4. As, however, we have thought fit to cause each ship to be fitted with a small steam-engine and a propeller, to be used only in pushing the ships through channels between masses of ice, when the wind is adverse, or in a calm, we trust the difficulty usually found in such cases will be much obviated, but as the supply of fuel to be taken in the ships is necessarily small you will use it only in cases of difficulty.

5. Lancaster Sound, and its continuation through Barrow's Strait, having been four

times navigated without any impediment by Sir Edward Parry, and since frequently by whaling ships, will probably be found without any obstacles from ice or islands; and Sir Edward Parry having also proceeded from the latter in a straight course to Melville Island, and returned without experiencing any, or very little, difficulty, it is hoped that the remaining portion of the passage, about 900 miles, to the Bhering's Strait may also be found equally free from obstruction; and in proceeding to the westward, therefore, you will not stop to examine any openings either to the northward or southward in that Strait, but continue to push to the westward without loss of time, in the latitude of about 74¼ degrees, till you have reached the longitude of that portion of the land on which Cape Walker is situated, or about 98 degrees west. From that point we desire that every effort be used to endeavour to penetrate to the southward and the westward in a course as direct towards Bhering's Strait as the position and extent of the ice, or the existence of land, at present unknown, may admit.

6. We direct you to this particular part of the Polar Sea as affording the best prospect of accomplishing the passage to the Pacific, in consequence of the unusual magnitude and apparently fixed state of the barrier of ice observed by the "Hecla" and the "Griper," in the year 1820, off Cape Dundas, the south-western extremity of Melville Island; and we, therefore, consider that loss of time would be incurred in renewing the attempt in that direction; but should your progress in the direction before ordered be arrested by ice of a permanent appearance, and that when passing the mouth of the Strait, between Devon and Cornwallis Islands, you had observed that it was open and clear of ice; we desire that you will duly consider, with reference to the time already consumed, as well as to the symptoms of a late or early close of the season, whether that channel might not offer a more practicable outlet from the Archipelago, and a more ready access to the open sea, where there would be neither islands nor banks to arrest and fix the floating masses of ice; and if you should have determined to winter in that neighbourhood, it will be a matter of your mature deliberation whether in the ensuing season you would proceed by the above-mentioned Strait, or whether you would persevere to the south-westward, according to the former directions.

7. You are well aware, having yourself been one of the intelligent travellers who have traversed the American shore of the Polar Sea, that the groups of islands that stretch from that shore to the northward to a distance not yet known, do not extend to the westward further than about the 120th degree of western longitude, and beyond this, and to Bhering's Strait, no land is visible from the American shore of the Polar Sea.

8. Should you be so fortunate as to accomplish a passage through Bhering's Strait, you are then to proceed to the Sandwich Islands, to refit the ships and refresh the crews, and if, during your stay at such place, a safe opportunity should occur of sending one of your officers or dispatches to England by Panama, you are to avail yourself of such opportunity to forward to us as full a detail of your proceedings and discoveries as the nature of the conveyance may admit of, and in the event of no such opportunity offering during your stay at the Sandwich Islands, you are on quitting them to proceed with the two ships under your command off Panama, there to land an officer with such dispatches, directing him to make the best of his way to England with them, in such a manner as our Consul at Panama shall advise, after which you are to lose no time in returning to England by way of Cape Horn.

9. If at any period of your voyage the season shall be so far advanced as to make it unsafe to navigate the ships, and the health of your crews, the state of the ships, and all concurrent circumstances should combine to induce you to form the resolution of wintering in those regions, you are to use your best endeavours to discover a sheltered and safe harbour, where the ships may be placed in security for the winter, taking such measures, for the health and comfort of the people committed to your charge as the materials with which you are provided for housing in the ships, may enable you to do—and if you should find it expedient to resort to this measure, and you should meet with any inhabitants, either Esquimaux or Indians, near the place where you winter, you are to endeavour by every means in your power to cultivate a friendship with them, by making them presents of such articles as you may be supplied with, and which may be useful or agreeable to them ; you will, however, take care not to suffer yourself to be surprized by them but use every precaution, and be constantly on your guard against any hostility ; you will, by offering rewards, to be paid in such a manner as you may think best, prevail on them to carry to any of the settlements of the Hudson's Bay Company, an account of your situation and proceedings, with an urgent request that it may be forwarded to England with the utmost possible dispatch.

10. In an undertaking of this description much must be always left to the discretion of the commanding officer, and, as the objects of this Expedition have been fully explained to you, and you may have already had much experience on service of this nature, we are convinced we cannot do better than leave it to your judgement, in the event of your not making the passage this season, either to winter on the coast, with the view of following up next season any hopes or expectations which your observations this year may lead you to entertain, or to return to England to report to us the result of such observations, always recollecting our anxiety for the health, comfort and safety of yourself, your officers and men ; and you will duly weigh how far the advantage of starting next season from an advanced position may be counterbalanced by what may be suffered during the winter, and by the want of such refreshment and refitting as would be afforded by your return to England.

11. We deem it right to caution you against suffering the two vessels placed under your orders to separate, except in the event of accident or unavoidable necessity, and we desire you to keep up the most unreserved communications with the commander of the "Terror," placing in him every proper confidence, and acquainting him with the general tenor of you orders, and with your views and intentions from time to time in the execution of them, that the service may have the full benefit of your united efforts in the prosecution of such a service ; and that, in the event of unavoidable separation, or of any accident to yourself, Captain Crozier may have the advantage of knowing, up to the latest practicable period, all your ideas and intentions relative to satisfactory completion of this interesting undertaking.

12. We also recommend, that as frequent an exchange take place as conveniently may be of the observations made in the two ships ; that any scientific discovery made by the one, be as quickly as possible communicated for the advantage and guidance of the other, in making their future observations, and to increase the probability of the observations of both being preserved.

13. We have caused a great variety of valuable instruments to be put on board the ships under your orders, of which you will be furnished with a list, and for the

Appendix III

return of which you will be held responsible ; among these, are instruments of the latest improvements for making a series of observations on terrestrial magnetism, which are at this time peculiarly desirable, and strongly recommended by the President and the Council of the Royal Society, that the important advantage be derived from observations taken in the North Polar Sea, in co-operation with the observers who are at present carrying on an uniform system at the magnetic observatories established by England in her distant territories, and, through her influence, in other parts of the world ; and the more desirable is this co-operation in the present year, when these splendid establishments, which do so much honour to the nations who have cheerfully erected them at a great expense, are to cease. The only magnetical observations that have been obtained very partially in the Arctic Regions, are now a quarter of a century old, and it is known that the phenomena are subject to considerable secular changes. It is also stated by Colonel Sabine, that the instruments and methods of observation have been so greatly improved, that the earlier observations are not to be named in point of precision with those which would now be made ; and he concludes by observing, that the passage through the Polar Sea would afford the most important service that now remains to be performed towards the completion of the magnetic survey of the globe.

14. Impressed with the importance of this subject, we have deemed it proper to request Lieut.-Colonel Sabine to allow Commander Fitzjames to profit by his valuable instructions, and we direct you, therefore, to place this important branch of science under the immediate charge of Commander Fitzjames; and as several other officers have also received similar instruction at Woolwich, you will therefore cause observations to be made daily on board each of the ships whilst at sea (and when not prevented by weather, and other circumstances) on the magnetic variation, dip and intensity, noting at the time the temperature of the air, and of the sea at the surface, and at different depths; and you will be careful that in harbour and on other favourable occasions those observations shall be attended to, by means of which the influence of the ship's iron on the result obtained to sea may be computed and allowed for.

15. In the possible event of the ships being detained during a winter in the high latitudes, the expedition has been supplied with a portable observatory, and with instruments similar to those which are employed in the fixed magnetical and meteorological observatories instituted by Her Majesty's Government in several of the British colonies.

16. It is our desire that, in case of such detention, observations should be made with these instruments, according to the system adopted in the aforesaid observatories, and detailed directions will be supplied for this purpose, which with the instruction received at Woolwich, will be found, as we confidently anticipate, to afford full and sufficient guidance for such observations, which will derive from their locality, peculiar interest, and high theoretical value.

17. We have also directed instruments to be specially provided for observations on atmospherical refraction at very low altitudes, in case of the expedition being detained during a winter in the high latitudes ; on this subject also particular directions will be supplied, and you will add any other meteorological observations that may occur to you of general utility ; you will also take occasions to try the depth of the sea and nature of the bottom, the rise, direction and strength of the tides, and the set and velocity of currents.

18. And you are to understand that although the effecting a passage from the Atlantic to the Pacific is the main object of this expedition, yet, that the ascertaining the true geological position of the different points of land near which you may pass, so far as can be effected without detention of the ships in their progress westward, as well as such other observations as you may have opportunities of making in natural history, must prove most valuable and interesting, and to call that of all the officers under your command to these points, as being objects of high interest and importance.

19. For the purpose, not only of ascertaining the set of the currents in the Arctic Seas, but also of affording more frequent chances of hearing your progress, we desire that you frequently, after you have passed the latitude of 65 degrees north, and once every day when you shall be in an ascertained current, throw overboard a bottle or copper cylinder closely sealed, and containing a paper stating the date and position at which it is launched, and you will give similar orders to the commander of the "Terror," to be executed in case of separation; and for this purpose, we have caused each ship to be supplied with papers, on which is printed, in several languages, a request that whoever may find it should take measures for transmitting it to this office.

20. You are to make use of every means in your power to collect and preserve specimens of animal, mineral and vegetable kingdoms, should circumstances place such within your reach without causing your detention, and of the larger animals you are to cause accurate drawings to be made, to accompany and elucidate the descriptions of them. In this, as well as in every other part of your scientific duty, we trust that you will receive material assistance from the officers under your command, several of whom are represented to us as well qualified in these respects.

21. In the event of any irreparable accident happening to either of the two ships, you are to cause the officers and crew of the disabled ship to be removed into the other, and with her singly to proceed in prosecution of the voyage, or return to England, according as circumstances shall appear to require, understanding that the officers and crews of both ships are hereby authorized and required to continue to perform the duties according to their respective ranks and stations on board either ship to which they may be removed, in the event of an occurrence of this nature. Should, unfortunately, your own ship be the one disabled, you are in that case to take command of the "Terror," and in the event of any fatal accident happening to yourself, Captain Crozier is hereby authorized to take command of the "Erebus," placing the officer of the expedition who may then be next in seniority to him in command of the "Terror." Also, in the event of your own inability, by sickness or otherwise, of any period of service, to continue to carry these instructions into execution, you are to transfer them to the officer next in command to you employed on the expedition, who is hereby required to execute them in the best manner he can for the attainment of the several objects herein set forth.

22. You are, while executing the service pointed out in these instructions, to take every opportunity that may offer of acquainting our secretary, for our information, with your progress, and on your arrival in England, you are immediately to repair to this office, in order to lay before us a full account of your proceedings in the whole course of your voyage, taking care before you leave the ship to demand from the officers, petty officers, and all other persons on board, the logs and journals they may have kept, together with any drawings or charts they may have made, which are all

sealed up, and you will issue similar directions to Captain Crozier and his officers. The said logs, journals or other documents to be thereafter disposed of as we may think proper to determine.

23. In the event of England becoming involved in hostilities with any other power during your absence, you are nevertheless clearly to understand that you are not on any account to commit any hostile act whatsoever, the expedition under your orders being only intended for the purpose of discovery and science, and it being the practice of all civilized nations to consider vessels so employed as excluded from the operations of war ; and, confiding in this feeling, we should trust that you would receive every assistance from the ships or subjects of any foreign power which you may fall in with ; but special application to that effect has been made to the respective governments.

> Given under our hands, this 5th day of May 1845.
> (signed) *Haddington.*
> G. *Cockburn*
> W. H. *Gage*
>
> Sir John Franklin, K.C.H. Captain of H.M.S. "Erebus," at Woolwich
> By command of their Lordships.
> (signed) *W.A.B. Hamilton*

Chapter Notes

Preface

1. Pete Seeger, "Waist Deep in the Big Muddy," 1967. Available at https://genius.com/Pete-seeger-waist-deep-in-the-big-muddy-lyrics.
2. Roland Huntford, *Scott and Amundsen*, London: Hodder and Stoughton, 1979.
3. Pierre Berton, *The Arctic Grail*, Toronto: McClelland & Stewart, 1988.

Chapter 1

1. Adrian Craciun, "The Franklin Mystery," *Literary Review of Canada*, May 2012. Available at http://reviewcanada.ca/magazine/2012/05/the-franklin-mystery/.
2. *The King's Mirror*, by an unknown Norseman, circa 1250, quoted in Edward Struzik, *Northwest Passage: The Quest for an Arctic Route to the East*, Toronto: KeyPorter Books, 1991. For more up-to-date sources, see Kirsten A. Seaver, *The Last Vikings: The Epic Story of the Great Norse Voyages*, London: Bloomsbury Academic, 2021, and Thomas H. McGovern, "The Archeology of the Norse North Atlantic," *Annual Review of Anthropology* 19 (1990): 331–51.
3. Richard J. Cyriax, *Sir John Franklin's Last Arctic Expedition: A Chapter in the History of the Royal Navy*, London: Methuen, 1939, reprint Sussex: Arctic Press, 1997, 3. There is an abundant literature on the Norse presence in North America. See, e.g., Hermann Pálsson, *The Vinland Sagas: The Norse Discovery of America*, New York: Penguin Classics, 1965; Helga Ingstad and Anne Stein Ingstad, *The Viking Discovery of America*, New York: Facts on File, 2001; and Erik Wahlgren, *The Vikings and America*, New York: Thames and Hudson, 1986.
4. See, among many other, often more skeptical, sources, Paul H. Chapman, *The Man Who Led Columbus to America*, Atlanta: Judson Press, 1973. There are no contemporary sources for Brendan's voyage; the first accounts appeared nearly two centuries after it supposedly occurred. See John D. Anderson, "The *Navigatio Brendani*, a Medieval Bestseller," *The Classical Journal* 83 (1988): 315–22.
5. Quoted in Pierre Berton, *Arctic Grail: The Quest for the Northwest Passage and the North Pole, 1818–1909*, Toronto: McClelland & Stewart, 1988, at 16.
6. Apart from a lone voyage by the Hudson's Bay Company's captain Christopher Middleton, who turned back at the accurately named Repulse Bay in the early 1740s. *Dictionary of Canadian Biography*, http://www.biographi.ca/en/bio/middleton_christopher_3E.html. Since Middleton was already in the Bay, as an HBC employee, this doesn't really count as an official English government expedition. For a brief popular summary of early English expeditions in Hudson's Bay and the Arctic archipelago, see Harry S. Anderson, *Exploring the Polar Regions*, New York: Facts on File, 2005, 30–42.
7. T.H. Levere, *Science and the Canadian Arctic: A Century of Exploration, 1818–1918*, Cambridge: Cambridge University Press, 1992. See also William Scoresby, *The Arctic Whaling Journals of William Scoresby*, London: Hakluyt Society, 2003.
8. On the key role of Murray as publisher of the various exploration accounts in the first half of the 19th century, and on the symbiotic relationship between Barrow and Murray, see Janice Cavell, "Making Books for Mr. Murray: The Case of Edward Parry's Third Arctic Narrative," *The Library* 14 (2013): 45–69.
9. John Barrow, *A Chronological History of Voyages into the Arctic Regions*, London: John Murray, 1818, reprint Newton Abbot, Devonshire: David & Charles, 1971.
10. Andrew Lambert, *The Gates of Hell: John Franklin's Tragic Quest for the Northwest Passage*, New Haven: Yale University Press, 2011, at 13.
11. See Bill Bell, "Authors in an Industrial Economy: The Case of John Murray's Travel Writers," *Romantic Textualities* 21 (2013): 9–29, and Ian MacLaren, "From Exploration to Publication: The Evolution of a 19th-Century Arctic Narrative," *Arctic* 47 (1994): 43–53.
12. On the many political and ideological uses of maps, see Mia Bennett, Wilfrid Greaves, Rudolf Riedlsperger and Alberic Botella, "Articulating the Arctic: Contrasting State and Inuit Maps of the Canadian North," *Polar Record* 52 (2016): 630–44. The mapping and claiming of putative "terra nullius" lands for the

Crown—despite the presence on those lands of people whose forebears had been there for thousands of years—was a significant part of the larger imperial project. On the use of "terra nullius" for British imperial purposes in Australia, see Stuart Banner, "Why *Terra Nullius*? Anthropology and Property Law in Early Australia," *Law and History Review* 23 (2005): 95–131.

13. Alexa Price, "'Our Proudest Heritage': Masculinity, Nostalgia, and the Sailing Navy on Display, 1820–1920," Ph.D. Dissertation, The George Washington University, 2019, at 86–87. See also Janice Cavell, *Tracing the Connected Narrative*, Toronto: University of Toronto Press, 2008, and Russell Potter, *Arctic Spectacles: The Frozen North in Visual Culture, 1818–1875*, Seattle: University of Washington Press, 2007.

14. See Douglas Wilkinson, *Arctic Fever: The Search for the Northwest Passage*, Toronto: Clark, Irwin & Co., 1971.

15. Mary Wollstonecraft Shelley, *Frankenstein, or the Modern Prometheus*, London: Penguin Classics, 1931, at 11 (originally published 1818).

16. Jules Verne, *The Voyages and Adventures of Captain Hatteras*, Boston: James R. Osgood & Co., 1876.

17. The development of the open Polar Sea theory is recounted in John K. Wright, "The Open Polar Sea," *Geographical Review* 43 (1953): 338–65. The remainder of this paragraph relies on that source.

18. See Alvyn A. Ruddock, "Columbus and Iceland: New Light on an Old Problem," *The Geographical Journal* 136 (1970): 177–89.

19. Wright, "The Open Polar Sea," 338–39.

20. *Id.* at 340–41.

21. Berton, *Arctic Grail*, 22. The narrative of that Spitzbergen voyage was not published for a quarter-century: Frederick W. Beechey, *A Voyage of Discovery Towards the North Pole, Performed in His Majesty's Ships* Dorothea *and* Trent *Under the Command of Captain David Buchan, R.N., 1818*, London: Richard Bentley, 1843.

22. William Scoresby, "On the Greenland or Polar Ice," *Memoirs of the Wernerian Society* 2 (1815): 268–338, cited in Constance Martin, "William Scoresby, Jr. (1789–1857) and the Open Polar Sea—Myth and Reality," *Arctic* 41 (1988): 39–47.

23. Michael Durey, "Exploration at the Edge: Reassessing the Fate of Sir John Franklin's Last Arctic Expedition," *The Great Circle: Journal of the Australian Association for Maritime History* 30 (2008): 3–40, at 8.

24. The First Secretary, a political appointee, was the spokesperson for the Admiralty in Parliament, while the Second Secretary, a career civil servant, basically ran the show on a day-to-day basis.

25. Parry was also the first of the 19th-century polar explorers to use a third ship as a transport, carrying extra supplies as far as Greenland, to lengthen the amount of time an expedition's two primary vessels could remain in the ice. May Fluhmann, *Second in Command: The Life of Francis R.M. Crozier*, Yellowknife: Department of Information, Government of the Northwest Territories, 1976, 14. Not everyone agrees, however, that Parry was a superior commander of exploration voyages. See, e.g., A.G.E. Jones, "Rear Admiral Sir William Parry: A Different View," *The Musk-Ox* 21 (1978): 3–10.

26. William Edward Parry, *Journal of a Voyage for the Discovery of a Northwest Passage from the Atlantic to the Pacific: Performed in the Years 1819–20 in His Majesty's Ships* Hecla *and* Griper, London: John Murray, 1821.

27. Anthony Brandt, *The Man Who Ate His Boots*, New York: Knopf, 2010.

28. Berton, *Arctic* Grail, at 63–75 has a succinct, if perhaps unnecessarily critical, summary of Franklin's 1819–22 expedition. Britain's 19th-century fascination with the Arctic is discussed in Kathryn Schulz, "Literature's Arctic Obsession," *New Yorker*, April 24, 2017, available at https://www.newyorker.com/magazine/2017/04/24/literatures-arctic-obsession.

29. John Ross, *Narrative of a Second Voyage in Search of the Northwest Passage and of a Residence in the Arctic Regions, 1829–33*, London: A.W. Webster, 1835.

30. George Back, *Narrative of an Expedition in HMS Terror Undertaken with a View to Geographical Discovery on the Arctic Shores in the Years 1836–7*, London: John Murray, 1838.

31. Cyriax, *Sir John Franklin's Last Arctic Expedition*, 25.

32. Lambert, *The Gates of Hell*, 149, 168.

33. Lambert, *The Gates of Hell*, 167–176.

34. See, e.g., Paul Nanton, *Arctic Breakthrough: Franklin's Expeditions 1819–1847*, Toronto: Clarke Irwin & Co., 1970, 225.

35. Lambert, *The Gates of Hell*, 144–50.

36. *Id.* at 145.

37. Fitzjames's biographer, William Battersby, raises the possibility that Fitzjames had lent John Barrow's son George money to help George out of some sort of scandal in 1841, recently enough to be in Barrow's mind as the Franklin expedition was being organized. William Battersby, *James Fitzjames: The Mystery Man of the Franklin Expedition*, Stroud: History Press, 2010, 135–36.

38. Michael Palin, *Erebus: One Ship, Two Epic Voyages, and the Greatest Naval Mystery of All Time*, Vancouver: Greystone Books, 2018, at 81.

39. In a letter to Dr. Adam Turnbull, Franklin wrote of Fitzjames, that "the Admiralty have appointed a Commander unsolicited on my part to the ship." George Mackannes, *Some Private Correspondence of Sir John and Lady Jane Franklin*, Sydney: D.S. Ford, 1947, cited by Allegra Rosenberg in the Facebook group "Remembering the Franklin Expedition," September 24, 2021.

40. Battersby, *James Fitzjames*, 152.

41. Frances Woodward, *Portrait of Jane: A Life of Lady Franklin*, London: Hodder & Stoughton, 1951, 257. Franklin's earnest attempts at modernization of the Tasmanian prison colony were resented by the landowners who benefited from convict labor and by the arguably corrupt government functionaries who did quite well under the old system and who convinced the Colonial Secretary in London to recall Franklin. See Kathleen Fitzpatrick, *Sir John Franklin in Tasmania, 1837–1843*, Melbourne: Melbourne University Press, 1949, and Kathleen Fitzpatrick, "Franklin, Sir John (1786–1847)," *Australian Dictionary of Biography*, vol. 1 (1966). Available at https://adb.anu.edu.au/biography/franklin-sir-john-2066.

42. Cyriax, *Sir John Franklin's Last Arctic Expedition*, 37. On Fitzjames's heroics in China, see Battersby, *James Fitzjames*, at 139–140.

43. Additional details of the crew's and officers' background and experience are in Chapter 4.

44. Life expectancy in England for someone born in 1796, as Franklin was, was about 38 years. https://www.statista.com/statistics/1040159/-life-expectancy-united-kingdom-all-time/. That number, though, included the many English babies who would die in childhood, and it probably underestimated the life expectancy of those born, as Franklin was, into reasonably well-off families. Still, 60 was a fairly advanced age for the rigors of Arctic exploration.

45. Quoted in Nanton, *Arctic Breakthrough*, 228.

46. See, e.g., Fitzjames's letter to his "sister," Elizabeth Coningham, July 1, 1845, referring to Franklin, "whose memory is as good as his judgment appears to be," in Russell Potter, Regina Koellner, Peter Carney and Mary Williamson, eds., *May Be Spared to Meet on Earth: Letters of the Lost Franklin Arctic Expedition*, Montreal: McGill-Queen's University Press, 2022, at 202.

47. *Erebus* was 372 tons, *Terror* 325. Cyriax, *Sir John Franklin's Last Arctic Expedition*, and their railway-locomotive steam engines produced about 20 horsepower apiece. [AU: something seems not quite right here; what "and their railway-locomotive steam engines…"?] Perhaps not very large by comparison to the tanker *Manhattan*, 115,000 tons and 43,000 horsepower, that made the first commercial voyage through the Passage in 1969, but considerably larger than some of the ships that had made important discoveries in the region. See https://www.maritime-executive.com/editorials/-photos-through-the-north-west-passage-with-the-manhattan-in-1969.

48. Lambert, *The Gates of Hell*, at 179. The much-ignored Admiralty gadfly Richard King had also tried to warn against a large-ship expedition. See Richard King, *The Franklin Expedition from First to Last*, London: John Churchill, 1855, reprint London: Forgotten Books, 2017, 8–9.

49. That small ships were better for Arctic exploration than larger, heavy-draft vessels was well known by the mid–19th century. In addition to King, see, e.g., John Shillinglaw, *Narrative of Arctic Discovery*, London: William Shoberl, 1851, at 51.

50. Lambert, *The Gates of Hell*, 313.

51. On Scott, see Roland Huntford, *Race to the South Pole: The Expedition Diaries of Scott and Amundsen*, London: Bloomsbury Academic, 2010; and Susan Solomon, *The Coldest March: Scott's Fatal Antarctic Expedition*, New Haven: Yale University Press, 2001.

52. Cyriax, *Sir John Franklin's Last Arctic Expedition*, 17–18. Beechey was sure that, with the addition of steam power, "no reasonable doubt can be entertained that the accomplishment of so desirable an object [i.e., completing the Northwest Passage] is practicable."

53. Cyriax, *Sir John Franklin's Last Voyage*, 54.

54. William Battersby and Peter Carney, "Equipping HM Ships Erebus and Terror, 1845," *International Journal for the History of Engineering and Technology* 81 (2011): 192–211.

55. It remains unclear just how many books they carried, but there were a lot. Martyn Beardsley, *Sir John Franklin: The Man Who Ate His Boots*, London: Short Books, 2005, 194, suggests "over a thousand," while at the high end, Arabella Edge's novel, *Fields of Ice*, London: Picador, 2011, 7, suggests some 17,000 volumes. Most other estimates are somewhere in between. Nanton, *Arctic Breakthrough*, 229, rather definitively states that *Erebus* carried 1,700 books and *Terror* 1,200. Aside from Bibles and prayer books, the only book that made it back from the Expedition was a copy of Oliver Goldsmith's 1766 novel, *Vicar of Wakefield*, retrieved by McClintock's search expedition in 1859, though Dickens and other popular authors of the time were apparently represented. Cyriax, *Sir John Franklin's Last Voyage*, 44.

56. Richard King (who had sailed in *Terror* on George Back's ill-fated 1836 expedition) had recommended a small, lightly manned overland journey, telling John Barrow that, under the Admiralty's plan for the Expedition, it was sending Franklin "to form the nucleus of an iceberg." Lambert, *The Gates of Hell*, 160; King, *The Franklin Expedition from First to Last*, 158–61. In addition to John Ross and George Back, there were other doubters. Some had argued that not even small ships could navigate the passage. See, e.g., Peter Heywood, *The Impracticability of a North-West Passage for Ships, Impartially Considered*, London: A.J. Valpy, 1824, reprint, Cambridge: Cambridge University Press, 2014.

57. Cyriax, *Sir John Franklin's Last Arctic Expedition*, 55.

58. Many of the letters are collected in Potter, et al., eds., *May We Be Spared to Meet on Earth*.

59. The sailing orders are reprinted in Appendix III and are available at https://www.canadianmysteries.ca/sites/franklin/archive/text/InstructionsToFranklin_en.htm.

60. Cyriax, *Sir John Franklin's Last Arctic Expedition*, 12–13.

61. See William Barr, "Searching for Franklin Where He Was Ordered to Go: Captain Erasmus Ommanney's Sledging Campaign to Cape Walker and Beyond, Spring 1851," *Polar Record* 52 (2016): 474–98.

62. Cyriax, *Sir John Franklin's Last Arctic Expedition*, 46–47.

63. *Id.* at 60.

64. See Stan Rogers, "Northwest Passage," 1981. Lyrics available at https://www.azlyrics.com/lyrics/stanrogers/northwestpassage.html.

65. Cyriax, *Sir John Franklin's Last Arctic Expedition*, 67–68.

66. Janice Cavell, "Who Discovered the Northwest Passage?" *Arctic* 71 (2018): 292–308, 296. Despite the complete lack of knowledge, especially before John Rae's and Francis McClintock's reports many years later, the Admiralty had no problem in granting promotions to many of the Expedition's officers. Fitzjames was raised from Commander to Captain at the end of 1845; the first lieutenants, Graham Gore in *Erebus* and Edward Little in *Terror*, were promoted to Commander in November 1846; and all the mates—Sargent, Des Voeux and Couch in *Erebus* and Hornby and Thomas in *Terror*—were made lieutenants before mid–1846. Richard J. Cyriax, *Sir John Franklin's Last Arctic Expedition*, at 28–30. And Franklin himself, still presumed to be alive even though he had died five years earlier, was promoted, by the workings of the Royal Navy's inexorable seniority system, to Rear Admiral in October 1852. *The London Gazette*, November 2, 1852, p. 2867.

67. W. Gillies Ross, "The Type and Number of Expeditions in the Franklin Search, 1847–1859," *Arctic* 55 (2002): 57–69. Ross categorizes the expeditions as being 20 direct search expeditions, 11 supply expeditions, one expedition in relief of lost searchers, four "bi-purpose" expeditions and three aborted search expeditions that never reached their intended search area.

68. A good reproduction of the Victory Point Record can be found at the Canadian Museum of History website, https://www.historymuseum.ca/blog/a-very-special-piece-of-paper/. The text is Transcribed in Appendix I.

69. Francis Leopold McClintock, *The Voyage of the "Fox" in the Arctic Seas: A Narrative of the Discovery of the Fate of Sir John Franklin and His Companions*, London: John Murray, 1859, 260.

70. David Woodman, *Unravelling the Franklin Mystery: Inuit Testimony*, 2nd ed., Montreal: McGill-Queen's University Press, 2015, 142, 298.

71. A video tour of the *Terror* wreck, posted by Parks Canada, is available at https://www.youtube.com/watch?v=OxyTZ3F7mkA.

72. Woodman, *Unravelling the Franklin Mystery*, 133, 152.

73. For a description of that exhibit, see Karen Ryan, *Death in the Ice: The Mystery of the Franklin Expedition*, Gatineau: Canadian Museum of History, 2018.

74. Kathleen Kasten-Mutkus, "Ghosts in the Archive: The Textual Lacunae of the Third Franklin Expedition," *Polar Record* 55 (2019): 417–24, at 418.

75. See Russell Potter, "The 'Peglar Papers' Revisited," *The Trafalgar Chronicle: The Yearbook of the 1805 Club*, 2014, 202–15. Available at http://www.ric.edu/faculty/rpotter/potter_peglar_trafchron.pdf.

76. Edward Parkinson, "'All Well': Narrating the Third Franklin Expedition," in John Moss, ed., *Echoing Silence: Essays on Arctic Narrative*, Ottawa: University of Ottawa Press, 1997, 43–52.

Chapter 2

1. Iris DeMent, "Let the Mystery Be" from the album *Infamous Angel*, 1992.

2. UNESCO estimate, available at http://www.unesco.org/new/en/culture/themes/underwater-cultural-heritage/underwater-cultural-heritage/wrecks/#:~:text=An%20estimated%203%20million%20shipwrecks,ocean%20floors%20around%20the%20planet.&text=A%20shipwreck%20by%20nature%20is,at%20the%20time%20of%.

3. Erika Behrisch Elce, "In Between Memory and Discovery: The Franklin Expedition Story Continues to Grow," *The Whig-Standard*, Kingston, Ontario, October 9, 2018, available at https://www.thewhig.com/opinion/columnists/-in-between-memory-and-discovery-franklin-expedition-story-continues-to-grow.

4. The group can be found at https://www.facebook.com/groups/11434844549/about/.

5. Genevieve LeMoine, Review of John V.H. Dippel, *To the Ends of the Earth: The Truth Behind the Glory of Polar Exploration*, Amherst: Prometheus, 2018; *Arctic* 72 (2019): 204–05.

6. See, e.g., Hal Arkowitz and Scott O. Lilienfeld, "Do the Eyes Have It? Why Science Tells Us Not to Rely on Eyewitness Accounts," *Scientific American Mind* 20 (2010): 68–69.

7. Dorothy Eber, *Encounters on the Passage: Inuit Meet the Explorers*, Toronto: University of Toronto Press, 2008; David C. Woodman, *Unravelling the Franklin Mystery: Inuit Testimony*, 2nd ed., Montreal: McGill-Queen's University Press, 2015; and David C. Woodman, *Strangers Among Us*, Montreal: McGill-Queen's University Press, 1996.

8. The Beck narrative is described, and effectively debunked, in Richard J. Cyriax, "Adam Beck and the Franklin Search," *Mariner's Mirror* 48 (1962): 35–51.

9. Dan Simmons, *The Terror*, Boston: Little, Brown, 2007.

10. Mordecai Richler, *Solomon Gursky Was Here*, New York: Knopf, 1990.

11. William T. Vollmann, *The Rifles*, New York: Viking, 1994.

12. John D. Trimmer, "The Present Situation in Quantum Mechanics: A Translation of Schrödinger's 'Cat Paradox' Paper," *Proceedings of the American Philosophical Society* 124 (1980): 323–38. Available at https://www.jstor.org/stable/986572#metadata_info_tab_contents.

13. For a non-physicist's explanation of the problem, see Leon M. Lederman and Christopher T. Hill, *Quantum Physics for Poets*, Amherst: Prometheus, 2011, 102–11. More generally, see Natalie Wolchover, "A Different Kind of Theory of Everything," *New Yorker*, February 19, 2019, available at https://www.newyorker.com/science/elements/a-different-kind-of-theory-of-everything.

14. Wolchover, "A Different Kind of Theory of Everything." A recent literary attempt to transfer this sense of uncertainty to the macroscopic sphere of human relations is Simon Stephens's play *Heisenberg*, London: Bloomsbury Methuen Drama, 2015.

15. Ed O'Loughlin, *Minds of Winter*, London: Riverrun, 2016, a recent novel that ropes in the Franklin mystery with other Arctic arcana, also proposes a kind of quantum history. Well worth reading.

16. Although a rigorous recent analysis of the bones found in the wake of the Franklin Expedition concludes that at most only 85 of the 105 officers and men known to have left the ships in April 1848 can be accounted for by the bones that have been discovered over the years. Douglas R. Stenton, "Finding the Dead Bodies: Bones and Burials from the 1845 Franklin Northwest Passage Expedition," *Polar Record* 54 (2018): 197–212.

17. It wasn't unheard of for women to disguise themselves as men and sail with the Royal Navy, though women's presence has been documented more frequently on pirate ships and other non-Establishment voyages. See, e.g., David Cordingly, *Women Sailors and Sailors' Women*, New York: Random House, 2001; Linda Grant DePauw, *Seafaring Women*, Boston: Houghton Mifflin, 1982; Suzanne Stark, *Female Tars: Women Aboard Ship in the Age of Sail*, Annapolis: Naval Institute Press, 2017; Mary Ann Talbot, *The Life and Surprising Adventures of Mary Ann Talbot*, London: Robert S. Kirby, 1809, reprinted in Mary Ann Talbot (Paul Royster, ed.), "The Life and Surprising Adventures of Mary Ann Talbot, in the Name of John Taylor (1809)," Lincoln: Faculty Publications, University of Nebraska Libraries 32, available at https://digitalcommons.unl.edu/libraryscience/32. An archaeological examination of bones from sites associated with the Franklin Expedition, in fact, found at least four bone fragments that appeared to be female, though these findings may have resulted from the natural degradation of the DNA over time. Douglas R. Stenton, Anne Keenleyside, Stephen Fratpietro and Robert Park, "DNA Analysis of Human Skeletal Remains from the 1845 Franklin Expedition," *Journal of Archaeological Sciences: Reports* 16 (2017): 409–17.

18. See Arthur N. Gilbert, "Buggery and the British Navy, 1700–1861," *Journal of Social History* 10 (1976): 72–98 for a concise summary of the gap between theory and practice on this issue.

19. Frankie Witzenburg, "The Lost Franklin Expedition," *Naval History Magazine*, October 2020, available at https://www.usni.org/magazines/naval-history-magazine/2020/october/lost-franklin-expedition, is a good recent summary of what the latest discoveries add to our knowledge.

Chapter 3

1. *Erebus* has received her own biography, Michael Palin, *Erebus: One Ship, Two Epic Voyages and the Greatest Naval Mystery of All Time*, Vancouver: Greystone Books, 2018, albeit a biography criticized by some reviewers for the occasional factual lapse. This chapter merely recounts some of the highlights of the ship's long, storied career.

2. Like *Erebus*, *Terror* now has her own biography: Matthew Betts, *HMS Terror: The Design, Fitting and Voyages of the Polar Discovery Ship*, Barnsley: Seaforth, 2022. Less romantic than Palin's paean to *Erebus*, but a lot more precise when it comes to describing the ship and its fittings.

3. A useful discussion of this type of ship is Chris Ware, *The Bomb Vessel: Shore Bombardment Ships of the Age of Sail*, London: Conway Maritime Press, 1994.

4. Palin, *Erebus*, at 8, says 104 feet.

5. Richard Cyriax, in *Sir John Franklin's Last Arctic Expedition: The Franklin Expedition, a Chapter in the History of the Royal Navy*, London: Methuen, 1939, reprint Sussex: Arctic Press, 1997, at 39, gives *Erebus*'s burthen as 370 tons, slightly less than other sources, but by no means a material difference.

6. One knot is equivalent to 1.15 miles per hour, or roughly 1.89 kilometers per hour.

7. These southern winter sojourns are described at some length in James Clark Ross, *A Voyage of Discovery and Research in the Southern and Antarctic Regions During the Years 1839–1843*, London: John Murray, 1847, reprint Cambridge: Cambridge University Press, 2011. See also E.A. (Ted) Michener, *Ice in the Rigging*, Hobart: National Museum of Tasmania, 2015.

8. Palin, *Erebus*, at 10–11.

9. *Id.* at 40–41.

10. Michael Smith, *Polar Crusader: Sir James Wordle—Exploring the Arctic and Antarctic*, Edinburgh: Birlinn, 2004, at 38.

11. See, e.g., Lorne K. Kriwoken and John W. Williamson, "Hobart, Tasmania: Antarctic and Southern Ocean Connections," *Polar Record* 29 (1993): 93–102, at 95.

12. *Id.*

13. Palin, *Erebus*, at 126.

14. J.E. Davis, *A Letter from the Antarctic*, London: William Clowes and Sons, 1901, at 19.

15. At the time, the *Illustrated London News* said the engines were from the Greenwich Railway; recent evidence suggests, however, that they came from the London and Croydon Railway. William Battersby and Peter Carney, "Equipping HM Ships *Erebus* and *Terror*, 1845," *International Journal for the History of Engineering & Technology* 81 (2011): 192–211.

16. Cyriax, *Sir John Franklin's Last Arctic Expedition*, at 61.

17. Betts, *HMS Terror*. Betts's book, based on his years of research, his building of a scale model of *Terror* and his work as a consultant on the television adaptation of Dan Simmons's *The Terror*, is the definitive study. Unless otherwise indicated, the facts about *Terror* given here are from this admirable book.

Chapter 4

1. See, e.g., Martyn Beardsley, *Deadly Winter: The Life of Sir John Franklin*, Annapolis: Naval Institute Press, 2002; August Henry Beesly, *Sir John Franklin*, New York: Putnam, 1881; Richard J. Cyriax, *Sir John Franklin's Last Arctic Expedition: The Franklin Expedition, a Chapter in the History of the Royal Navy*, London: Methuen, 1939, reprint Sussex: Arctic Press, 1997; Andrew Lambert, *Franklin: Tragic Hero of Polar Navigation*, London: Faber & Faber, 2009; Albert Hastings Markham (with the aid of Sophia Cracroft), *Life of Sir John Franklin and the North-West Passage*, London: George Philip & Son, 1891; Sherrard Osborn, *The Career, Last Voyage and Fate of Sir John Franklin*, London: Bradbury & Evans, 1860; B.J. Rule, *Polar Knight: The Mystery of Sir John Franklin*, New Smyrna Beach: Luthers, 1998; and H. D. Traill, *Life of Sir John Franklin, RN*, London: John Murray, 1896.

2. The most famous poem about Franklin, carved on his monuments at Westminster Abbey and in Tasmania, is by his relation-by-marriage, Alfred Lord Tennyson:

> Not here! The white north hath thy bones and thou
> Heroic sailor soul
> Art passing on thine happier voyage now
> Toward no earthly pole.

See also, e.g., Chandos Hoskyns Abrahall, *Arctic Enterprise: A Poem in Seven Parts*, London: Hope, 1856; Algernon Charles Swinburne, *The Death of Sir John Franklin*, in E. Gosse and T.J. Wise, eds., *The Posthumous Poems of Algernon Charles Swinburne*, London: William Heineman, 1917, at 79; Joseph Addison Turner, *The Discovery of Sir John Franklin and Other Poems*, Mobile, n.p., 1858; and Owen Alexander Vidal, *A Poem Upon the Life and Character of Sir John Franklin*, Oxford: T. & G. Shrimpton, 1860.

3. Sten Nadolny, *The Discovery of Slowness*, New York: Viking, 1987.

4. May Fluhmann, *Second in Command: The Life of Francis R. M. Crozier*, Yellowknife: Department of Information, Government of the Northwest Territories, 1976, available at https://drive.google.com/file/d/1sW0Fn-p3F4w-xmfDCL1--wj_dERUsM9E/view, and Michael Smith, *Captain Francis Crozier: Last Man Standing?*, Cork: Collins, 2006, updated and republished as *Icebound in the Arctic: The Mystery of Francis Crozier and the Franklin Expedition*, Dublin: The O'Brien Press, 2014.

5. Factual: William Battersby, *James Fitzjames: the Mystery Man of the Franklin Expedition*, Stroud: History Press, 2010. Fictional: John Wilson, *North with Franklin: The Lost Journals of James Fitzjames*, Markham: Fitzhenry & Whiteside, 1999.

6. On Franklin see, e.g., Martyn Beardsley, *Deadly Winter*; Cyriax, *Sir John Franklin's Last Arctic Expedition*; and Geoffrey. F. Lamb, *Franklin, Happy Voyager*, London: Ernest Benn, 1956.

7. Shane McCorristine, "A Manuscript History of the Franklin Family by Sophia Cracroft (1853)," *Polar Record* 51 (2015): 72–90.

8. Franklin's service record, at least since 1804, can be found at https://www.nationalarchives.gov.uk/trafalgarancestors/details.asp?id=3093.

9. John Franklin, *Narrative of a Journey to the Shores of the Polar Sea in the Years 1819, 20, 21 and 22*, London: John Murray, 1823, reprint Edmonton: M. G. Hurtig, 1969.

10. A phrase used by several of Franklin's biographers. See, e.g., Martyn Beardsley, *Sir John Franklin: The Man Who Ate His Boots*, London: Short Books, 2005; Anthony Brandt, *The Man Who Ate His Boots*, Knopf: New York, 2010; and Russell Potter, "Introduction: Exploration and Sacrifice: The Cultural Logic of Arctic Discovery," in Frédéric Regard, ed., *The Quest for the Northwest Passage: British Narratives of Arctic Exploration, 1576–1874*, London: Pickering & Chatto, 2013, at 1–17.

11. Adeline Johns-Putra, "Historicizing the Networks of Ecology and Culture: Eleanor Anne Porden and Nineteenth-Century Climate Change," *ISLE Interdisciplinary Studies in Literature and Environment* 21 (2015): 27–46.

12. Janice Cavell, "Lady Lucy Barry and Evangelical Reading on the First Franklin Expedition," *Arctic* 63 (2010): 131–40.

13. *Id.* at 135.

14. John Franklin, *Narrative of a Second Expedition to the Shores of the Polar Sea*, London: John Murray: 1828, reprint, New York: Greenwood Press, 1969.

15. Lady Jane is the subject of a great many biographies and commentaries. See, e.g., Alison Alexander, *The Ambitions of Jane Franklin: Victorian Lady Adventurer*, Sydney: Allen & Unwin, 2013; Adrienne Eberhard, *Lady Jane Franklin*,

Melbourne: Black Pepper, 2004; Erika Behrisch Elce, ed., *As Affecting the Fate of My Absent Husband: Selected Letters of Lady Franklin Concerning the Search for the Lost Franklin Expedition*, Montreal: McGill-Queen's University Press, 2009; Penny Russell, ed., *This Errant Lady: Jane Franklin's Overland Journey to Port Philip and Sydney, 1839*, Canberra: National Library of Australia, 2002; Penny Russell, "Wife Stories: Narrating Marriage and Self in the Life of Jane Franklin," *Victorian Studies* 48 (2005): 35–57; and Frances Woodward, *Portrait of Jane: A Life of Lady Franklin*, London: Hodder & Stoughton, 1951.

16. From the importation of George I in 1714 until the ascension of Queen Victoria, British monarchs were simultaneously rulers of Britain and of the German princely state of Hanover. By the time of William IV, however, they generally spoke English. Franklin seems to have been particularly proud of this honor, and the discovery of his medal, recovered from the Inuit at Repulse Bay by John Rae in 1854, was an important element in convincing the public that the Expedition had indeed been lost.

17. From a pre-contact population of anywhere from 3,000 to 15,000, the Aboriginal population of Van Diemen's Land had diminished, primarily succumbing to European-introduced diseases, to just a few hundred by the time Franklin arrived, confined on Flinders Island northeast of Hobart. The last full-blooded Tasmanian Aboriginal died in 1876, though there are some thousands of people alive today with Aboriginal ancestry. See Benjamin Madley, "From Terror to Genocide: Britain's Tasmanian Penal Colony and Australia's History Wars," *Journal of British Studies* 47 (2008): 77–106, and Josephine Flood, *The Original Australians: Story of the Aboriginal People*, Longford: Allen & Unwin, 2006.

18. See, e.g., Kathleen Fitzpatrick, *Sir John Franklin in Tasmania, 1837–1843*, Melbourne: Melbourne University Press, 1949; and Sir John's own, albeit self-serving, narrative, *Narrative of Some Passages in the History of Van Diemen's Land*, London: Richard and John E. Taylor, 1845, reprint, Cambridge: Cambridge University Press, 2012.

19. McCorristine, "A Manuscript History of the Franklin Family," at 75.

20. Unless otherwise indicated, the facts of Crozier's life are from Fluhmann, *Second in Command*, and Smith, *Icebound in the Arctic*.

21. R.J. Campbell, "Crozier's Date of Birth," *Polar Record* 45 (2009): 83–84.

22. Alison Long, "A Note Relating to the Birthdate of Captain Francis Rawdon Moira Crozier, R.N., F.R.S., F.R.A.S. *Polar Record* 58(e24) (2022): 1–5, available at https://doi.org/10.1017/S0032247422000225.

23. Smith, *Icebound in the Arctic*, at 25–26, states that Francis's father pulled strings with Lord Downshire, a local grandee, to assure Francis a place in the Navy, but whether this was Francis's wish or his father's is unclear.

24. *Id.* at 30–31.

25. Smith, *Icebound in the Arctic*, at 99–100, speculates that Crozier wooed the twenty-something young poet Jean Ingelow while he was back in England in 1837–39; other than one cryptic, ambiguous poem of Ingelow's, there is no evidence to back up the assertion.

26. James Clark Ross, *A Voyage of Discovery and Research in the Southern and Antarctic Regions During the Years 1839–1843*, London: John Murray, 1847, reprint Cambridge: Cambridge University Press, 2011.

27. Smith, *Icebound in the Arctic*, at 153.

28. Michael Palin, *Erebus: One Ship, Two Epic Voyages, and the Greatest Naval Mystery of All Time*, Vancouver: Greystone Books, 2018, at 175.

29. *Id.* at 182.

30. Smith, *Icebound in the Arctic*, at 157.

31. On the role of "interest," patronage and nepotism at the Admiralty in the first half of the 19th century, see Charles I. Hamilton, "John Wilson Croker: Patronage and Clientage at the Admiralty, 1809–1857," *The Historical Journal* 43 (2000): 49–77.

32. Matthew Betts, *HMS Terror: The Design, Fitting and Voyages of the Polar Discovery Ship*, Barnsley: Seaforth, 2022, at 57.

33. Palin, *HMS Erebus*, at 229.

34. William Battersby, *James Fitzjames: The Mystery Man of the Franklin Expedition*, Stroud: History Press, 2010. Otherwise unattributed factual statements in this section are from Battersby's work.

35. Smith, *Icebound in the Arctic*, at 165.

36. On the role of the opium trade in sustaining the broader British imperial effort, see Carl A. Trocki, *Opium, Empire and the Global Political Economy: A Study of the Asian Opium Trade, 1750–1950*, London: Routledge, 1999.

37. Battersby, *James Fitzjames*, at 136.

38. Much of the information on Gore comes from the excellent blog post: William Battersby, "The Remarkable Background of Lt. Graham Gore," *Hidden Tracks* blog, November 29, 2009. Available at https://hidden-tracks-book.blogspot.com/2009/11/remarkable-background-of-lt-graham-gore.html.

39. Potter, et al., eds., *May We Be Spared to Meet on Earth*, at 75.

40. *Id.* at 82.

41. S. Mays, A. Ogden, J. Montgomery, S. Vincent, W. Battersby and G.M. Taylor, "New Light on the Personal Identification of a Skeleton of a Member of Sir John Franklin's Last Expedition to the Arctic, 1845," *Journal of Archaeological Science* 38 (2011): 1571–82.

42. Harry Goodsir says that Fairholme was Lord Forbes's "nephew." Potter, et al., eds., *May We Be Spared to Meet on Earth*, at 144. I have not independently researched the genealogy.

43. *Id.* at 219.

44. Ian R. Stone, "Fairholme v Fairholme's Trustees: Presumption of Death on the 1845 Franklin Expedition," *Polar Record* 32 (1996): 239–41.
45. Potter, et al., eds., *May We Be Spared to Meet on Earth*.
46. *Id.* at 211.
47. John Goodsir and Harry Goodsir, *Anatomical and Pathological Observations*, Edinburgh: Myles MacPhail, 1845.
48. David C. Woodman, *Unravelling the Franklin Mystery: Inuit Testimony*, 2nd ed., Montreal: McGill-Queen's University Press, 2nd ed., 2015.
49. *Id.* at 118–19.
50. Russell Potter, "The Grave of Lieutenant Irving?" *Visions of the North* blog, January 3, 2010, available at https://visionsnorth.blogspot.com/search?q=John+Irving+Grave.
51. Details of Lieutenant Irving's life and career can be found in Ralph Lloyd-Jones, "An Evangelical Christian on Franklin's Last Expedition: Lieutenant John Irving of HMS *Terror*," *Polar Record* 33 (1997): 327–32. An earlier biography, in the typical 19th-century hagiographic style, is Benjamin Bell, *Lieut. John Irving, R.N.: A Memorial Sketch with Letters in Sir John Franklin's Last Expedition to the Arctic Regions*, Edinburgh: David Douglas, 1881, reprint Cambridge: Cambridge University Press, 2014.
52. Janice Cavell, "Lady Lucy Barry and Evangelical Reading on the First Franklin Expedition," *Arctic* 63 (2010): 131–40.
53. These facts, and most of the following concerning McDonald, are from Ian Barrie, "Alexander M'Donald L.R.S.C.E. (1817–c. 1848)," *Arctic* 62 (2009): 239–40. See also W. Gillies Ross, "William Penny (1809–1892)," *Arctic* 36 (1983): 380–81.
54. Alexander McDonald, *A Narrative of Some Passages in the History of Enoolooapik, a Young Eskimo, Who Was Brought to Britain in 1839, in the Ship "Neptune" of Aberdeen*, Edinburgh: Fraser & Co., 1841. Enoolooapik remains of interest into the 21st century; see H.G. Jones, "The Inuit as Geographers: The Case of Eenoolooapik," *Inuit Studies* 28 (2004): 58–72.
55. Potter, et al., eds., *May We Be Spared to Meet on Earth*, at 284–5.
56. Mordecai Richler, *Solomon Gursky Was Here*, New York: Knopf, 1990.
57. Blanky is mentioned briefly in Richard J. Cyriax, *Sir John Franklin's Last Arctic Expedition: The Franklin Expedition, a Chapter in the History of the Royal Navy*. London: Methuen, 1939, reprint Sussex: Arctic Press, 1997. A more complete version of his story is Andrés Paredes, "Thomas Blanky: A Life in Hell," *Kabloonas* blog, June 26, 2015, available at https://kabloonas.blogspot.com/2015/06/thomas-blanky-live-in-hell.html.
58. Potter, et al., eds., *May We Be Spared to Meet on Earth*, at 297.
59. *Id.* at 40–41. Alas, the particular candidate, one Abraham Rose Bradford of HMS *Actaeon*, was unsuitable, in Fitzjames's view, because "he is no 'ologist—he can't stuff birds—give long names to shiny things—or put moss in blotting paper." *Id.* at 41. Bradford's lack of experience in the ice would have been no problem, though.
60. Betts, *HMS Terror*, at 77.
61. Ralph Lloyd-Jones, "The Royal Marines on Franklin's Last Expedition," *Polar Record* 40 (2004): 319–26; Ralph Lloyd-Jones, "The Men Who Sailed with Franklin," *Polar Record* 41 (2005): 311–18; Ralph Lloyd-Jones, "Further Light on Franklin's Men," *Polar Record* 47 (2011): 379–82; and Ralph Lloyd-Jones, "Franklin's Men and Their Families: New Evidence from the Allotment Books," *Polar Record* 54 (2018): 267–74.
62. Though it's possible that Evans was only 17 but was listed as 18 so he could be paid as a "boy 1st class." Lloyd-Jones, "The Men Who Sailed with Franklin," at 316.
63. Glenn M. Stein, "Scattered Memories and Frozen Bones: Revealing a Sailor of the Franklin Expedition, 1845–1848," *Journal of the Orders and Medals Research Society* 46 (2007): 224–32.
64. Lloyd-Jones, "Further Light on Franklin's Men," at 380.
65. Lloyd-Jones, "The Men Who Sailed with Franklin," at 312.
66. Lloyd-Jones, "Franklin's Men and Their Families," at 268.
67. *Id.*
68. Lloyd-Jones, "The Men Who Sailed with Franklin," at 318.
69. Owen Beattie and James M. Savelle, "Discovery of Human Remains from Sir John Franklin's Last Expedition," *Historical Archaeology* 17 (1983): 100–05; Owen Beattie, Eric Damkjar, Walter Kowal and Roger Amy, "Anatomy of an Arctic Autopsy," *Medical Post* 20 (1985): 1–2; and Roger Amy, Rakesh Bhatnagar, Eric Damkjar and Owen Beattie, "The Last Franklin Expedition: Report of a Postmortem Examination of a Crew Member," *Canadian Medical Association Journal* 135 (1986): 115–17.
70. "Accounting for the 105 Survivors," *Starvation Cove* blog, January 13, 2018. Available at starvationcove.blogspot.com.
71. S. Mays, A. Ogden, J. Montgomery, S. Vincent, W. Battersby and G.M. Taylor, "New Light on the Personal Identification of a Skeleton of a Member of Sir John Franklin's Last Expedition to the Arctic, 1845," *Journal of Archaeological Science* 38 (2011): 1571–82.
72. Russell Potter has done the heroic work of transcribing the entire collection of papers, available at https://w3.ric.edu/faculty/rpotter/aglooka/peglar-fulltext-rev_2000.pdf.
73. Richard J. Cyriax and A.G.E. Jones, "The Papers in the Possession of Harry Peglar, Captain of the Foretop, HMS Terror, 1845," *Mariners' Mirror* 40 (1954): 186–95.

74. Stein, "Scattered Memories and Frozen Bones."

75. Douglas R. Stenton, Stephen Fratpietro, Anne Keenleyside and Robert W. Park, "DNA Identification of a Sailor from the 1845 Franklin Northwest Passage Expedition," *Polar Record* 57 (2021): e14. Available at doi:10.1017/S0032247421000061.

76. See, e.g., Alexander, *The Ambitions of Jane Franklin*; Eberhard, *Lady Jane Franklin*; Elce, ed., *As Affecting the Fate of My Errant Husband*; Russell, "Wife Stories"; and Woodward, *Portrait of Jane*.

77. Potter, et al., eds., *May We Be Spared to Meet on Earth*.

78. Erika Behrisch Elce, "Widows' Men: The Admiralty Board, Precedent, and Pensions for the Widows of the Lost Franklin Expedition," *The Journal of Imperial and Commonwealth History*, 47 (2019): 28–50.

Chapter 5

1. W. Gillies Ross, "Nineteenth-Century Explorations of the Arctic," in John Logan Allen, ed., *North American Explorations, Vol. 3*, Lincoln, University of Nebraska Press, 1997, 244–331, 249.

2. Stan Rogers, "Northwest Passage," 1981, available at https://genius.com/Stan-rogers-northwest-passage-lyrics.

3. There are many versions of "Northwest Passage" available online. Here's one: https://www.youtube.com/watch?v=I3eBUUoNN8A. And here's the closest I could find to the cruise ship lounge version, the Adventure Canada expedition staff singing the tune: https://www.google.com/search?q=stan+rogers+northwest+passage&tbm=vid&sxsrf=AJOqlzXq8F92_oBmO1hwoQJiCajGhbAAnQ:1673293897670&ei=SXC8Y-nIKMuVwbkPpaK8kAU&start=20&sa=N&ved=2ahUKEwipkYDDobv8AhXLSjABHSURD1I4ChDw0wN6BAgPEBk&biw=1257&bih=634&dpr=2#fpstate=ive&vld=cid:73327a6c,vid:BgFS9Q8r7KA.

4. Lars Pilø, "Buried in the Ice: The Franklin Expedition Cemetery," https://secretsoftheice.com/news/2019/10/28/franklin-expedition/, posted October 28, 2019.

5. Alison Freebairn, "Lost and Found: the Beechey Island Papers," *Visions of the North* blog, September 10, 2019, available at https://visionsnorth.blogspot.com/2019/09/lost-and-found-beechey-island-papers.html.

6. C.S. MacKinnon, "The Wintering-over of Royal Navy Ships in the Canadian Arctic, 1819–1876," *The Beaver*, Winter 1984/85, 12–21, at 14.

7. Id.

8. William Battersby and Peter Carney, "Equipping HM Ships *Erebus* and *Terror*, 1845," *International Journal for the History of Engineering & Technology* 81 (2011): 192–211.

9. MacKinnon, "Wintering-over," at 17.

10. On Goodsir, see Matthew H. Kaufman, "Harry Goodsir and the Franklin Expedition, of 1845," *Journal of Medical Biography* 12 (2004): 82–89.

11. Andrew Lambert, *The Gates of Hell: John Franklin's Tragic Quest for the Northwest Passage*, New Haven: Yale University Press, 2011, at 12.

12. On the worldwide network of scientists who collaborated on magnetic observations throughout much of the 19th century, see S.R.C. Malin and D.R. Barraclough, "Humboldt and the Earth's Magnetic Field," *Quarterly Journal of the Royal Astronomical Society* 32 (1991): 279–93.

13. A good, readable account of how the dipping needle worked, and the skill required for effective use of the instrument, is Edward J. Gillin, "The Instruments of Expeditionary Science and the Reworking of Nineteenth Century Magnetic Experiment," *Royal Society Publishing*, March 30, 2022. Available at https://royalsocietypublishing.org/doi/10.1098/rsnr.2022.0002

14. Shane McCorristine, "Cross-Dressing, Feasts and Fun," *UpHere*, December 2017, available at https://uphere.ca/articles/cross-dressing-feasts-and-fun. For a more complete description of the "Parry System" for keeping the crew's health and morale up during the long, cold, dark Arctic winters, see MacKinnon, "Wintering-over" and Shane McCorristine and Jane S.P. Mocellin, "Christmas at the Poles: Emotion, Food and Festivities on Polar Expeditions, 1818–1912," *Polar Record* 52 (2016): 562–77. A very recent study of how Royal Navy crews entertained themselves through the winter is Eavan O'Dochartaigh, *Visual Culture and Arctic Voyages: Personal and Public Art and Literature of the Franklin Search Expeditions*, Cambridge: Cambridge University Press, 2022.

15. Patrick O'Neill, "Theatre in the North: Staging Practices of the British Navy in the Canadian Arctic," *Dalhousie Review* 73 (1994): 356–84, and Mike Pearson, "'No Joke in Petticoats': British Polar Expeditions and Their Theatrical Presentations," *TDR (The Drama Review)* 48 (2004): 48–59.

16. Elaine Hoag, "Caxtons of the North: Mid-Nineteenth-Century Arctic Shipboard Printing," *Book History* 4 (2001): 81–114. See also Ian S. MacLaren, "The Poetry of the 'New Georgia Gazette' or 'Winter Chronicle' 1819–1820," *Canadian Poetry* 30 (1992): 41–73.

17. A useful summary of where things were on Beechey Island is Todd Hansen, "Documents and Survey re Beechey Island," *Polar Record* 46 (2010): 193–99; see also Todd Hansen, "Additional Documents and Survey on the Franklin Sites of Beechey Island, Nunavut, Canada," *Polar Record* 48 (2012): 195–96.

18. May Fluhmann, *Second in Command: the Life of Francis R. M. Crozier*, Yellowknife: Department of Information, Government of the Northwest Territories, 1976, at 96–97, available at

https://drive.google.com/file/d/1sW0Fn-p3F4w-xmfDCL1--wj_dERUsM9E/view?fbclid=IwAR1AOlOlq791X3xm23yGiDK4hXyrSHI92PQTgeOIyEP_tNV1gVdqFS5w6_E2-18.

19. Stan Rogers, "Northwest Passage."

20. Iron Maiden, "Stranger in a Strange Land," lyrics by Adrian Smith, available at https://genius.com/Iron-maiden-stranger-in-a-strange-land-lyrics. The lyrics also reference Robert Heinlein's science-fiction novel of the same name and Aldous Huxley's *Brave New World*.

21. Though the article is now more than half a century old, the definitive source for descriptions of Franklin-era sledge travel by the Royal Navy is still Richard J. Cyriax, "Arctic Sledge Travelling by Officers of the Royal Navy, 1819–1849," *Mariner's Mirror* 49 (1963): 127–42. See also Francis Leopold McClintock, "On Arctic Sledge Travelling," *Proceedings of the Royal Geographical Society of London* 19 (1874–75): 464–79, and C.S. Mackinnon, "The British Man-Hauled Sledging Tradition," in Patricia D. Sutherland, ed., *The Franklin Era in Canadian Arctic History 1845–1849*, Ottawa: National Museums of Canada, Mercury Series, Archaeological Survey of Canada Paper No. 131, 1985, 129–40. Although modern dictionaries find little nor no difference between "sled" and "sledge," British authors up to and including Cyriax tended to use "sledge," while North American writers more often use "sled." The two terms are used interchangeably here, and no significance should be attached to any particular use of the terms.

22. For example, on William Parry's North Pole attempt in 1827, he brought along large sleds fitted with both runners and wheels, for hauling boats that were to be used when the expedition met open water. But on their first try, the heavy wheeled sledges proved absolutely immovable. Parry's next attempt, which did reach a record furthest north at 82°45'34" fell back on the Royal Navy's default option: plucky British sailors hauling their own sledges. See Cyriax, "Arctic Sledge Travelling," at 132.

23. Cyriax, "Arctic Sledge Traveling."

24. Richard Collinson, *Journal of HMS Enterprise on the Expedition in Search of Sir John Franklin's Ships by Behring Strait, 1850–1855*, London: Sampson Low, Marston, Searle, & Rivington, 1889, at 368.

25. Mackinnon, "The British Man-Hauled Sledging Tradition," at 134.

26. William E. Parry, *Journal of a Voyage for the Discovery of a Northwest Passage from the Atlantic to the Pacific: Performed in the Years 1819–20 in His Majesty's Ships* Hecla *and* Griper, London: John Murray, 1821, at 298.

27. William E. Parry, *Journal of a Second Voyage of Discovery of a Northwest Passage*, London: John Murray, 1824, reprint New York: Greenwood Press, 1968.

28. Lambert, *The Gates of Hell*, at 200.

29. *Id.* at 201.

30. Pierre Berton, *The Arctic Grail*, Toronto: McClelland & Stewart, 1988, at 187, cited in William Barr, "The Use of Dog Sledges During the British Search for the Missing Franklin Expedition in the North American Arctic Islands, 1848–59," *Arctic* 62 (2009): 257–72.

31. Barr, "The Use of Dog Sledges" at 269.

32. *Id.* at 269–72.

33. Mackinnon, "The British Man-Hauled Sledging Tradition," at 130.

34. E.E. Rich, *Hudson's Bay Company, 1763–1870*, London: Hudson's Bay Record Society, 1959, quoted in Paul Nanton, *Arctic Breakthrough: Franklin's Expeditions 1819–1847*, Toronto: Clarke Irwin & Co., 1970, at 226.

35. See, e.g., the otherwise absurd B J. Rule, *Polar Knight: The Mystery of Sir John Franklin*, New Smyrna Beach: Luthers, 1998, at 98 (messages supposedly delivered to the author via telepathy from Sir John!).

36. Richard J. Cyriax, *Sir John Franklin's Last Arctic Expedition: The Franklin Expedition, a Chapter in the History of the Royal Navy*, London: Methuen, 1939, reprint Sussex: Arctic Press, 1997, at 110.

37. Dip circles, which were intended to measure both the direction and the intensity of the earth's magnetic field, were idiosyncratic and somewhat unreliable instruments, but the scientific establishment in England believed strongly in their efficacy. See Matthew Goodman, "Proving Instruments Credible in the Early Nineteenth Century: The British Magnetic Survey and Site-Specific Experimentation," *Notes and Records of the Royal Society* 70 (2016): 251–68. Available at https://www.ncbi.nlm.nih.gov/pmc/articles/PMC4978730/.

38. Memorialized in Gustavus Mercator's world map, the open polar sea theory continued to entrance educated Englishmen long after its existence had been effectively disproved. See Kathryn Schulz, "Literature's Arctic Obsession," *New Yorker*, April 24, 2017, available at https://www.newyorker.com/magazine/2017/04/24/literatures-arctic-obsession.

Chapter 6

1. Franklin's orders are reproduced in Appendix III. They are available online at https://www.canadianmysteries.ca/sites/franklin/archive/text/InstructionsToFranklin_en.htm.

2. Janice Cavell, "Who Discovered the Northwest Passage?" *Arctic* 71 (2018): 292–308, at 296.

3. McClure's narrative of that journey is Robert L.M. McClure, *The Discovery of the Northwest Passage by HMS 'Investigator,'* London: Longmans, Brown, Green, Longmans and Roberts, 1857, reprint, ed. Sherrard Osborn, Rutland: Charles Tuttle & Co., 1969. A more modern examination is Glenn M. Stein, *Discovering the Northwest Passage: The Four-Year Arctic Odyssey*

of H.M.S. Investigator and the McClure Expedition, Jefferson: McFarland, 2015.

4. William H. Browne, "Report of the Proceedings of the Sledge 'Enterprise,' in charge of Lieut. W. H. Browne, Her Majesty's Ship 'Resolute,' commencing 15th April, ending 28th May 1851," in *Additional Papers Relative to the Arctic Expedition under the Orders of Captain Austin and Mister William Penny,* London: G.E. Eyre and W. Spottiswode, 1851, at 65–75.

5. A.G. Findlay, "On the Probable Course Pursued by Sir John Franklin's Expedition," *Journal of the Royal Geographical Society* 26 (1856): 26–35.

6. William E. Parry, *Journal of a Voyage for the Discovery of a Northwest Passage from the Atlantic to the Pacific: Performed in the Years 1819–20 in His Majesty's Ships* Hecla *and* Griper, London: John Murray, 1821, at 298.

7. Henry Larson, *The North-West Passage, 1940–42 and 1944,* Vancouver: City Archives, 1954, at 39–45, cited in Cavell, "Who Discovered the Northwest Passage," at 304.

8. Bea Alt, Roy M. Koerner, David A. Fisher, and J.C. Bourgeois, "Arctic Climate During the Franklin Era, as Deduced from Ice Cores," in Patricia Sutherland, ed., *The Franklin Era in Canadian Arctic History 1845–1859,* Ottawa: National Museums of Canada, Mercury Series, Archaeological Survey of Canada Paper No. 131, 1985, 69–92.

9. J.L. Robinson, "Conquest of the Northwest Passage by R.C.M.P. Schooner *St. Roch,*" *Canadian Geographical Journal* 30 (1945): 52–73, at 73.

10. Barrow, convinced of his own superior analytical abilities, rejected Parry's advice to navigate via the coastlines and believed that, west and south of Cape Walker, there was only open sea as far as Banks Land (later confirmed to be Banks Island), Wollaston Land and Victoria Land (the latter two eventually becoming known as parts of the much larger—a bit larger, in fact, than Great Britain—Victoria Island). See John Barrow, et al., "Communications on a North-West Passage, and Further Survey of the Northern Coast of America," *Journal of the Royal Geographical Society of London* 6 (1836): 34–50.

11. John Franklin, Letter to James Clark Ross, July 9, 1845, Scott Polar Research Institute, MS 248/316/25, cited in Cavell, "Who Discovered the Northwest Passage?" at 296.

12. Cavell, "Who Discovered the Northwest Passage?" at 296.

13. *Id.* at 301.

14. Francis Leopold McClintock, *The Voyage of the "Fox" in the Arctic Seas: A Narrative of the Discovery of the Fate of Sir John Franklin and His Companions,* London: John Murray, 1859, at 341.

15. Francis Leopold McClintock, "Discoveries by the Late Expedition in Search of Sir John Franklin and His Party," *Journal of the Royal Dublin Society* 1 (1857): 183–250, at 195.

16. Moira Dunbar, "The Effect of Sea Ice Conditions on Maritime Arctic Expeditions during the Franklin Era," in Patricia Sutherland, ed., *The Franklin Era in Canadian Arctic History 1845–1859,* Ottawa: National Museums of Canada, Mercury Series, Archaeological Survey of Canada Paper No. 131, 1985, 114–121.

17. Robert K. Headland, et al., *Transits of the Northwest Passage to End of the 2022 Navigation Season,* Cambridge: Scott Polar Research Center, 2022, available at https://www.spri.cam.ac.uk/resources/infosheets/northwestpassage.pdf.

18. Clifford G. Hickey, James M. Savelle and George B. Hobson, "The Route of Sir John Franklin's Third Arctic Expedition: An Evaluation and Test of an Alternate Hypothesis," *Arctic* 46 (1993): 78–81.

19. See Brian Payton, *The Ice Passage,* Toronto: Doubleday Canada, 2016, at 284.

Chapter 7

1. See Bea Alt, Roy M. Koerner, David A. Fisher and J.C. Bourgeois, "Arctic Climate During the Franklin Era, as Deduced from Ice Cores," in Patricia Sutherland, ed. *The Franklin Era in Canadian Arctic History 1845–1859,* Ottawa: National Museums of Canada, Mercury Series, Archaeological Survey of Canada Paper No. 131, 1985, 69–92.

2. Rajmund Przybylak, "Air Temperature in the Canadian Arctic in the Mid-Nineteenth Century Based on Data from Expeditions," *Prace Geograficzne* 107 (2000): 251–58.

3. John Rae, *Narrative of an Expedition to the Shores of the Arctic Sea, in 1846 and 1847,* London: T. W. Boone & Co., 1850, 5–6. Available at http://www.gutenberg.org/ebooks/39917.

4. R.K. Headland, et al., "Transits of the Northwest Passage to End of the 2022 Navigation Season," Scott Polar Research Institute Infosheet, November 18, 2022, available at https://www.spri.cam.ac.uk/resources/infosheets/northwestpassage.pdf. The statistics that follow are from this publication.

5. See Kaila Jefferd-Moore, "Canadian Submarines Not Part of International Under-Ice Exercise," *CBC News,* June 11, 2018. Available at https://www.cbc.ca/news/canada/north/-canadian-submarines-not-part-of-international-arctic-under-ice-exercise-1.4699208.

6. Richard J. Cyriax, *Sir John Franklin's Last Arctic Expedition: The Franklin Expedition, a Chapter in the History of the Royal Navy,* London: Methuen, 1939, reprint, Sussex: Arctic Press, 1997, 125–26. Or perhaps it was "Poets Bay," or even "Poetess Bay." Franklin scholar Peter Carney discusses the name problem in his blog: Peter Carney, "Return to Poctes Bay," *Erebus & Terror Files* blog, August 1, 2014, available at https://erebusandterrorfiles.

blogspot.com/2014/08/return-to-poctes-bay.html.

7. For details of Amundsen's trip, see Roald Amundsen, *The Northwest Passage*, London: Archibald Constable & Co., 1908.

8. Donat Pharand and Leonard H. Legault, *The Northwest Passage: Arctic Straits*, Leyden: Brill/Nijhoff, 1984.

9. Headland, et al., "Transits of the Northwest Passage."

10. For a more detailed discussion of the navigational issues on this eastern and southern route, see Russell Potter, "A Navigable Northwest Passage," *Visions of the North* blog, September 29, 2012, available at http://visionsnorth.blogspot.com/2012/09/a-navigable-northwest-passage.html.

11. Amundsen, *The Northwest Passage*.

12. Potter, "A Navigable Northwest Passage."

13. There is an abundant scholarly literature on the place of exploration in general, and polar exploration in particular, in British culture, both elite and popular, in the 19th century. Even the explorers themselves, by dint of exposure to the narratives, engravings, panoramas and other popular representations of their exploits, would have absorbed some of the heroic zeitgeist. See, e.g., Stephanie Barczewski, *Myth and National Identity in Nineteenth-Century Britain: The Legends of King Arthur and Robin Hood*, New York: Oxford University Press, 2000; Barczewski, *Heroic Failure and the British*, New Haven: Yale University Press, 2016; Janice Cavell, "Miss Porden, Mrs. Franklin, and the Arctic Expeditions: Eleanor Anne Porden and the Construction of Arctic Heroism (1818–1825)," in Frédéric Regard, ed., *Arctic Exploration in the Nineteenth Century: Discovering the Northwest Passage*, London: Pickering & Chatto, 2013, 79–94; Heather Davis-Fisch, *Loss and Cultural Remains in Performance: The Ghosts of the Franklin Expedition*, New York: Palgrave Macmillan, 2012; Erika Behrisch Elce, "Voices of Silence, Texts of Truth: Imperial Discourse and Cultural Negotiations in Nineteenth-Century British Arctic Exploration" (Ph.D. Thesis, Queen's University, Kingston, Ontario, 2002); Russell Potter, *Arctic Spectacles: The Frozen North in Visual Culture*, New York: University of Washington Press, 2007; Potter, "Introduction: Exploration and Sacrifice: The Cultural Logic of Arctic Discovery," in Frédéric Regard, ed., *The Quest for the Northwest Passage: British Narratives of Arctic Exploration, 1576–1874*, London: Pickering & Chatto, 2013, 1–18; Kathryn Schulz, "Literature's Arctic Obsession," *New Yorker*, April 24, 2017; and Francis Spufford, *I May Be Some Time: Ice and the English Imagination*, New York: St. Martin's Press, 1996.

14. A.G.E. Jones, "Sir James Clark Ross and the Voyage of the *Enterprise* and *Investigator*, 1848–49," *Geographical Journal* 137 (1971): 165–79.

Chapter 8

1. The official narrative of John Ross's voyage, including some of his nephew James Clark Ross's exploits, is John Ross, *Narrative of a Second Voyage in Search of the Northwest Passage and of a Residence in the Arctic Regions, 1829–33*, London: A.W. Webster, 1835. See also Ernest Dodge, *The Polar Rosses: John and James Clark Ross and Their Explorations*, London: Faber & Faber, 1973 and Maurice J. Ross, *Polar Pioneers: John Ross and James Clark Ross*, Montreal: McGill-Queen's University Press, 1994.

2. Richard J. Cyriax, *Sir John Franklin's Last Arctic Expedition: The Franklin Expedition, a Chapter in the History of the Royal Navy*, London: Methuen, 1939, reprint Sussex: Arctic Press, 1997, 132–133.

3. Ian S. MacLaren, "The Poetry of the 'New Georgia Gazette' or 'Winter Chronicle' 1819–1820," *Canadian Poetry* 30 (1992): 41–73.

4. Sammy Hudes, "Accidents, Not Illness, May Have Killed Many Officers in Franklin Expedition, Says Scottish Study," *Toronto Star*, September 21, 2016. Available at https://www.thestar.com/news/canada/2016/09/21/accidents-not-illness-may-have-killed-many-officers-in-franklin-expedition-says-scottish-study.html1.

5. A.G.E. Jones, "Sir James Clark Ross and the Voyage of the *Enterprise* and *Investigator*, 1848–49, *Geographical Journal* 137 (1971): 165–79, at 175.

Chapter 9

1. Communication in Facebook group "Remembering the Franklin Expedition," March 31, 2020. https://www.facebook.com/groups/11434844549/10157940372034550/?comment_id=10157942584354550&reply_comment_id=10157950348164550&notif_id=1585519848273654&notif_t=group_comment_mention.

2. John Ross, *Narrative of a Second Voyage in Search of the Northwest Passage and of a Residence in the Arctic Regions, 1829–33*, London: A.W. Webster, 1835.

3. George Back, *Narrative of an Expedition in HMS* Terror *Undertaken with a View to Geographical Discovery on the Arctic Shores in the Years 1836–7*, London: John Murray, 1838.

4. Richard J. Cyriax, *Sir John Franklin's Last Arctic Expedition: The Franklin Expedition, a Chapter in the History of the Royal Navy*, London: Methuen, 1939, reprint, Sussex: Arctic Press, 1997, at 196, concludes that everyone on the expedition died fairly soon after leaving the ship, with the possible exception of a few men who may have returned to the ships and survived a while longer.

5. See, e.g., David C. Woodman, *Unravelling the Franklin Mystery*, 2nd ed., Montreal:

McGill-Queen's University Press, 2015, 306–09. Charles Francis Hall, in his multi-year sojourn with the Inuit, collected several tales of survivors who were said to have lived on into the 1850s. For a closer look at the possibility of survivors, see Chapter 15 below.

6. Thomas Simpson, *Narrative of the Discoveries on the North Coast of America*, London: Richard Bentley, 1843, ch. 15.

7. Jonathan Schaeffer, ed., *Toward No Earthly Pole: Letters from John Franklin's Last Expedition*, self-published, Amazon, 2019.

8. For a history and map of the north magnetic pole's gyrations over the years, see Philip W. Livermore, Christopher C. Finlay and Matthew Bayliff, "Recent North Magnetic Pole Acceleration toward Siberia Caused by Flux Lobe Elongation," *Nature GeoScience* 13 (2020): 387–91. A more readily understandable version is Jennifer Lemann, "The Magnetic North Pole Is Rapidly Moving Because of Some Blobs," *Popular Mechanics*, March 7, 2022, available at https://www.popularmechanics.com/science/environment/a32496561/why-magnetic-north-pole-moving/.

Chapter 10

1. Richard J. Cyriax, *Sir John Franklin's Last Arctic Expedition: The Franklin Expedition, a Chapter in the History of the Royal Navy*, London: Methuen, 1939, reprint, Sussex: Arctic Press, 1997, at 75.

2. Claudio Aporta, "The Trail as Home: Inuit and Their Pan-Arctic Network of Routes," *Human Ecology* 37 (2009): 131–46, at 137.

3. Claudio Aporta, "Life on the Ice: Understanding the Codes of an Environment," *Polar Record* 38 (2002): 341–54.

4. Aporta, "The Trail as Home."

5. Michael Durey, "Exploration at the Edge: Reassessing the Fate of Sir John Franklin's Last Arctic Expedition," *The Great Circle: Journal of the Australian Association for Maritime History* 30 (2008): 3–40, at 29.

6. Paul Watson, *Ice Ghosts: The Epic Hunt for the Lost Franklin Expedition*, New York: W.W. Norton, 2017, at 149.

7. *Id.* at 158.

8. Cyriax, *Sir John Franklin's Last Arctic Expedition*, 144–46.

9. William. R. Hobson, *Report of Sledge Journey, April-June 1859*, reprinted in Douglas R. Stenton, "A Most Inhospitable Coast: The Report of Lieutenant William Hobson's 1859 Search for the Franklin Expedition on King William Island," *Arctic* 67 (2014): 511–22; Francis Leopold McClintock, *The Voyage of the "Fox" in the Arctic Seas: A Narrative of the Discovery of the fate of Sir John Franklin and his Companions*, London: John Murray, 1860.

10. William H. Gilder, *Schwatka's Search: Sledging in the Arctic in Search of the Franklin Records*, New York: Scribner's, 1881; Heinrich Klutschak, *Overland to Starvation Cove: With the Inuit in Search of Franklin, 1878–1880*, translated and edited by William Barr, Toronto: University of Toronto Press, 1993; Frederick Schwatka, *The Long Arctic Search: The Narrative of Lt. Frederick Schwatka, USA, 1878–1880, Seeking the Records of the Lost Franklin Expedition*, edited by Edouard Stackpole, Mystic, CT, Marine Historical Association, 1965; Douglas R. Stenton, Anne Keenleyside and Robert W. Park, "The 'Boat Place' Burial: Skeletal Evidence from the 1845 Franklin Expedition," *Arctic* 68 (2015): 32–44.

11. J.E. Nourse, ed., *Narrative of the Second Arctic Expedition Made by Charles Francis Hall: His Voyage to Repulse Bay, Sledge Journeys to the Straits of Fury and Hecla and to King William's Land, and Residence among the Eskimos during the Years 1864-'69*, Washington, D.C.: Government Printing Office, 1879, at 419–42.

12. On this last point, see Anne Keenleyside, Margaret Bertulli and Henry C. Fricke, "The Final Days of the Franklin Expedition: New Skeletal Evidence," *Arctic* 50 (1997): 36–46.

13. Douglas R. Stenton and Robert W. Park, "History, Oral History and Archaeology: Reinterpreting the 'Boat Places' of Erebus Bay," *Arctic* 70 (2017): 203–18.

14. Robin M. Rondeau, "The Wrecks of Franklin's Ships *Erebus* and *Terror*; Their Likely Location and the Cause of Failure of Previous Search Expeditions," *Journal of the Hakluyt Society*, March 2010: 1–11.

15. *Id.* at 6. Rondeau, writing in 2010, also presciently predicted that the wrecks of *Erebus* and *Terror* would likely be found intact, and not, as reported in some of the Inuit narratives, crushed by the ice while still on the surface. His interpretation of the Inuit stories is that those narratives, which spoke of seeing *umiak* (large, by Inuit standards, boats) crushed by ice, actually referred to ships' boats and not to the two Expedition ships themselves. *Id.* at 8. Rondeau was wrong, though, about where the ships would eventually be found, predicting that they would be quite near the position referred to in the Victory Point Record, rather than where they actually were discovered, much further south. *Id.* at 11.

16. William Gibson, "Some Further Traces of the Franklin Retreat," *Geographical Journal* 79 (1932): 402–08.

17. May Fluhmann, *Second in Command: The Life of Francis R. M. Crozier*, Yellowknife: Department of Information, Government of the Northwest Territories, 1976, quoting *Journal of the Royal Geographical Society* (1857): 323.

18. James Anderson, "Chief Factor James Anderson's Back River Journal of 1855," *Canadian Field Naturalist*, 54 (1940): 63–67, 84–89, 125–26, 134–36 and 167–69, and 55 (1941): 9–11 and 20–26.

19. George Back, *Narrative of the Arctic Land*

Expedition to the Mouth of the Great Fish River, London: John Murray, 1836, reprinted Dartmouth: CD-Academic Book Co., 1999.

20. Fluhmann, *Second in Command*, 118–19.

21. Douglas R. Stenton, "Finding the Dead: Bodies, Bones and Burials from the 1845 Franklin Northwest Passage Expedition," *Polar Record* 54 (2018): 197–212.

22. Richard J. Cyriax, "Recently Discovered Traces of the Franklin Expedition," *The Geographical Journal* 117 (1951): 211–14.

Chapter 11

1. L.T. Burwash, "The Franklin Search," *Canadian Geographical Journal* 1, no. 7 (1930).

2. Joseph Frey, "Inside the Terror—Life on Board Sir John Franklin's Lost Ships," *Geographical Magazine* 27 (2019). Available at http://geographical.co.uk/people/explorers/item/3525-inside-terror.

3. Harry Wilson, "Sir John Franklin's HMS Terror Believed Found in Arctic," *Canadian Geographic*, September 12, 2016. Available at https://www.canadiangeographic.ca/article/-sir-john-franklins-hms-terror-believed-found-arctic.

4. Elaine Anselmi, "Closing in on Franklin," *UpHere*, January/February 2018. Available at https://uphere.ca/articles/closing-franklin.

5. Charles Dagneau, "Interpretative Essay," *The Franklin Mystery: Life and Death in the Arctic* website. Available at https://www.canadianmysteries.ca/sites/franklin/interpretation/experts/interpretationDagneau_en.htm.

6. See Russell Potter, *Finding Franklin*, Montreal: McGill-Queen's University Press, 2016, at 38. The argument for a return to the ships fairly soon after the initial abandonment is made by David Woodman, *Unravelling the Franklin Mystery: Inuit Testimony*, 2nd ed., Montreal: McGill-Queen's University Press, 2015, at 110–12.

7. Russell Potter, "The Grave of Lieutenant Irving?" *Visions North* blog, January 3, 2010. Available at https://visionsnorth.blogspot.com/2010/01/grave-of-lieutenant-irving.html.

8. Richard J. Cyriax, "The Two Franklin Expedition Records Found on King William Island," *Mariner's Mirror* 44 (1958): 178–89.

9. See the discussion of these medical issues in Chapter 13.

10. See, e.g., Episode 7 of the AMC television series *The Terror*, where a full-fledged mutiny, led by Cornelius Hickey, splits the crew into two warring camps.

11. On this point, see Leonard F. Guttridge, *Mutiny: A History of Naval Insurrection*, Annapolis: Naval Institute Press, 1992, at 1.

12. The *Bounty* story is perhaps best known through fiction: Charles Nordhoff and James Norman Hall, *Mutiny on the Bounty*, New York: Grosset and Dunlap, 1932, and especially the 1945 edition, with illustrations by N.C. Wyeth. The story has also received numerous non-fiction treatments and is the subject of half a dozen films. The best recent non-fiction account is Caroline Alexander, *The Bounty: The True Story of the Mutiny on the Bounty*, New York: Viking Penguin, 2003. See also Owen Rutter, ed., *The Court Martial of the Bounty Mutineers*, London: Hodge, 1931; Alexander McKee, *HMS Bounty*, New York: William Morrow, 1962; and Richard Hough, *Captain Bligh and Mr. Christian*, Annapolis: Naval Institute Press, 2000.

13. On Bligh's tenure in New South Wales, which to some extent echoes that of Franklin in Van Diemen's Land, see Russell Earls Davis, *Bligh in Australia: A New Appraisal of William Bligh and the Rum Rebellion*, Warriewood: Woodslane Press, 2010.

14. David Grann, *The Wager: A Tale of Shipwreck, Mutiny and Murder*, New York: Doubleday, 2023; Stanley W.C. Pack, *The Wager Mutiny*, London: Alvin Redman, 1952; C.H. Layman, *The Wager Disaster: Mayhem, Mutiny and Murder in the South Seas*, London: Unicorn Press, 2015.

15. British Admiral Richard Kempenfelt, quoted in Guttridge, *Mutiny*, at 7.

16. *Id*.

17. Niklas Frykman, "Connections Between Mutinies in European Navies," *International Review of Social History* 58 (2013): 87–107; Jonathan Sayles Neale, "Forecastle and Quarterdeck: Protest, Discipline and Mutiny in the Royal Navy, 1793–1814," Ph.D. thesis, University of Warwick, 1990.

18. Frykman, "Connections Between Mutinies," at 87.

19. N.A.M. Rodger, "Shipboard Life in the Old Navy: The Decline of the Old Order," in Lewis R. Fischer, et al., eds., *The North Sea: Twelve Essays on Social History of Maritime Labour*, Stavanger: Maritime Museum, 1992, 29–39, at 32.

20. Wartime crews in the British, French and Dutch navies typically contained anywhere from 20 percent to as much as 70 percent foreigners. Niklas Frykman, "Seamen on Late Eighteenth-Century European Warships," *International Review of Social History* 54 (2009): 69–93, at 71–73.

21. Niklas Frykman, "The Mutiny on the *Hermione*: Warfare, Revolution and Treason in the Royal Navy," *Journal of Social History* 44 (2010): 158–87; Dudley Pope, *The Black Ship*, London: Weidenfeld and Nicolson, 1963.

22. Frykman, "Seamen on Late Eighteenth-Century Warships, 71-73.

23. Guttridge, *Mutiny*, at 45. See also Lawrence James, *Mutiny in the British and Commonwealth Forces, 1797–1956*, London: Buchan & Enright, 1987, at 36.

24. David C. Woodman, *Strangers Among Us*, Montreal: McGill-Queen's University Press, 1996 and Woodman, *Unravelling the Franklin Mystery: Inuit Testimony*.

25. See Douglas R. Stenton, Anne Keenleyside and Robert W. Park, "The 'Boat Place' Burial: Skeletal Evidence from the 1845 Franklin Expedition," *Arctic* 68 (2015): 32–44, and Douglas R. Stenton and Robert W. Park, "History, Oral History and Archaeology: Reinterpreting the 'Boat Places' of Erebus Bay," *Arctic* 70 (2017): 203–18. There is a helpful diagram of the boat that Lieutenant Hobson found at Erebus Bay in 1859 in Scott Cookman, *Ice Blink: The Tragic Fate of Sir John Franklin's Last Polar Expedition*, New York: John Wiley & Sons, 2000, at 15.

26. Stenton, Keenleyside and Park, "The 'Boat Place' Burial," at 42.

27. Stenton and Park, "History, Oral History and Archaeology," at 216–17.

Chapter 12

1. The authoritative source on Franklin search and rescue missions is W. Gillies Ross, "The Type and Number of Expeditions in the Franklin Search 1847–1859," *Arctic* 55 (2002): 57–69, from which most of the facts in this paragraph have been taken.

2. See David C. Woodman, *Unravelling the Franklin Mystery*, 2nd ed., Montreal: McGill-Queen's University Press, 2015.

3. The most complete account of McClure's voyage is Glenn M. Stein, *Discovering the North-West Passage: The Four-Year Arctic Odyssey of H.M.S. Investigator and the McClure Expedition*, Jefferson: McFarland, 2015.

4. Details of Kennedy's 1853 voyage are in William Barr, "'The Cold of Valparaiso': The Disintegration of William Kennedy's Second Franklin Search Expedition, 1853–54," *Polar Record* 34 (1998): 203–18. Facts in this discussion are taken from that account.

5. Richard J. Cyriax, *Sir John Franklin's Last Arctic Expedition: The Franklin Expedition, a Chapter in the History of the Royal Navy*, London, Methuen, 1939, reprint, Sussex Arctic Press, 1997, at 51.

6. Kenn Harper, "The Travels of an Inuit Story," *Nunatsiaq News* online, July 7, 2019, https://nunatsiaq.com/stories/article/taissumani-july-5/.

7. William Barr, "Franklin in Siberia?—Lieutenant Bedford Pim's Proposal to Search the Arctic Coast of Siberia. 1851–52," *Arctic* 45 (1992): 36–46. The story of searching efforts from the Pacific through the Bering Strait is told at some length in John R. Bockstoce, "The Search for Sir John Franklin," in John R. Bockstoce, *Furs and Frontiers in the Far North*, New Haven: Yale University Press, 2009, 227–59, and John R. Bockstoce, "The Search for Sir John Franklin in Alaska," in Patricia D. Sutherland, ed., *The Franklin Era in Canadian Arctic History, 1845–1859*, Ottawa: National Museums of Canada, Mercury Series, Archaeological Survey of Canada Paper No. 131, 1985, 93–113.

8. Fridtjof Nansen, *Farthest North: Being the Record of a Voyage of Exploration of the Ship "Fram," 1893–96 and of a Fifteen Months' Sleigh Journey by Dr. Nansen and Lieut. Johansen*, London: Harper & Bros., 1897.

9. See, e.g., Bob Bartlett, *The Karluk's Last Voyage: An Epic of Death and Survival in the Arctic, 1913–1916*, New York: Cooper Square Press, 2001.

10. Paul Watson, *Ice Ghosts: The Epic Hunt for the Lost Franklin Expedition*, New York: W.W. Norton, 2017, at 173.

11. David Woodman, *Strangers Among Us*, Montreal: McGill-Queen's University Press, 1995.

12. Woodman, *Unravelling the Franklin Mystery*.

13. For a summary of the argument, see Woodman, *Strangers Among Us*, at 130–36.

Chapter 13

1. The Admiralty's official action at the end of March had been prefigured by an announcement on January 20, 1854, in the *London Gazette*. Erika Behrisch Elce, "Widows' Men: The Admiralty Board, Precedent, and Pensions for the Widows of the Lost Franklin Expedition," *Journal of Imperial and Commonwealth History* 47 (2019): 28–50, at 30.

2. "Notice," *The Times* (London), January 21, 1854.

3. Elce, "Widows' Men."

4. Andrew Lambert, *The Gates of Hell: John Franklin's Tragic Quest for the Northwest Passage*, New Haven: Yale University Press, 2011, at 243, and Trevor H. Levere, "Science and the Canadian Arctic, 1818–76, from Sir John Ross to Sir George Strong Nares," *Arctic* 41 (1988): 127–37. By 1860, the total cost of all search expeditions, both official and private, was on the order of £2 million, or roughly a quarter of a billion U.S. dollars in today's values. Hugh N. Wallace, "Private Expeditions in the Search for Sir John Franklin," in Patricia D. Sutherland, ed., *The Franklin Era in Canadian Arctic History, 1845–1859*, Ottawa: National Museums of Canada, Mercury Series, Archaeological Survey of Canada Paper No. 131, 1985, 42–53. On American Elisha Kent Kane's efforts, see Ken McGoogan, *Race to the Polar Sea, the Heroic Adventures of Elisha Kent Kane*, Berkeley: Counterpoint, 2009.

5. Lambert, *The Gates of Hell*, at 257; David Lyon & Rif Winfield, *The Sail and Steam Navy List*, London: Chatham, 2004, at 241–42.

6. S. Mays, G.J.R. Maat and H.H. de Boer, "Scurvy as a Factor in the Loss of the 1845 Franklin

Expedition to the Arctic: A Reconsideration," *International Journal of Osteoarchaeology* 25 (2015): 334–44. See also Scott Cookman, *Ice Blink: The Tragic Fate of Sir John Franklin's Last Polar Expedition,* New York: John Wiley & Sons, 2000, and Maurice J. Ross, *Polar Pioneers: John Ross and James Clark Ross,* Kingston: McGill-Queen's University Press, 1994.

7. Jonathan M. Karpoff, "Public versus Private Initiative in Arctic Exploration: The Effect of Incentives and Organizational Structure," *Journal of Political Economy* 109 (2001): 38–78. True to his Chicago School ideology, Karpoff finds that publicly funded expeditions "(1) had poorly motivated and prepared leaders; (2) wrongly separated the initiating and implementing functions of leadership [e.g., John Barrow was the initiator, Sir John Franklin the mere implementer]; and (3) were slow to exploit new information about clothing, diet, shelter, modes of Arctic travel, organizational structure and optimal party size." Crew size was also markedly different; publicly funded expeditions averaged 69 men, private ones only 15. Fifteen men might have had a chance of surviving in 1848; 105 certainly did not.

8. Douglas Stenton, "Finding the Dead: Bodies, Bones and Burials from the 1845 Franklin Northwest Passage Expedition," *Polar Record* 54 (2018): 197–212.

9. Bea Uusma, *The Expedition: A Love Story,* translated by Agnes Broom, London: Head of Zeus, 2014, 244–70. Possible causes of death for the Andrée expedition include carbon dioxide poisoning, suffocation due to oxygen depletion in their tent, algae poisoning, scurvy, trichinosis, Vitamin A poisoning, botulism from tinned food, hypothermia, lead poisoning (again from tinned food), gunshot wounds, polar bear attacks and morphine overdoses.

10. John Geiger and Owen Beattie, *Frozen in Time: The Fate of the Franklin Expedition,* New York: Dutton, 1988. For a brief review of the Beechey Island deaths and their possible causes, see Lars Pilø, "Buried in Ice—the Franklin Expedition Cemetery," available at https://secretsoftheice.com/news/2019/10/28/franklin-expedition/.

11. Roger Amy, Rakesh Bhatnagar, Eric Damkjar and Owen Beattie, "The Last Franklin Expedition: Report of a Post-Mortem Examination of a Crew Member," *Canadian Medical Association Journal* 135 (1986): 115–17 and Roger Amy, "Report on the Exhumation of John Torrington, Deceased January 4, 1846," available at https://lookaside.fbsbx.com/file/autopsy-torrington.pdf?token=AWwypZ30XtfyutPV2dFzbfI8Fr_5oT8BUyjUfpTeDvitfbWyxRAeM4DK_4GHfXUAciBnEjUQSsQTDgVrTIr0vmrGqqMPN4TL1wz33Kjp8eMAdym163DMDFAbSnRa4fqk1KT46d0kQpLl26uYWSLd4_46.

12. Paul Watson, *Ice Ghosts: The Epic Hunt for the Lost Franklin Expedition,* New York: W.W. Norton, 2017, at 47. See also Jannine Forst and Terence A. Brown, "A Case Study: Was Private William Braine of the 1845 Franklin Expedition a Victim of Tuberculosis?" *Arctic* 70 (2017): 381–88 (casting doubt on tuberculosis as Braine's cause of death). Braine's body showed evidence that it had been attacked by rats before burial, which would be consistent with death away from the ship and a delay between death and burial. Gillian Hutchinson, *Sir John Franklin's Erebus and Terror Expedition Lost and Found,* London: Bloomsbury, 2017, at 82.

13. Steve Inskeep, "Remembering John Glenn," *Morning Edition,* NPR, December 9, 2016, https://www.npr.org/2016/12/09/504930256/-remembering-john-glenn.

14. Robert E. Peary, *Secrets of Polar Travel,* New York: The Century Co., 1917, at 58.

15. Jean-Marie Ananquaq and Guy Mary-Rousseliere, OMI, "The Schwatka Expedition as Seen by the Inuit," n.d., available at https://lookaside.fbsbx.com/file/The%20Schwatka%20Expedition%20as%20seen%20by%20the%20Inuit%20-%20Eskimo%2C%20n.s.%2038-39%2C%201990%2C%20pp%-205-11%2C%20.pdf?token=AWzpLpNdDIW_nu001I1tR3wBgHbddA0fL7V0IpmuVUlQMcAN7xesn1ruIsYh2ex42kiFuB-HSO_g7U8UiuYHTzPMuPTjRsNmhYNVPexZlz4zZdZZ1_XGafvct3QPpdf78sazRpVFA-W3kY1FFRKMUcHI.

16. Keith Millar, A.W. Bowman, William Battersby and R.R. Welbury, "The Health of Nine Royal Naval Arctic Crews, 1848 to 1854: Implications for the Lost Franklin Expedition," *Polar Record* 52 (2016): 423–41. For the short version, see Sammy Hudes, "Accidents, Not Illness, May Have Killed Many Officers in Franklin Expedition, Says Scottish Study," *Toronto Star,* September 21, 2016, available at https://www.thestar.com/news/canada/2016/09/21/accidents-not-illness-may-have-killed-many-officers-in-franklin-expedition-says-scottish-study.html.

17. Cluny MacPherson, "The First Recognition of Beri-Beri in Canada," Letter to the Editor, *Canadian Medical Association Journal* 95 (1966): 278–79.

18. Available at https://www.pbs.org/wgbh/nova/arctic/provisions.html.

19. Millar, et al., "The Health of Nine Royal Navy Arctic Crews," at 434.

20. The best study on botulism as a possible cause of death on the Franklin Expedition is B. Zane Horowitz, "Polar Poisons: Did Botulism Doom the Franklin Expedition?" *Journal of Toxicology: Clinical Toxicology* (2003): 841–47, though Horowitz rejects the likelihood that botulism, if it occurred, was from Goldner's tinned food.

21. J.C. Drummond and W.R. Lewis, "Historical Introduction," in International Tin Research and Development Council, *Historic Tinned Foods,* Greenford: International Tin Research and Development Council Publication No. 85,

1939, at 26. Available at https://archive.org/stream/in.ernet.dli.2015.211975/2015.211975.-Historic-Tinned_djvu.txt.

22. Scott Cookman, *Ice Blink: The Tragic Fate of Sir John Franklin's Last Polar Expedition,* New York: John Wiley & Sons, 2000.

23. Though not so inedible that a substantial portion of the cans rejected for Royal Navy use were not then given to the London poor.

24. Constantine Ardeleanu, "A British Meat Cannery in Moldavia (1844–52)," *The Slavonic and East European Review* 90, (2012): 671–704.

25. Peter Carney, "J'Accuse!—The Case of Stephan Goldner—Britain's Dreyfus," *Erebus & Terror Files* blog, April 30, 2020, available at https://erebusandterrorfiles.blogspot.com/2020/04/-jaccuse-case-of-stephan-goldner.html.

26. Drummond and Lewis, "Historical Introduction," at 24–25.

27. Christopher Lloyd and J.S.L. Coulter, *Medicine in the Navy 1200–1900*, vol. IV, 1815–1900, London: E. & S. Livingstone, 1969.

28. Carney, "J'Accuse—the Case of Stephan Goldner."

29. Drummond and Lewis, "Historical Introduction," at 12–13.

30. *Id.* at 20.

31. May Fluhmann, *Second in Command: The Life of Francis R. M. Crozier,* Yellowknife: Department of Information, Government of the Northwest Territories, 1976, at 99.

32. Horowitz, "Polar Poisons," at 842.

33. *Id.* at 845–46.

34. Michael Durey, "Exploration at the Edge: Reassessing the Fate of Sir John Franklin's Last Arctic Expedition," *The Great Circle: Journal of the Australian Association for Maritime History* 30 (2008): 3–40, at 26.

35. Millar, et al., "The Health of Nine Royal Navy Arctic Crews," at 436.

36. See, e.g., Joseph B. Mclaughlin, Jeremy Sobel, Tracey Lynn, Elizabeth Funk and John P. Middaugh, "Botulism Type E Outbreak Associated with Eating a Beached Whale, Alaska," *Emerging Infectious Diseases* 10 (2004): 1685–87, summarizing the various types of botulism associated with eating wild-caught game in the Arctic.

37. Russell Potter, Regina Koellner, Peter Carney and Mary Williamson, eds., *May We Be Spared to Meet on Earth: Letters of the Lost Franklin Arctic Expedition,* Montreal: McGill-Queen's University Press, 2022, at 88 (letter from Harry Goodsir).

38. B. Alt, R.M. Koerner, D.A. Fisher, and J. C. Bourgeois, "Arctic Climate During the Franklin Era, as Deduced from Ice Cores," in Sutherland, ed., *The Franklin Era in Canadian Arctic History,* 69–92.

39. Peter Marchand, *Life in the Cold: An Introduction to Winter Ecology,* Hanover: University Press of New England, 1987, at 186.

40. *Id.* at 200.

41. *Id.* at 190–91.

42. For a serious, non-sensational look at the story of the Donner Party, see Ethan Rarick, *Desperate Passage: The Donner Party's Perilous Journey West,* New York: Oxford University Press, 2008.

43. Leonard Guttridge, *Ghosts of Cape Sabine: The Harrowing True Story of the Greeley Expedition,* New York: Putnam, 2000, cited in Lawrence A. Palinkas and Peter Suedfeld, "Psychological Effects of Polar Expeditions," *The Lancet* 371 (2008): 153–63, at 154.

44. Palinkas and Suedfeld, "Psychological Effects of Polar Expeditions," at 155.

45. H.L. Reed, K.R. Reedy and Lawrence A. Palinkas, "Impairment in Cognitive and Exercise Performance during Prolonged Antarctic Residence: Effect of Thyroxine Supplementation in the Polar Triiodothyronine Syndrome," *Journal of Clinical Endocrinology and Metabolism* 86 (2001): 110–16.

46. Lawrence A. Palinkas, "Going to Extremes: The Cultural Context of Stress, Illness and Coping in Antarctica," *Social Science and Medicine* 35 (1992): 651–64.

47. Jack W. Stuster, "Bold Endeavors: Behavioral Lessons from Polar and Space Exploration," *Gravitational and Space Biology Bulletin* 13 (2000): 49–57 (referring, in particular, to the *Belgica* Antarctic expedition at the end of the 19th century and to the 1957 International Geophysical Year Antarctic expedition). A first-hand narrative of the problems on the *Belgica* journey is Frederick A. Cook, *Through the First Antarctic Night, 1899–1900: A Narrative of the Voyage of the Belgica Through Newly Discovered Lands and over an Unknown Sea about the South Pole,* Montreal: McGill-Queen's University Press, 1980 (originally published 1900).

48. J.N. Norman, "A Comparison of the Patterns of Illness and Injury Occurring in Offshore Structures in the Northern North Sea and the Stations of the British Antarctic Survey," *Arctic Medical Research* 50 (1991): 719–21; L. Rosen, K.H. Knudson and P. Fancher, "Prevalence of Seasonal Affective Disorder among U.S. Army Soldiers in Alaska," *Military Medicine* 167 (2002): 581–84.

49. J.M. Haggarty, Z. Cernovsky, M. Husni, P. Kermeen and H. Merskey, "Seasonal Affective Disorder in an Arctic Community," *Acta Psychiatrica Scandinavica* 105 (2002): 378–84.

50. Stuster, "Bold Endeavors." Stuster's findings are explained in much greater detail in his book: Jack W. Stuster, *Bold Endeavors: Lessons from Polar and Space Exploration,* Annapolis: Naval Institute Press, 1996.

51. See, inter alia, William Battersby, "Identification of the Probable Source of the Lead Poisoning Observed in Members of the Franklin Expedition," *Journal of the Hakluyt Society,* September 2008: 1–6; Owen Beattie, "Elevated Bone

Lead Levels in a Crewman from the Last Arctic Expedition of Sir John Franklin," in Patricia D. Sutherland, ed., *The Franklin Era in Canadian Arctic History*, 131; Peter Carney, "Further Light on the Source of Lead in Human Remains from the 1845 Franklin Expedition," *Journal of the Hakluyt Society*, September 2016: 3–17; Anne Keenleyside, X. Song, D.R. Chettle and C.E. Webber, "The Lead Content of Human Bones from the 1845 Franklin Expedition," *Journal of Archaeological Science* 23 (1996): 461–65; Walter Kowal, P.M. Krahn and Owen Beattie, "Lead Levels in Human Tissues from the Franklin Forensic Project," *International Journal of Environmental Analytical Chemistry* 35 (1989): 119–26; Walter Kowal, Owen Beattie, H. Baadsgaard and P.M. Krahn, "Source Identification of Lead Found in Tissues of Sailors from the Franklin Arctic Expedition of 1845," *Journal of Archaeological Science* 18 (1991): 193–203; R. Martin, S. Naftel, S. Macfie, K. Jones and A. Nelson, "Pb Distribution in Bones from the Franklin Expedition: Synchrotron X-ray Fluorescence and Laser Ablation/Mass Spectroscopy," *Applied Physics A* 111 (2013): 23–29; Keith Millar, Adrian W. Bowman and William Battersby, "A Re-Analysis of the Supposed Role of Lead Poisoning in Sir John Franklin's Last Expedition, 1845–1848," *Polar Record* 51 (2015): 224–38; and Treena Swanston, et al., "Franklin Expedition Lead Exposure: New Insights from High Resolution Confocal X-ray Fluorescence Imaging of Skeletal Microstructure," *PLOS One* 13, no. 8 (2018), available at https://journals.plos.org/plosone/article?id=10.1371/journal.pone.0202983.

52. See, e.g., Beattie, "Elevated Bone Lead Levels."

53. Battersby, "Identification of the Probable Source of the Lead Poisoning."

54. See, e.g., the sources cited in Millar and Bowman, "Cognitive Archaeology."

55. See, e.g., Ann Shirley, "Lead Poisoning and the Franklin Expedition," *Polar Record* 28 (1992): 73; Derek Fordham, "Lead Poisoning and the Franklin Expedition," *Polar Record* 27 (1991): 371; and Swanston, et al., "Franklin Expedition Lead Exposure," all arguing that lead poisoning was, at most, a minor factor in the deaths of the Expedition's crew.

56. Lori D'Ortenzio, Michael Inskip, William Manton and Simon Mays, "The Franklin Expedition: What Sequential Analysis of Hair Reveals about Lead Exposure Prior to Death," *Journal of Archaeological Sciences: Reports* 21 (2018): 401–05; Kiona Smith, "Strands of Hair Shed Light on Doomed 19th-Century Arctic Expedition," *Ars Technica Online*, September 30, 2018, available at https://arstechnica.com/science/2018/09/did-lead-poisoning-finish-off-a-doomed-arctic-expedition/.

57. Smith, "Strands of Hair."

58. Millar and Bowman, "Cognitive Archaeology."

59. Millar, Bowman and Battersby, "A Reappraisal of the Supposed Role of Lead Poisoning."

60. *Id.*

61. Durey, "Exploration at the Edge," at 26.

62. Millar, Bowman, Battersby and Welbury, "The Health of Nine Royal Navy Arctic Crews," at 434.

63. K.T.H. Farrer, "Lead and the Last Franklin Expedition," *Journal of Archaeological Science* 20 (1993): 399–409; Carney, "Further Light on the Source of Lead."

64. Battersby, "Identification of the Probable Source of the Lead"; William Battersby and Peter Carney, "Equipping HM Ships Erebus and Terror, 1845," *International Journal for the History of Engineering & Technology* 81 (2011): 192–211. The basic Fraser's Patent Stove had been used in polar voyages since Parry's expeditions 25 years earlier, but without the modifications employed for the Franklin Expedition. See Fluhmann, *Second in Command*, at 14. See also Matthew Betts, *HMS Terror: The Design, Fitting and Voyages of the Polar Discovery Ship*, Barnsley: Seaforth, 2022.

65. Carney, "Further Light on the Source of Lead."

66. Millar, Bowman, Battersby and Welbury, "The Health of Nine Royal Navy Arctic Crews," at 434.

67. Scurvy was first suggested as a factor in the Franklin disaster in Richard J. Cyriax, *Sir John Franklin's Last Arctic Expedition: The Franklin Expedition, a Chapter in the History of the Royal Navy*, London: Methuen, 1939, reprint, Sussex: Arctic Press, 1997.

68. Durey, "Exploration at the Edge," at 27.

69. Jeremy Hugh Baron, "Sailors' Scurvy Before and After James Lind—a Reassessment," *Nutrition Reviews* 67 (2009): 315–32, at 316–17.

70. *Id.* at 317.

71. *Id.* at 318–24.

72. James Lind, *A Treatise on the Scurvy*, London: A. Millar, 1753. See also Robert E. Feeney, *Polar Journeys: The Role of Food and Nutrition in Early Exploration*, Fairbanks: American Chemical Society and University of Alaska Press, 1998, at 8.

73. Michael Bartholomew, "James Lind and Scurvy: A Revaluation," *Journal for Maritime Research* 4 (2002): 1–14, and Betts, *HMS Terror*. See also Mark Harrison, "Scurvy on Sea and Land: Political Economy and Natural History, c. 1780–c. 1850," *Journal for Maritime Research* 15 (2013): 7–25.

74. Feeney, *Polar Journeys: The Role of Food and Nutrition in Early Exploration*, at 8.

75. *Id.* at 25.

76. Watson, *Ice Ghosts* at 25. See also Mays, et al., "Scurvy as a Factor."

77. Millar, Bowman, Battersby and Welbury, "The Health of Nine Royal Navy Arctic Crews," at 431.

78. *Id.*

79. Mays et al., "Scurvy as a Factor."

80. See, e.g., R.L. Richards, "Rae of the Arctic," *Medical History* 19 (1975): 176–93. John Rae describes how he and his small party survived in John Rae, *Narrative of an Expedition to the Shores of the Arctic Sea, in 1846 and 1847*, London: T.W. Boone & Co., 1850, available at http://www.gutenberg.org/ebooks/39917.

81. Millar, Bowman, Battersby and Welbury, "The Health of Nine Royal Navy Arctic Crews," at 430.

82. *Id.* at 431.

83. Richard J. Cyriax, "A Historic Medicine Chest," *Canadian Medical Association Journal* 57 (1947): 295–300.

84. Mays, et al., "Scurvy as a Factor."

85. *Id.*

86. Much of the following discussion relies on Mechtild Opel, "Chocolate in the Arctic," *Trimaris* blog, April 24, 2019. Available at https://www.trimaris.de/?s=Chocolate.

87. Janet MacDonald, *Feeding Nelson's Navy: The True Story of Food at Sea in the Georgian Era*, London: Chatham, 2004, at 9–10. Although MacDonald focuses on a slightly earlier period than that of the Expedition, the ration she describes remained the Royal Navy standard through 1847.

88. Vilhjalmur Steffansson, *The Friendly Arctic*, New York: Macmillan, 1921, at 323.

89. John Rae, quoted in Alexa Price, "'Our Proudest Heritage': Masculinity, Nostalgia, and the Sailing Navy on Display, 1820–1920," Ph.D. Dissertation, The George Washington University, 2019, at 100.

90. Recounted in Liz Cruwys, "Henry Grinnell and the American Franklin Searches," *Polar Record* 26 (1990): 211–16.

91. Chauncey Loomis, *Weird and Tragic Shores: The Story of Charles Francis Hall, Explorer*, New York: Knopf, 1991.

92. Ken McGoogan, "Mystery Solved!!! Polar Bears Explain the Fate of the Franklin Expedition," September 7, 2018, available at https://kenmcgoogan.com/tag/canadian-mysteries/. See also Delbert Young, "Was There an Unsuspected Killer Aboard the *Unicorn*?" *The Beaver* 304, no. 3 (1973): 4–15.

93. Frank H. Connell, "Trichinosis in the Arctic," *Arctic* 2 (1949): 98–107.

94. See Russell H. Taichman, Tom Gross and Mark P. MacEachern, "A Critical Assessment of the Oral Condition of the Crew of the Franklin Expedition," *Arctic* 70, no. 1 (2017): 25–36.

95. See, e.g., Amy Bhatnagar, et al., "The Last Franklin Expedition"; D.N.H. Notman, L. Anderson, Owen Beattie & Roger Amy, "Arctic Paleology: Portable Radiographic Examination of Two Frozen Sailors from the Franklin Expedition (1845–1848)," *American Journal of Roentgenology* 149, no. 2 (1987): 347–50.

96. Janine Forst and Terence A. Brown, "A Case Study: Was Private William Braine of the 1845 Franklin Expedition a Victim of Tuberculosis?" *Arctic* 70 (2017): 381–88.

97. See Jenny R. Christensen, Joyce M. McBeth, Nicole J. Sylvain, Jody Spence, and Hing Man Chan, "Hartnell's Time Machine: 170-Year-Old Nails Reveal Severe Zinc Deficiency Played a Greater Role Than Lead in the Demise of the Franklin Expedition," *Journal of Archaeological Science: Reports* 16 (2017): 430–44; Keith Millar and Adrian W. Bowman, "Hartnell's Time Machine" Reprise: Further Implications of Zinc, Lead and Copper in the Thumbnail of a Franklin Expedition Crewmember," *Journal of Archaeological Science: Reports* 13 (2017): 286–90, available at https://eprints.gla.ac.uk/141446/1/141446.pdf.

98. Christensen, et al., "Hartnell's Time Machine."

Chapter 14

1. Ann Gibbons, "Archaeologists Rediscover Cannibalism," *Science* 277 no. 5326 (August 1, 1997): 635–37.

2. Adriana Craciun, "Writing the Disaster: Franklin and *Frankenstein*," *Nineteenth Century Literature* 65 (2011): 433–80, at 461–62. Craciun does point out, however, that Mary Shelley's monster was, in contrast to some, at least, of Franklin's sailors, a vegan.

3. Charles Dickens, "The Lost Arctic Voyagers," *Household Words*, December 2 and 9, 1854: 362–65 and 387–93.

4. For a particularly clinical, unsentimental and graphic description of how it might have happened, see Andrew Lambert, *The Gates of Hell: John Franklin's Tragic Quest for the Northwest Passage*, New Haven: Yale University Press, 2011, at 1.

5. Alexa Price, "'Our Proudest Heritage': Masculinity, Nostalgia, and the Sailing Navy on Display, 1820–1920," Ph.D. Dissertation, The George Washington University, 2019, at 90.

6. Martyn Beardsley's relatively recent retelling of the story, Martyn Beardsley, *Deadly Winter: The Life of Sir John Franklin*, Annapolis: Naval Institute Press, 2002, returns a verdict of "not proven," based on some ambiguities in the research published by Anne Keenleyside and colleagues (see below, this chapter). A most unscientific internet posting by E.C. Coleman, "The Franklin Expedition: Cannibalism?" 1998, available at *Roger Parsons Lincolnshire World website*, http://www.rogerparsons.info/franklin.html, argues against cannibalism, as does Coleman's more recent book, Ernest C. Coleman, *No Earthly Pole: The Truth About the Franklin Expedition 1845*, Stroud: Amberley, 2020, but that's pretty much the extent of current denialism.

7. "The Arctic Expedition," *The Times* (London), October 23, 1854.

8. Edgar Allan Poe, *The Narrative of Arthur Gordon Pym of Nantucket*, New York: Hill & Wang, 1960 (originally published 1838).

9. Herman Melville, *Moby-Dick: An Authoritative Text, Context, Criticism*, New York: W.W. Norton, 2018 (first published 1851).

10. H.L. Malchow, *Gothic Images of Race in Nineteenth Century Britain*, Palo Alto: Stanford University Press, 1996, at 105–10. Even the supposedly authoritative *Encyclopaedia Britannica* discussed cannibalism (listed under "Anthropophagia") as something done only by savages, though that category included, perhaps, Britons' ancestors the Picts as well.

11. See Ibram X. Kendi, *Stamped from the Beginning: The Definitive History of Racist Ideas in America*, New York: Nation Books, 2016, 79–91.

12. See generally Patrick Brantlinger, *Taming Cannibalism*, Ithaca: Cornell University Press, 2011.

13. Dickens, "The Lost Arctic Voyagers."

14. J.E. Nourse, ed., *Narrative of the Second Arctic Expedition Made by Charles Francis Hall: His Voyage to Repulse Bay, Sledge Journeys to the Straits of Fury and Hecla and to King William's land, and Residence among the Eskimos During the Years 1864-'69*, Washington, D.C.: Government Printing Office, 1879, at 107–09, 405, 419–20 and 608. Nourse's published version of Hall's journals omits many additional references to cannibalism, which may be found in the Hall archive at the Smithsonian Institution in Washington. Some of these notes he had, in fact, supplied to Lady Jane Franklin at the time, on condition that they not be published. See Janice Cavell, "Publishing Sir John Franklin's Fate: Cannibalism, Journalism and the 1881 Edition of Leopold McClintock's *The Voyage of the "Fox" in the Arctic Seas*," *Book History* 16 (2013): 155–84, at 166.

15. G. Richard Scott and Sean McMurry, "The Delicate Question: Cannibalism in Prehistoric and Historic Times," in Kelly Dixon, Julie Schablitsky and Shannon Novak, eds., *An Archaeology of Desperation: Exploring the Donner Party's Alder Creek Camp*, Norman: University of Oklahoma Press, 2011, 211–243, at 232.

16. For an example of Dickens's overt racism, see, e.g., Charles Dickens, "The Noble Savage," n.d. Available at www.ric.edu/faculty/rpotter/temp/noblesav.html.

17. Cavell, "*Publishing Sir John Franklin's Fate.*"

18. Fairholme Papers, S. Cracroft to Elizabeth Murray, February 21, 1873, Public Archives, Ottawa, Canada, cited in Anne Keenleyside, Margaret Bertulli and Henry C. Fricke, "The Final Days of the Franklin Expedition: New Skeletal Evidence," *Arctic* 50 (1997): 36–46, at 37.

19. Keenleyside, Bertulli and Fricke, "Final Days of the Franklin Expedition"; Anne Keenleyside, "The Last Resort: Cannibalism in the Arctic," *The Explorers Journal* 72 (1995): 138–43. See also the earlier work by Owen Beattie and colleagues: Owen Beattie, "A Report on Newly Discovered Skeletal Remains from Sir John Franklin's Last Expedition," *The Musk-Ox* 33 (1983): 68–77; Owen Beattie and James M. Savelle, "Discovery of Human Remains from Sir John Franklin's Last Expedition," *Historical Archaeology* 17 (1983): 100–05; Owen Beattie and John Geiger, *Frozen in Time: The Fate of the Franklin Expedition*, Vancouver: Douglas & McIntyre, 1987 and Walter Kowal, P.M. Krahn and Owen Beattie, "Lead Levels in Human Tissues from the Franklin Forensic Project," *International Journal of Environmental Analytical Chemistry* 35 (1989): 119–26.

20. Keenleyside, et al., "The Final Days of the Franklin Expedition" 41–42.

21. See, e.g., Paola Villa, "Cannibalism in Prehistoric Europe," *Evolutionary Anthropology* 1 (1992): 93–104; Y. Fernandez-Jalvo, J.C. Diez, I. Caceres and J. Rosell, "Human Cannibalism in the Early Pleistocene of Europe (Gran Dolina, Sierra de Atapuerca, Burgos, Spain)," *Journal of Human Evolution* 37 (1999): 591–622.

22. See Lawrence Goldman, ed., *The Anthropology of Cannibalism*, Westport: Bergin & Garvey, 1999, and Paula Brown and Donald Tuzin, eds., *The Ethnography of Cannibalism*, Washington, D.C.: Society for Psychological Anthropology, 1983.

23. Laurel L. Fox, "Cannibalism in Natural Populations," *Annual Review of Ecology and Systematics* 6 (1975): 87–106.

24. Mariko Hiraiwa-Hasigawa, "Cannibalism among Non-Human Primates," in Mark A. Elger and Bernard J. Crespi, eds., *Cannibalism: Ecology and Evolution Among Diverse Taxa*, Oxford: Oxford University Press, 1992, 323–38; J.D. Bygott, "Cannibalism Among Wild Chimpanzees," *Nature* 238 (1976): 410–11; Jane Goodall, "Infant-Killing and Cannibalism in Free-Ranging Chimpanzees," *Folia Primatologica* 28 (1977): 259–82.

25. P. Saladié and A. Rodriguez-Hidalgo, "Archaeological Evidence for Cannibalism in Pre-Historic Western Europe from Homo Antecessor to the Bronze Age," *Journal of Archaeological Method and Theory* (2016), available at doi: 10.1007/s10816-016-9306-y.

26. Christy G. Turner II and Jacqueline Turner, *Man Corn: Cannibalism and Violence in the Pre-Historic American Southwest*, Salt Lake City: University of Utah Press, 1999, and Turner and Turner, "Cannibalism in Chaco Canyon: The Charnel Pit Excavated in 1926 at Small House Ruin by Frank H. H. Roberts, Jr.," *American Journal of Physical Anthropology* 91 (1993): 421–39.

27. Jerry Melbye and Scott I. Fairgrieve, "A Massacre and Possible Cannibalism in the Canadian Arctic: New Evidence for the Saunatuk Site (NgTn1)," *Arctic Anthropology* 31 (1994): 57–77.

28. Dirk H. R. Spenneman, "Cannibalism in

Fiji: the Analysis of Butchering Marks on Human Bones and the Historical Record," *Domodomo* 2 (1997): 29–46; David DeGusta, "Fijian Cannibalism: Osteological Evidence from Navatu," *American Journal of Physical Anthropology* 110 (1999): 215–41; David W. Steadman, Susan C. Antón and Patrick V. Kirch, "Ana Manuku: A Prehistoric Ritual Site on Mangaia, Cook Islands," *Antiquity* 74 (2000): 873–83; Patrick Brantlinger, "Missionaries and Cannibals in Nineteenth-Century Fiji," *History and Anthropology* 17 (2006): 21–38; Ross Bowden, "Maori Cannibalism: An Interpretation," *Oceania* 55 (1984): 81–99; and David Wetherell, "Accounts of Fighting and Cannibalism in Eastern New Guinea During the Missionary Contact Period, 1877–1888, as Told to Charles Abel," *Pacific Studies* 26 (2003): 37–52.

29. Carole A. Travis-Henikoff, *Dinner with a Cannibal*, Santa Monica: Santa Monica Press, 2008.

30. Bill Schutt, *Cannibalism: A Perfectly Natural History*, Chapel Hill: Algonquin Books, 2018, at 133–73. See also Roger Rosenberg, "The Donner Party: Trapped in the Mountains," *Academia Letters*, 2021, available at https://doi.org/10.20935/AL2482.

31. Nando Parrado, *Miracle in the Andes: 72 Days on the Mountain and My Long Trek Home*, New York: Broadway Books, 2006.

32. See, e.g., Jerome T. Whitfield, Wandagi H. Pako, John Collinge and Michael P. Alpers, "Mortuary Rites of the South Fore and Kuru," *Philosophical Transactions of the Royal Society B: Biological Sciences* 363 (2008): 3721–24.

33. Brantlinger, *Taming Cannibalism*, at 30; see also the discussions of "compassionate cannibalism" in Beth A. Conklin, *Consuming Grief: Cannibalism in an Amazonian Society*, Austin: University of Texas Press, 2001, and I.M. Lewis, *Religion in Context: Cults and Charisma*, Cambridge: Cambridge University Press, 2001.

34. Brantlinger, *Taming Cannibalism*, at 27–45; see also R.D. Shaw, "Three-Day Visitors: The Samo Response to Colonialism in Western Province, Papua New Guinea," in N.M. McPherson, ed., *Colonial New Guinea: Anthropological Perspectives*, Pittsburgh: University of Pittsburgh Press, 2001, 171–93.

35. Scott and McMurry, "The Delicate Question," at 227. Others offer slightly different typologies; see, e.g., Tim D. White, *Prehistoric Cannibalism at Mancos—5MTUMR 2346*, Princeton: Princeton University Press, 1992 (survival, funerary, gastronomic); Lynn Flinn, Christy G. Turner II, and Alan Brew, "Additional Evidence for Cannibalism in the Southwest: the Case of LA-4528," *American Antiquity* 41 (1976): 308–18 (famine, dietetic, magical, pietistic); and Turner and Turner, *Man Corn* (emergency, sociobiology, social contract, ritual sacrifice and social pathology).

36. Scott and McMurray, "The Delicate Question," at 216.

37. Stanley Garn, "The Noneconomic Nature of Eating People," *American Anthropologist* 81 (1979): 902–03; Stanley Garn and W.D. Block, "The Limited Nutritional Value of Cannibalism," *American Anthropologist* 70 (1970): 106.

38. James Cole, "Assessing the Calorific Significance of Episodes of Human Cannibalism in the Paleolithic," *Scientific Reports* 7, 44707 (2017), available at https://doi.org/10.1038/srep44707.

39. *Id.* Cole estimates that a human thigh averages 13,350 calories, a calf 4,490 calories, and an upper arm 7,450 calories. The most nutritious internal organs are the liver (2,750 calories) and the heart (650). Hard-working crew members pulling sledges might have required at least 4,000 calories a day, though a far lesser number of calories would have at least fended off starvation, if not allowing much work to be done.

40. Charles Dickens, *Great Expectations*, London: Chapman & Hall, 1861, at 3.

41. Quoted in Brantlinger, *Taming Cannibalism*, at 68.

42. See, e.g., Christine Kinealy, *The Great Calamity*, Dublin: Gill & Macmillan, 1994, and James S. Donnelly, Jr., *The Great Irish Potato Famine*, Stroud: Sutton, 2001.

43. Ronan McGreevey, "The Role of 'Survivor Cannibalism' During the Great Famine," *The Irish Times*, November 30, 2020, available at https://www.irishtimes.com/news/ireland/irish-news/role-of-survivor-cannibalism-during-great-famine-detailed-in-new-tv-documentary-1.4423323.

44. *Id.* at 70–71. For a more complete picture of the Pakeha Maori, see Trevor Bentley, *Pakeha Maori: The Extraordinary Story of the Europeans Who Lived as Maori in Early New Zealand*, London: Penguin, 1999.

45. Pitirim Sorokin, *Man and Society in Calamity: The Effects of War, Revolution, Famine, Pestilence Upon the Human Mind, Behavior, Social Organization and Cultural Life*, New York: E.P. Dutton, 1942; and Sorokin, *Hunger as a Factor in Human Affairs*, Gainesville: University Presses of Florida, 1975.

46. Scott and McMurray, "The Delicate Question," at 236.

47. Jasper Becker, *Hungry Ghosts: Mao's Secret Famine*, New York: Free Press, 1996.

48. Alexis Peri, *The War Within: Diaries from the Siege of Leningrad*, Cambridge: Harvard University Press, 2017.

49. *Id.*

50. *Id.*; see also Mark P. Donnelly and Daniel Diehl, *Eat Thy Neighbor: A History of Cannibalism*, Phoenix: Mill Sutton, 2006.

51. Yuki Tanaka, *Hidden Horrors: Japanese War Crimes in World War II*, 2nd ed., Lanham: Rowman & Littlefield, 2017.

52. See, e.g., Jisheng Yang, *Tombstone: The Great Chinese Famine, 1958–1962*, New York: Farrar, Strauss & Giroux, 2012, and, more broadly,

Key Rae Chong, *Cannibalism in China,* Wakefield: Longwood Academic, 1990.

53. Lewis Petrinovich, *The Cannibal Within,* New York: Aldine de Gruyter, 2006.

54. *Id.,* see also Edward E. Leslie, *Desperate Journeys, Abandoned Souls: True Stories of Castaways and Other Survivors,* Boston: Houghton Mifflin, 1988, and A.W. Brian Simpson, *Cannibalism and the Common Law: The Story of the Tragic Last Voyage of the* Mignonette *and the Strange Legal Proceedings to Which It Gave Rise,* Chicago: University of Chicago Press, 1984.

55. 14 QBD 273 DC (1884). For a full account of the case and the circumstances that led to it, see Simpson, *Cannibalism and the Common Law,* and M.G. Mallin, "In Warm Blood: Some Historical and Procedural Aspects of *Regina v. Dudley and Stephens,*" *University of Chicago Law Review* 34 (1967): 387–407.

56. See, e.g., the early 17th-century case of sailors from St. Christopher's Island in the Caribbean, mentioned in Simpson, *Cannibalism and the Common Law,* and the 1875 case of James Archer in Singapore, described in A.W.B. Simpson, "Regina v. Archer and Muller (1875): The Leading Case That Never Was," *Oxford Journal of Legal Studies* 2 (1982): 181–96.

57. Films: Iradj Azimi, dir., *Le Radeau de la Méduse* (1994); Peter Webber, dir., *The Medusa* (2018). Novels: Peter Weiss, *The Aesthetics of Resistance,* vol. 2, Durham: Duke University Press, 2020 (originally published in German 1978); Julian Barnes, *A History of the World in 10½ Chapters,* London: Jonathan Cape, 1989; and Alessandro Baricco, *Ocean Sea,* Rome: Rizzoli Editore, 1999 (originally published in Italian 1993).

58. Scott and McMurry, "The Delicate Question," at 229–30.

59. Leonard F. Guttridge, *Ghosts of Cape Sabine: The Harrowing True Story of the Greeley Expedition,* New York: G.P. Putnam's Sons, 2000; and Buddy Levy, *Labyrinth of Ice: The Triumphant and Tragic Greeley Polar Expedition,* New York: St. Martin's Press, 2019.

60. Scott and McMurry, "The Delicate Question," at 234–235; See also Paul H. Gantt, *The Case of Alfred Packer: The Man-Eater,* Denver: University of Colorado Press, 1952, and Ervan F. Kushner, *Alfred Packer, Cannibal! Victim?* Frederick: Platte 'N Press, 1980. Packer is also immortalized in song, Phil Ochs, "The Ballad of Alferd [sic] Packer," 1964, lyrics available at http://web.cecs.pdx.edu/~trent/ochs/lyrics/ballad-alferd-packer.html, and in the names of cafeterias at the University of Colorado in Boulder and at the Philadelphia Folk Festival (the "Alfred E. Packer Memorial Dining Hall ... serving humanity since 1874").

61. Piers Paul Read, *Alive: The Story of the Andes Survivors,* Philadelphia: J.B. Lippincott, 1974.

62. *Id.* at 238.

63. Schutt, *Cannibalism: A Perfectly Natural History.*

64. The bones were originally discovered by amateur explorer Barry Ranford. See Barry Ranford, "Bones of Contention," *Equinox* 74 (1994): 69–87.

65. Keenleyside, et al., "The Final Days of the Franklin Expedition," at 40.

66. Christy G. Turner II and Jacqueline A. Turner, "Cannibalism in the Prehistoric American Southwest: Occurrence, Taphonomy, Explanation and Suggestions for Standardized World Definition," *Anthropological Science* 103 (1995): 1–22 and Keenleyside, et al., "The Final Days of the Franklin Expedition," at 41–42.

67. Keenleyside, et al., "The Final Days of the Franklin Expedition," at 42.

68. Simon Mays and Owen Beattie, "Evidence for End-Stage Cannibalism on Sir John Franklin's Last Expedition to the Arctic, 1845," *International Journal of Osteoarchaeology* 26 (2015): 778–86.

69. Dan Simmons, *The Terror,* Boston: Little, Brown, 2007.

70. "The C, the C, the Open C," *The Terror,* Season 1, Episode 9, AMC, 2018.

Chapter 15

1. Edmund Falconer, "Last of the Crew," in Edmund Falconer, *Bequest of My Boyhood: Poems,* London: Tinsley Brothers, 1863, 40–41.

2. Mordecai Richler, *Solomon Gursky Was Here,* New York: Knopf, 1990.

3. William Vollmann, *The Rifles,* New York: Viking Books, 1994.

4. See David C. Woodman, *Unravelling the Franklin Mystery: Inuit Testimony,* 2nd ed., Montreal: McGill-Queen's University Press, 2015, at 306–07.

5. Woodman: *Unravelling the Franklin Mystery: Inuit Testimony,* at 227 and 257–58.

6. Janice Cavell, "Publishing Sir John Franklin's Fate: Cannibalism, Journalism and the 1881 Edition of Leopold McClintock's *The Voyage of the "Fox" in the Arctic Seas,*" *Book History* 16 (2013), 155–84, at 165.

7. *Id.* at 170.

8. Farley Mowat, *Ordeal by Ice,* Boston: Atlantic/Little, Brown, 1960, 323.

9. See generally David C. Woodman, *Strangers Among Us,* Montreal: McGill-Queen's University Press, 1995.

10. M. John Roobol, "Status of the History of the Lost Franklin Expedition of 1845," *Academia Letters* 1461, July 2021, available at https://www.academia.edu/49826555/Status_of_the_History_of_the_Lost_Franklin_Expedition_of_1845.

11. Woodman, *Unravelling the Franklin Mystery,* and Woodman, *Strangers Among Us,*. John Roobol has expanded on his views in two 2019 books, M. John Roobol, *Franklin's Fate: An*

Investigation into What Happened to the Lost 1845 Expedition of Sir John Franklin, Canterbury: The Conrad Press, 2019, and the fictionalized account, M. John Roobol, *Trapped,* Canterbury: The Conrad Press, 2019.

12. With one exception. Erasmus Ommanney led a sledging journey in the spring of 1851 that headed southwest from Cape Walker, following the line of Franklin's Admiralty orders, though without finding any trace of the Expedition. William Barr, "Searching for Franklin Where He Was Ordered to Go: Erasmus Ommanney's Sledging Campaign to Cape Walker and Beyond, Spring, 1851," *Polar Record* 52 (2016): 474–98.

13. Woodman, *Strangers Among Us.*

14. *Id.* at 131.

15. *Id.* at 135.

16. "It seems a pity, but I do not think I can write more. R. Scott . . . for God's sake look after our people." Robin McKie, "Tired, Frozen, Beaten: Image of Captain Scott's Expedition that Foretold a Tragedy," *The Observer,* November 11, 2017, available online at https://www.theguardian.com/uk-news/2017/nov/12/-captain-scott-antarctic-image-foretold-national-tragedy#:~:text=His%20final%20words%20were%20written,sake%20look%20after%20our%20people.%E2%80%9D.

17. H.M. Ami, "Notes on the Adelaide Peninsula Skull from the Canadian Arctic Regions," *Transactions of the Royal Society of Canada,* Sec. II (1928): 318–23.

18. Alexa Price, "'Our Proudest Heritage': Masculinity, Nostalgia, and the Sailing Navy on Display, 1820–1920," Ph.D. Dissertation, The George Washington University, 2019, at 111.

Chapter 16

1. Edward A. Inglefield, *A Summer Search for Sir John Franklin, with a Peep into the Polar Basin,* London: Thomas Harrison, 1853, at 186.

2. The two stories are described in detail in Tom Gross & Russell S. Taichman, "A Comparative Analysis of the Su-pung-er and Bayne Testimonies Related to the Franklin Expedition," *Polar Record* 53 (2017): 561–79.

3. For one story of a search based on the Bayne narrative, see Richard Finnie, *Lure of the North,* Philadelphia: David McKay Co., 1940, at 51–66.

4. Richard J. Cyriax, "The Unsolved Problem of the Franklin Expedition Records Supposedly Buried on King William Island," *Mariner's Mirror* 55 (1969): 23–32.

5. Richard J. Cyriax, "Sir John Franklin: A Note on the Absence of Records on the Shores Which He Sailed Past During His Last Voyage," *Scottish Geographical Magazine* 75 (1959): 30–40, at 30.

6. For a more detailed description of the 1967 Canadian Army expedition, see Paul Watson, *Ice Ghosts: The Epic Hunt for the Lost Franklin Expedition,* New York, W.W. Norton, 2017, at 231–47.

7. *Id.* at 246.

8. May Fluhmann, *Second in Command: The Life of Francis R. M. Crozier,* Yellowknife: Department of Information, Government of the Northwest Territories, 1976, at 113.

9. Jonathan Pryor, "Interment Without Earth: A Study of Sea Burials During the Age of Sail," *NewEnglandBurialsatSea.com*, November 26, 2010, available at https://www.newenglandburialsatsea.com/interment-without-earth-a-study-of-sea-burials-during-the-age-of-sail/.

10. Herman Melville, *White-Jacket, or, The World From a Man O' War,* New York: Harper & Brothers, 1850.

11. See David C. Woodman, *Unravelling the Franklin Mystery: Inuit Testimony,* 2nd ed., Montreal: McGill-Queen's University Press, 2015, at 29.

12. This last possibility was suggested by noted Franklin researcher Ralph Lloyd-Jones in a post on the Facebook group, "Remembering the Franklin Expedition," March 8, 2020. For details of the preservation of Admiral Nelson's body, see William Beatty, *The Death of Lord Nelson,* 2nd ed., Westminster: Constable, 1895.

13. David J. Stewart, "'Rocks and Storms I'll Fear No More': Anglo-American Maritime Memorialization, 1700–1940," Ph.D. dissertation, Texas A&M University, 2004.

14. Beatty, *The Death of Lord Nelson,* at 61.

15. Nelson, as it happens, insisted on being buried at St. Paul's Cathedral, well up on a hill overlooking London, rather than at Westminster, since he believed that Westminster would one day revert to the swamp from which it came. *Id.* at 70.

16. Pryor, "Interment Without Earth."

17. Russell Potter, Regina Koellner, Peter Carney and Mary Williamson, eds., *May We Be Spared to Meet on Earth: Letters of the Lost Franklin Arctic Expedition,* Montreal: McGill-Queen's University Press, 2022, at 89.

18. Matt Bendoris, "Scots Grave Could Hold Clues to Franklin's Doomed Arctic Expedition," *Scottish Sun,* March 23, 2021, available at https://www.thescottishsun.co.uk/tv/6864064/scots-grave-could-hold-clue-to-franklins-doomed-expedition/.

19. Russell Potter, "The Grave of Lieutenant Irving?" *Visions of the North blog,* January 3, 2010, available at https://visionsnorth.blogspot.com/search?q=John+Irving+Grave.

Chapter 17

1. Ken Coates, "Very Magnetic North," *Literary Review of Canada,* April 2017, available at http://reviewcanada.ca/magazine/2017/04/very-magnetic-north/.

2. Michael Durey, "Exploration at the Edge: Reassessing the Fate of Sir John Franklin's Last Arctic Expedition," *The Great Circle (Australian Association for Maritime History)* 30 (2008): 3–40, at 4.

3. Richard C. Davis, ed., *Sir John Franklin's Journals and Correspondence: The First Arctic Land Expedition, 1819–1822*, Toronto: The Champlain Society, 1995, at xciii.

4. Andrew Lambert, *The Gates of Hell: John Franklin's Tragic Quest for the Northwest Passage*, New Haven: Yale University Press, 2011, at 22.

5. See, e.g., H.D. Traill, *Life f Sir John Franklin, RN*, London: John Murray, 1896.

6. *Id.* at 5–6.

7. See, e.g., Owen Alexander Vidal, *A Poem Upon the Life and Character of Sir John Franklin*, Oxford: T. & G. Shrimpton, 1860; Joseph Addison Turner, *A Poem Upon the Life and Character of Sir John Franklin*, Mobile, 1858. The same hero-worship is evident in the early biographies of Franklin, such as H.D. Traill, *Life of Sir John Franklin, RN*, London: John Murray, 1896.

8. See, e.g., Pierre Berton, *The Arctic Grail*, Toronto: McClelland and Stewart, 1988 and Richard C. Davis, "'Once Bitten, Twice Shy': Cultural Arrogance and the Final Franklin Expedition," *Polar Geography* 26 (2002): 21–38.

9. Ken McGoogan, *Fatal Passage: The Untold Story of Scotsman John Rae, the Arctic Adventurer Who Discovered the Fate of Franklin*, New York: Basic Books, 2002, at 221.

10. Martyn Beardsley, *Deadly Winter: The Life of Sir John Franklin*, Annapolis: Naval Institute Press, 2002, at 236.

11. See, e.g., Fred Bodsworth, "Franklin's Folly," *Maclean's* 64, no. 6 (March 15, 1951): 19–20, 47–50, portraying Franklin as a useless aristocrat—something he decidedly was not—and concluding that "white gloves, swords and polished buttons were only shortcuts to death in the Arctic."

12. Pierre Berton, *The Arctic Grail*, Toronto: McClelland & Stewart, 1988.

13. Richard King, *The Franklin Expedition, From First to Last*, London: T. Brettell, 1855, at 179; reprint, London: Forgotten Books, 2017, cited in Michael Durey, "Exploration at the Edge: Reassessing the Fate of Sir John Franklin's Last Arctic Expedition," *The Great Circle: Journal of the Australian Association for Maritime History* 30 (2008): 3–40 and 186–95, at 3.

14. On the use of Arthurian and chivalric imagery in Victorian literature, see, e.g., Stephanie Barczewski, *Myth and National Identity in Nineteenth-Century Britain: The Legends of King Arthur and Robin Hood*, New York: Oxford University Press, 2000 and Mark Girouard, *Return to Camelot: Chivalry and the English Gentleman*, New Haven: Yale University Press, 1981.

15. Scott's quest, and the attendant English tradition of the noble but doomed hero, are explored in detail in Francis Spufford, *I May Be Some Time: Ice and the English Imagination*, New York: St. Martin's Press, 1997.

16. May Fluhmann, *Second in Command: The Life of Francis R. M. Crozier*, Yellowknife: Department of Information, Government of the Northwest Territories, 1976, at 91. Available at https://drive.google.com/file/d/-1sW0Fn-p3F4w-xmfDCL1--wj_dERUsM9E/view?fbclid=IwAR1AOlOlq791X3xm23yGiDK4hXyrSHI92PQTgeOIyEP_tNV1gVdqFS5w6_E2-18. The letter as printed in a recent collection merely says: "I cannot bear going on board *Erebus*—Sir John is very kind and would have me there dining every day if I would go." Russell Potter, Regina Koellner, Peter Carney and Mary Williamson, eds., *May We Be Spared to Meet on Earth: Letters of the Lost Franklin Arctic Expedition*, Montreal: McGill-Queen's University Press, 2022, at 297.

17. Sherrill Grace, "Reconfiguring North: Canadian Identity in the 21st Century," in Marc Maufort and Franca Bellarsi, eds., *Reconfigurations: Canadian Literatures and Post-Colonial Identities*, Brussels: P.I.E./Pieter Lang, 2002: 215–30.

18. Janice Cavell, "Going Native in the North: Reconsidering British Attitudes During the Franklin Search, 1848–1859," *Polar Record* 45 (2009): 25–35.

19. *Id.*

20. See Richard C. Davis, "'Which an Affectionate Heart Would Say,' John Franklin's Personal Correspondence, 1819–1844," *Polar Record* 31 (1997): 189–212.

21. See Janice Cavell, "Lady Lucy Barry and Evangelical Reading on the First Franklin Expedition," *Arctic* 63 (2010): 131–40.

22. *Id.* at 139.

Chapter 18

1. Lisa M. Hodgetts, "The Rediscovery of HMS *Investigator*: Archaeology, Sovereignty and the Colonial Legacy in Canada's Arctic," *Journal of Social Archaeology* 13 (2012): 80–100, at 81.

2. Ian S. MacLaren, "Tracing One Discontinuous Line Through Poetry of the Northwest Passage," *Canadian Poetry* 39 (1996): 7–48.

3. On this point, see David McRobert, *Arctic Sovereignty Initiatives in the Canadian North: A Historical Review, 1700–1980*. Report for the President's Advisory Committee on Northern Studies, York University, 1982 and Rosanna White, "Ceremonies of Possession: Performing Sovereignty in the Canadian Arctic," Ph.D. Dissertation, University of London, 2019.

4. Adriana Craciun, "Franklin's Sobering True Legacy," *Ottawa Citizen*, September 10, 2014, available at htpps://www.Ottawacitizen.com/news/national/Adriana-craciun-franklins-sobering-true-legacy.

5. See Klaus Dodds, "'We Are a Northern

Country': Stephen Harper and the Canadian Arctic." *Polar Record* 47 (2011): 371–82.

6. The most complete review of the international law position regarding the Northwest Passage is Donat Pharand and Leonard H. Legault. *The Northwest Passage: Arctic Straits,* Leyden: Brill/Nijhoff, 1984. See also Donat Pharand, "Canada's Sovereignty over the Northwest Passage." *Michigan Journal of International Law* 10 (1989): 653–78. And Michael Byers and Suzanne Lalonde, "Who Controls the Northwest Passage?" *Vanderbilt Journal of Transnational Law* 42 (2009): 1133–1210.

7. Noted in Richard C. Davis, "Introduction," in Richard C. Davis, ed., *Sir John Franklin's Journals and Correspondence: The First Arctic Land Expedition, 1819–1822,* Toronto, Champlain Society, 1995.

8. *See* Janice Cavell, "Representing Akaitcho: European Vision and Revision in the Writing of Sir John Franklin's Narrative of a Journey to the Shores of the Polar Sea," *Polar Record* 44 (2008): 25–34

9. "'Hallowed Space': Divers Pull 275 Artifacts from 2022 Excavation of Franklin Ship," *Lethbridge News Now*, December 18, 2022, available at https://lethbridgenewsnow.com/2022/12/18/-hallowed-space-divers-pull-275-artifacts-from-2022-excavation-of-franklin-ship/.

10. Facebook group "Remembering the Franklin Expedition," December 18, 2022, available at https://www.facebook.com/groups/11434844549/?hoisted_section_header_type=recently_seen&multi_permalinks=10160173565169550.

11. David Woodman, *Unravelling the Franklin Mystery: Inuit Testimony*, 2nd ed., Montreal and Kingston: McGill-Queen's University Press, 2015 and Woodman, *Strangers Among Us*, Montreal: McGill-Queen's University Press, 1995.

12. Dorothy Eber, *Encounters on the Passage: Inuit Meet the Explorers*. Toronto: University of Toronto Press, 2008.

Bibliography

Books and Articles

Abrahall, Chandos Hoskyns. *Arctic Enterprise: A Poem in Seven Parts*. London: Hope, 1856.

Albanov, Valerian. *In the Land of the White Death*. Translated by William Barr. New York: Modern Library, 2000.

Alexander, Alison. *The Ambitions of Jane Franklin: Victorian Lady Adventurer*. Sydney: Allen & Unwin, 2013.

Alexander, Caroline. *The Bounty: The True Story of the Mutiny on the Bounty*. New York: Viking Penguin, 2003.

Alt, Bea, Roy M. Koerner, David A. Fisher, and J.C. Bourgeois. "Arctic Climate During the Franklin Era, as Deduced from Ice Cores." In Patricia Sutherland, ed., *The Franklin Era in Canadian Arctic History 1845–1859*. Ottawa: National Museums of Canada, Mercury Series, Archaeological Survey of Canada Paper No. 131, 1985, 69–92.

Ami, H.M. "Notes on the Adelaide Peninsula Skull from the Canadian Arctic Regions." *Transactions of the Royal Society of Canada*, Sec. II (1928): 318–23.

Amundsen, Roald. *The Northwest Passage*. London: Archibald Constable & Co., 1908.

Amy, Roger, Rakesh Bhatnagar, Eric Damkjar, and Owen Beattie. "The Last Franklin Expedition: Report of a Post-Mortem Examination of a Crew Member." *Canadian Medical Association Journal* 135 (1986): 115–17.

Anderson, Harry S. *Exploring the Polar Regions*. New York: Facts on File, 2005.

Anderson, James. "Chief Factor James Anderson's Back River Journal of 1855." *Canadian Field Naturalist* 54 (1940): 63–67, 84–89, 125–26, 134–36 and 167–69 and 55 (1941): 9–11 and 20–26.

_____. *The Hudson Bay Expedition in Search of Sir John Franklin*. Toronto: Canadiana House, 1969.

Anderson, John D. "The 'Navigatio Brendoni,' A Medieval Bestseller." *The Classical Journal* 83 (1988): 315–22.

Anselmi, Elaine. "Closing in on Franklin." *UpHere*, January/February 2018. Available at https://uphere.ca/articles/closing-franklin.

Aporta, Claudio. "Life on the Ice: Understanding the Codes of an Environment." *Polar Record* 38 (2002): 341–54.

_____. "The Trail as Home: Inuit and Their Pan-Arctic Network of Routes." *Human Ecology* 37 (2009): 131–46.

Arctic Rewards and Their Claimants. London: T. Hatchard, 1856.

Ardeleanu, Constantine. "A British Meat Cannery in Moldavia (1844–52)." *The Slavonic and East European Review* 90 (2012): 671–704.

Arens, William. *The Man-Eating Myth: Anthropology and Anthropophagy*. New York: Oxford University Press, 1979.

Arkowitz, Hal, and Scott O. Lilienfeld. "Do the Eyes Have It? Why Science Tells us Not to Rely on Eyewitness Accounts." *Scientific American Mind* 20 (2010): 68–69.

Armston-Sheret, Edward. "Tainted Bodies: Scurvy, Bad Food and the Reputation of the British National Antarctic Expedition, 1901–1904." *Journal of Historical Geography* 65 (2019): 19–28.

Ashcroft, Frances. *Life at the Extremes: The Science of Survival*. Berkeley: University of California Press, 2000.

Atwood, Margaret. "The Age of Lead," In Margaret Atwood, *Wilderness Tips*. New York: Random House, 1991, 157–75.

_____. "Concerning Franklin and His Gallant Crew." In Margaret Atwood, ed., *Strange Things: The Malevolent North in Canadian Literature*. Oxford: Clarendon Press, 1995, 7–34.

Back, George. *Narrative of an Expedition in* HMS Terror *Undertaken with a View to Geographical Discovery on the Arctic Shores in the Years 1836–7*. London: John Murray, 1838.

_____. *Narrative of the Arctic Land Expedition to the Mouth of the Great Fish River*. London: John Murray, 1836; reprint, Dartmouth: CD-Academic Book Co., 1999.

Bankes, Nigel. "Her Majesty's Ships *Erebus* and *Terror* and the Intersection of Legal Norms." *The Northern Review* 50 (2020): 47–81.

Banner, Stuart. "Why Terra Nullius? Anthropology and Property Law in Early Australia." *Law and History Review* 23 (2005): 95–131.

Barczewski, Stephanie. *Heroic Failure and the British*. New Haven: Yale University Press, 2016.

_____. *Myth and National Identity in Nineteenth-Century Britain: The Legends of King Arthur and Robin Hood.* New York: Oxford University Press, 2000.

Baricco, Alessandro. *Ocean Sea.* Rome: Rizzoli Editore, 1999 (originally published in Italian 1993).

Barker, Francis, Peter Hulme, and Margaret Iverson, eds. *Cannibalism and the Colonial World.* Cambridge: Cambridge University Press, 1998.

Barnes, Julian. *A History of the World in 10½ Chapters.* London: Jonathan Cape, 1989.

Baron, Jeremy Hugh. "Sailors' Scurvy Before and After James Lind—a Reassessment." *Nutrition Reviews* 67 (2009): 315–32.

Barr, William. *Arctic Hell-Ship: The Voyage of HMS Enterprise.* Calgary: University of Alberta Press, 2007.

_____. "'The Cold of Valparaiso'; The Disintegration of William Kennedy's Second Franklin Search Expedition, 1853–1854." *Polar Record* 34 (1998): 203–18.

_____. "Franklin in Siberia?—Lieutenant Bedford Pim's Proposal to Search the Arctic Coast of Siberia, 1851–52." *Arctic* 45 (1992): 36–46.

_____. *A Frenchman in Search of Franklin: de Bray's Arctic Journal 1852–54.* Toronto: University of Toronto Press, 1992.

_____. "Identification of the Franklin Expedition Wreck." *Polar Record* 51 (2015): 218.

_____. "Klutschak: An Artist in Search of Franklin." *The Beaver,* June-July 1991: 12–25.

_____. "Misinterpretation and Obfuscation." *Polar Record* 51 (2015): 222–23.

_____. "Searching for Franklin from Australia: William Parker Snow's Initiative of 1853." *Polar Record* 33 (1977): 145–50.

_____. "Searching for Franklin Where He Was Ordered to Go: Erasmus Ommanney's Sledging Campaign to Cape Walker and Beyond, Spring 1851." *Polar Record* 52 (2016): 474–98.

_____. "The Use of Dog Sledges During the British Search for the Missing Franklin Expedition in the North American Arctic Islands, 1848–59." *Arctic* 62 (2009): 257–72.

_____. "William Robert Hobson (1831–1880)." *Arctic* 39 (1986): 184–85.

_____, Nadine Forestier Blazart, and Jean-Claude Forestier-Blazart. "'The Last Duty of an Officer': Lieutenant de Vaisseau Joseph-René Bellot, 1826–1853, in the Franklin Search." *Polar Record* 50 (2014): 1–30.

_____, and Glenn M. Stein. "Frederick J. Krabbe: Last Man to See HMS *Investigator* Afloat, May 1854." *Journal of the Hakluyt Society,* January 2017: 1–34. Available at https://www.hakluyt.com/downloadable_files/Journal/krabbe.pdf.

_____, ed. *Searching for Franklin: The Land Arctic Searching Expedition 1855: James Anderson's and James Stewart's Expedition via the Back River.* London: Hakluyt Society, 1999.

Barrett, Andrea. *Archangel.* New York: W.W. Norton, 2013.

_____. *Voyage of the Narwhal.* New York: W.W. Norton, 1998.

Barrie, Ian. "Alexander M'Donald L.R.C.S.E. (1817-c. 1848)." *Arctic* 62 (2009): 239–40.

Barrow, John. *An Auto-Biographical Memoir of Sir John Barrow, Bart.* London: John Murray, 1847.

_____. "Burney-Behring's Strait and the Polar Basin." *Quarterly Review* 18 (1818): 431–59.

_____. *A Chronological History of Voyages into the Arctic Regions.* London: John Murray, 1818; reprint, Newton Abbot: David & Charles, 1971.

_____. "Communications on a North-West Passage, and Further Survey of the Northern Coast of America." *Journal of the Royal Geographical Society of London* 6 (1836): 34–50.

_____. *The Eventful History of the Mutiny and Piratical Seizure of HMS 'Bounty' Its Causes and Consequences.* London: John Murray, 1831.

_____. "Lord Selkirk and the North-West Company/Lord Selkirk and the North-West Passage." *Quarterly Review* 16 (1816): 129–72.

_____. "On the Polar Ice and Northern Passage into the Pacific." *Quarterly Review* 18 (1817): 199–223.

_____. *Voyages of Discovery and Research Within the Arctic Regions.* London: John Murray, 1846.

Bartholomew, Michael. "James Lind and Scurvy: A Revaluation." *Journal for Maritime Research* 4 (2002): 1–14.

Bartlett, Bob. *The Karluk's Last Voyage: An Epic of Death and Survival in the Arctic, 1913–1916.* New York: Cooper Square Press, 2001.

Battersby, William. "Identification of the Probable Source of the Lead Poisoning Observed in Members of the Franklin Expedition." *Journal of the Hakluyt Society* (2008): 1–6.

_____. *James Fitzjames: The Mystery Man of the Franklin Expedition.* Stroud: History Press, 2010.

_____. "The Remarkable Background of Lt. Graham Gore." *Hidden Tracks* blog, November 29, 2009. Available at https://hidden-tracks-book.blogspot.com/2009/11/remarkable-background-of-lt-graham-gore.html.

_____, and Peter Carney. "Equipping HM Ships *Erebus* and *Terror,* 1845." *International Journal for the History of Engineering & Technology* 81 (2011): 192–211.

Bayliss, Richard. "Sir John Franklin's Last Arctic Expedition: A Medical Disaster." *Journal of the Royal Society of Medicine* 95 (2002): 151–53.

Beardsley, Martyn. *Deadly Winter: The Life of Sir John Franklin.* Annapolis: Naval Institute Press, 2002.

_____. *Sir John Franklin: The Man Who Ate His Boots.* London: Short Books, 2005.

Beattie, Owen. "Elevated Bone Lead Levels in a Crewman from the Last Arctic Expedition of Sir John Franklin." In Patricia D. Sutherland, ed., *The Franklin Era in Canadian Arctic History, 1845–1859.* Ottawa: National Museums of Canada, Mercury Series, Archaeological Survey of Canada Paper No. 131, 1985, 131.

_____. "A Report on Newly Discovered Skeletal Remains from Sir John Franklin's Last Expedition." *The Musk-Ox* 33 (1983): 68–77.

_____, Eric Damkjar, Walter Kowal, and Roger Amy. "Anatomy of an Arctic Autopsy." *Medical Post* 20 (1985): 1–2.

_____, and John Geiger. *Dead Silence: The Greatest Mystery in Arctic Exploration*. London: Bloomsbury, 1993.

_____, and John Geiger. *Frozen in Time: The Fate of the Franklin Expedition*. Vancouver: Douglas & McIntyre, 1987.

_____, Walter Kowal, and Halfdan Baadsgard. "Did Solder Kill Franklin's Men?" *Nature* 343 (1990): 319–20.

_____, and James M. Savelle. "Discovery of Human Remains from Sir John Franklin's Last Expedition." *Historical Archaeology* 17 (1983): 100–105.

Beatty, Rachel. *The Jill Tars: Seven Remarkable Accounts of Female Sailors who Served and Fought as Men*. London: Leonaur, 2015.

Beatty, William. *The Death of Lord Nelson*, 2nd ed. Westminster: Constable, 1895.

Becker, Jasper. *Hungry Ghosts: Mao's Secret Famine*. New York: Free Press, 1996.

Beechey, Frederick W. *A Voyage of Discovery Towards the North Pole, Performed in His Majesty's Ships* Dorothea *and* Trent *Under the Command of Captain David Buchan, R.N., 1818*. London: Richard Bentley, 1843.

Beesly, Augustus Henry. *Sir John Franklin*. Dublin: Marcus Ward and Co., 1881.

Belcher, Edward. *The Last of the Arctic Voyages*. London: Lovell Reeve, 1855.

Bell, Benjamin. *Lieut. John Irving, R.N.: A Memorial Sketch with Letters in Sir John Franklin's Last Expedition to the Arctic Regions*. Edinburgh: David Douglas, 1881, reprint, Cambridge: Cambridge University Press, 2014.

Bell, Bill. "Authors in an Industrial Economy: The Case of John Murray's Travel Writers." *Romantic Textualities* 21 (2013): 2–9.

Bender, A.E. "The History and Implications of Processed Food." In J. Watt, E.J. Freeman and W.F. Bynum, eds., *Starving Sailors: The Influence of Nutrition upon Naval and Maritime History*. Greenwich: National Maritime Museum, 1981, 117–22.

Bendoris, Matt. "Scots Grave Could Hold Clues to Franklin's Doomed Arctic Expedition." *Scottish Sun*, March 23, 2021. Available at https://www.thescottishsun.co.uk/tv/6864064/scots-grave-could-hold-clue-to-franklins-doomed-expedition/16-15.

Bennett, Mia M., Wilfred Greaves, Rudolf Riedlsperger and Alberic Botella. "Articulating the Arctic: Contrasting State and Inuit Maps of the Canadian North." *Polar Record* 52 (2016): 630–44.

Bentley, Trevor. *Pakeha Maori: The Extraordinary Story of the Europeans Who Lived as Maori in Early New Zealand*. London: Penguin, 1999.

Berton, Pierre. *The Arctic Grail*. Toronto: McClelland & Stewart, 1988.

Betts, Matthew. *HMS Terror: The Design, Fitting and Voyages of the Polar Discovery Ship*. Barnsley: Seaforth, 2022.

Blackmore, Richard Doddridge. *The Fate of Franklin*. London: R. Hardwicke, 1860.

Boas, Franz. *The Central Eskimo*. Washington, D.C.: Bureau of Ethnology, 1888.

Bockstoce, John R. *Furs and Frontiers in the Far North*. New Haven: Yale University Press, 2009.

_____. "The Search for Sir John Franklin in Alaska." In Patricia Sutherland, ed., *The Franklin Era in Canadian Arctic History 1845–1859*. Ottawa: National Museums of Canada, Mercury Series, Archaeological Survey of Canada Paper No. 131, 1985, 93–113.

_____, ed. *The Journal of Rochfort Maguire, 1852–1854: Two Years at Point Barrow, Alaska, Aboard* HMS Plover *in the Search for Sir John Franklin*. London: Hakluyt Society, 1988.

Bodsworth, Fred. "Franklin's Folly." *Maclean's* 64, no. 6 (March 15, 1951): 19–20, 47–50.

Boucher, Ellen. "Arctic Mysteries and Imperial Ambitions: The Hunt for Sir John Franklin and the Victorian Culture of Survival." *The Journal of Modern History* 90 (2018): 40–75.

Bowden, Ross. "Maori Cannibalism: An Interpretation." *Oceania* 55 (1984): 81–99.

Bown, Stephen R. *Scurvy*. New York: St. Martin's Press, 2004.

Brandt, Anthony. *The Man Who Ate His Boots*. New York: Knopf, 2010.

Brantlinger, Patrick. "Missionaries and Cannibals in Nineteenth-Century Fiji." *History and Anthropology* 17 (2006): 21–38.

_____. *Taming Cannibalism*. Ithaca: Cornell University Press, 2011.

Bravo, Michael T. "Geographies of Exploration and Improvement: William Scoresby and Arctic Whaling, 1782–1822." *Journal of Historical Geography* 32 (2006): 512–538.

_____. "Science and Discovery in the Admiralty Voyages to the Arctic Regions in Search of a North-West Passage (1815–1825)." Ph.D. dissertation, Cambridge University, 1992.

Brazzelli, Nicoletta. "In Search of John Franklin: The 'Arctic Adventures' of Lady Franklin." In Nicholas Brownlees, ed., *The Language of Discovery, Exploration and Settlement*. Newcastle upon Tyne: Cambridge Scholars, 2020, 145–59.

Brodie, Erasmus Henry. *Euthanasia: A Poem in Four Cantos of Spenserian Metre on the Discovery of the North-West Passage by Sir John Franklin*. London: Longmans & Co., 1866.

Brown, Paula, and Donald Tuzin, eds. *The Ethnography of Cannibalism*. Washington, D.C.: Society for Psychological Anthropology, 1983.

Browne, William H. "Report of the Proceedings of the Sledge 'Enterprise,' in Charge of Lieut.

W.H. Browne, Her Majesty's Ship 'Resolute,' Commencing 15th April, Ending 28th May 1851." In *Additional Papers Relative to the Arctic Expedition under the Orders of Captain Austin and Mister William Penny*. London: G.E. Eyre and W. Spottiswode, 1851.

Bullocke, John G. *Sailors' Rebellion: A Century of Naval Mutinies*. London: Eyre and Spottiswode, 1938.

Bunyan, Ian, Jenni Calder, Dale Idiens, and Bryce Wilson. *No Ordinary Journey: John Rae, Arctic Explorer, 1813–1893*. Edinburgh: National Museums of Scotland, and Montreal: McGill-Queen's University Press, 1993.

Burns, Fiona Hamilton. "H.M.S. *Herald* in Search of Franklin." *The Beaver* 294 (Autumn 1963): 3–13.

Burwash, Lachlin T. *Canada's Western Arctic*. Ottawa: King's Printer, 1931.

_____. "The Franklin Search." *Canadian Geographical Journal* 1, no. 7 (November 1930).

Byers, Michael, and Suzanne Lalonde. "Who Controls the Northwest Passage?" *Vanderbilt Journal of Transnational Law* 42 (2009): 1133–1210.

Bygott, J.D. "Cannibalism Among Wild Chimpanzees." *Nature* 238 (1976): 410–11.

Calef, George. *Caribou and the Barren Lands*. Ottawa: Canadian Arctic Resources Committee, 1981, reprint, Willowdale: Firefly Books, 1995.

Campbell, I.T. "Energy Balance Under Polar Conditions." *Human Nutrition: Applied Nutrition* 36A (1982): 165–78.

Campbell, Richard J. "Crozier's Date of Birth." *Polar Record* 45 (2009): 83–84.

_____. "The Journal of Sergeant William K. Cunningham, R.M. of HMS Terror." *Journal of the Hakluyt Society* 2 (2009): 82.

Carpenter, Kenneth J. *The History of Scurvy and Vitamin C*. Cambridge: Cambridge University Press, 1986.

Carney, Peter. "Further Light on the Source of Lead in Human Remains from the 1845 Franklin Expedition." *Journal of the Hakluyt Society* (September 2016): 3–17.

_____. "J'Accuse!—The Case of Stephan Goldner—Britain's Dreyfus." *Erebus & Terror Files* blog, April 30, 2020. Available at https://erebusandterrorfiles.blogspot.com/2020/04/-jaccuse-case-of-stephan-goldner.html.

_____. "'Return to Poctes Bay.'" *Erebus & Terror Files* blog, August 1, 2014. Available at https://erebusandterrorfiles.blogspot.com/2014/08/-return-to-poctes-bay.html.

Cato, Nancy. *North-West by South*. London: Heinemann, 1965.

Cavell, Janice. "Arctic Exploration in Canadian Print Culture, 1890–1930." *Papers of the Bibliographic Society of Canada* 44 (2006): 7–43.

_____. "Comparing Mythologizers: Twentieth-Century Canadian Constructions of Sir John Franklin." In Norman Hillmer and Adam Chapnick, eds., *Canadas of the Mind: The Making and Unmaking of Canadian Nationalisms in the Twentieth Century*. Montreal: McGill-Queen's University Press, 2007, 15–45.

_____. "Going Native in the North: Reconsidering British Attitudes During the Franklin Search, 1848–1859." *Polar Record* 45 (2009): 25–35.

_____. "Lady Lucy Barry and Evangelical Reading on the First Franklin Expedition." *Arctic* 63 (2010): 131–40.

_____. "Making Books for Mr. Murray: The Case of Edward Parry's Third Arctic Narrative." *The Library* 14 (2013): 45–69.

_____. "Miss Porden, Mrs. Franklin, and the Arctic Expeditions: Eleanor Anne Porden and the Construction of Arctic Heroism (1818–25)." In Frédéric Regard, ed., *Arctic Exploration in the Nineteenth Century: Discovering the Northwest Passage*. London: Pickering & Chatto, 2013, 79–94.

_____. "Publishing Sir John Franklin's Fate: Cannibalism, Journalism and the 1881 Edition of Leopold McClintock's *The Voyage of the "Fox" in the Arctic Seas*." *Book History* 16 (2013): 155–84.

_____. "Representing Akaitcho: European Vision and Revision in the Writing of Sir John Franklin's Narrative of a Journey to the Shores of the Polar Sea." *Polar Record* 44 (2008): 25–34.

_____. *Tracing the Connected Narrative: Arctic Explorations in British Print Culture, 1818–1860*. Toronto: University of Toronto Press, 2008.

_____. "Who Discovered the Northwest Passage?" *Arctic* 71 (2018): 292–308.

Cavell, Samantha. "A Social History of Midshipmen and Quarterdeck Boys in the Royal Navy, 1761–1831." Ph.D. Dissertation, University of Exeter, 2010.

Cherry-Garrard, Apsley. *The Worst Journey in the World, Antarctic, 1910–1913*. London: Constable & Co., 1922.

Chong, Key Rae. *Cannibalism in China*. Wakefield: Longwood Academic, 1990.

Christensen, Jenny R., Joyce M. McBeth, Nicole J. Sylvain, Jody Spence, and Hing Man Chan. "Hartnell's Time Machine: 170-Year-Old Nails Reveal Severe Zinc Deficiency Played a Greater Role Than Lead in the Demise of the Franklin Expedition." *Journal of Archaeological Science: Reports* 16 (2017): 430–44.

Coates, Ken. "Very Magnetic North" (review of Paul Watson, *Ice Ghosts*). *Literary Review of Canada* (April 2017). Available at http://reviewcanada.ca/magazine/2017/04/very-magnetic-north/.

Cole, James. "Assessing the Calorific Significance of Episodes of Human Cannibalism in the Paleolithic." *Scientific Reports* 7, 44707 (2017). Available at https://doi.org/10.1038/srep44707.

Coleman, Ernest C. *No Earthly Pole: The Truth About the Franklin Expedition 1845*. Stroud: Amberley, 2020.

Collinson, Richard. *The Journal of HMS*

Enterprise on the Expedition in Search of Sir John Franklin's Ships by Behring Strait, 1850–1855. London: Sampson Low, Marston, Searle & Rivington, 1889.

Conklin, Beth A. *Consuming Grief: Cannibalism in an Amazonian Society*. Austin: University of Texas Press, 2001.

Connell, Frank H. "Trichinosis in the Arctic." *Arctic* 2 (1949): 98–107.

Conrad, Joseph. "Geography and Some Explorers." *National Geographic*, March 1924, reprinted in Richard Curle, ed., *Last Essays*. London: J.M. Dent & Sons, 1926, 10–17.

Cook, Frederick A. *Through the First Antarctic Night, 1899–1900: A Narrative of the Voyage of the Belgica Through Newly Discovered Lands and over an Unknown Sea about the South Pole*. Montreal: McGill-Queen's University Press, 1980 (originally published 1900).

Cooke, Alan. "A Bibliographical Introduction to Sir John Franklin's Expeditions and the Franklin Search." In Patricia D. Sutherland, ed., *The Franklin Era in Canadian Arctic History, 1845–1859*. Ottawa: National Museums of Canada, Mercury Series, Archaeological Survey of Canada Paper No. 131, 1985, 12–20.

_____, and Clive Holland. *The Exploration of Northern Canada*. Toronto: Arctic History Press, 1978.

Cookman, Scott. *Ice Blink: The Tragic Fate of Sir John Franklin's Last Polar Expedition*. New York: John Wiley & Sons, 2000.

Cordingly, David. *Women Sailors and Sailors' Women*. New York: Random House, 2001.

Craciun, Adriana. "The Disaster of Franklin: Victorian Exploration in the Twenty-First-Century Arctic." In Heidi Hansson and Anka Ryall, eds., *Arctic Modernities: The Environmental, the Exotic and the Everyday*. Newcastle upon Tyne: Cambridge Scholars, 2017, 191–212.

_____. "The Franklin Mystery." *Literary Review of Canada*. May 2012. Available at http://reviewcanada.ca/magazine/2012/05/the-franklin-mystery/.

_____. "The Franklin Relics in the Arctic Archive." *Victorian Literature and Culture* 42 (2014): 1–30.

_____. "Franklin's Sobering True Legacy." *Ottawa Citizen*, September 10, 2014. Available at htpps://www.Ottawacitizen.com/news/national/Adriana-craciun-franklins-sobering-true-legacy.

_____. "The Scramble for the Arctic." *Interventions: International Journal of Post-Colonial Studies* 11 (2009): 102–14.

_____. "What Is an Explorer?" *Eighteenth-Century Studies* 45 (2011): 29–51.

_____. *Writing Arctic Disaster: Authorship and Exploration*. Cambridge: Cambridge University Press, 2016.

_____. "Writing the Disaster: Franklin and *Frankenstein*." *Nineteenth Century Literature* 65 (2011): 433–80.

Crouse, Nellis Maynard. *In Search of the Western Ocean*. London: J.M. Dent & Sons, 1928.

_____. *Search for the Northwest Passage*. New York: Columbia University Press, 1934.

Cruwys, Liz. "Henry Grinnell and the American Franklin Searches." *Polar Record* 26 (1990): 211–16.

_____. "Profile: Henry Grinnell." *Polar Record* 27 (1991): 115–19.

Cyriax, Richard J. "Adam Beck and the Franklin Search." *Mariner's Mirror* 48 (1962): 35–51.

_____. "Arctic Sledge Travelling by Officers of the Royal Navy, 1819–1849." *Mariner's Mirror* 49 (1963): 127–42.

_____. "Captain Hall and the So-Called Survivors of the Franklin Expedition." *Polar Record* 4 (1944): 170–85.

_____. "A Historic Medicine Chest." *Canadian Medical Association Journal* 57 (1947): 295–300.

_____. "James Clark Ross and the Franklin Expedition." *Polar Record* 43 (1942): 528–40.

_____. "The Position of Victory Point, King William Island." *Polar Record* 6 (1952): 496–507.

_____. "Recently Discovered Traces of the Franklin Expedition." *Geographical Journal* 117 (1951): 211–14.

_____. "Sir James Clark Ross and the Franklin Expedition." *Polar Record* 3 (1942): 170–85.

_____. "Sir John Franklin: A Note on the Absence of Records on the Shores Which He Sailed Past During His Last Voyage." *Scottish Geographical Magazine* 75 (1959): 30–40.

_____. *Sir John Franklin's Last Arctic Expedition: The Franklin Expedition, a Chapter in the History of the Royal Navy*. London: Methuen, 1939, reprint, Sussex: Arctic Press, 1997.

_____. "The Two Franklin Expedition Records Found on King William Island." *Mariner's Mirror* 44 (1958): 178–89.

_____. "The Unsolved Problem of the Franklin Expedition Records Supposedly Buried on King William Island." *Mariner's Mirror* 55 (1969): 23–32.

_____. "The Voyage of H.M.S. *North Star*, 1849–50." *Mariner's Mirror* 50 (1964): 307–18.

_____, and A.G.E. Jones. "The Papers in the Possession of Harry Peglar, Captain of the Foretop, HMS Terror, 1845." *Mariner's Mirror* 40 (1954): 186–95.

Dagneau, Charles. "Interpretative Essay." *The Franklin Mystery: Life and Death in the Arctic* website. Available at https://www.canadianmysteries.ca/sites/franklin/interpretation/experts/interpretationDagneau_en.htm.

Davidson, Peter. *The Idea of North*. London: Reaktion Books, 2005.

Davis, J.E. *A Letter from the Antarctic*. London: William Clowes and Sons, 1901.

Davis, Richard C. "'Once Bitten, Twice Shy': Cultural Arrogance and the Final Franklin Expedition." *Polar Geography* 26 (2002): 21–38.

_____. "'Which an Affectionate Heart Would Say,' John Franklin's Personal Correspondence, 1819–1844." *Polar Record* 31 (1997): 189–212.

_____, ed. *Sir John Franklin's Journals and Correspondence: The First Arctic Land Expedition, 1819–1822.* Toronto: Champlain Society, 1995.

_____, ed. *Sir John Franklin's Journals and Correspondence: The Second Arctic Land Expedition, 1825–1827.* Toronto: Champlain Society, 1998.

Davis, Russell Earls. *Bligh in Australia: A New Appraisal of William Bligh and the Rum Rebellion.* Warriewood: Woodslane Press, 2010.

Davis-Fisch, Heather. *Loss and Cultural Remains in Performance: The Ghosts of the Franklin Expedition.* New York: Palgrave MacMillan, 2012.

Day, Allen Edwin. *Search for the Northwest Passage: An Annotated Bibliography.* New York: Garland, 1986.

Dease, Peter Warren. *From Barrow to Boothia: The Arctic Journal of Chief Factor Peter Warren Dease, 1836–1839.* Edited by William Barr. Montreal: McGill-Queen's University Press, 2002.

Debenham, Frank. "The *Erebus* and *Terror* at Hobart." *Polar Record* 3 (1942): 368–75.

deBray, Émile Frédéric. *A Frenchman in Search of Franklin: DeBray's Arctic Journal, 1852–1854.* Edited and translated by William Barr. Toronto: University of Toronto Press, 1992.

DeGusta, David. "Fijian Cannibalism: Osteological Evidence from Navatu." *American Journal of Physical Anthropology* 110 (1999): 215–41.

Delgado, James P. *Across the Top of the World: The Quest for the Northwest Passage.* New York: Checkmark Books and London: British Museum Press, 1999.

Dennett, J.F. *The Voyages and Travels of Captains Ross, Parry, Franklin and Mr. Belzoni.* London: William Wright, 1835.

De Pauw, Linda Grant. *Seafaring Women.* Boston: Houghton Mifflin, 1982.

Dickens, Charles. "Christmas in the Frozen Regions." *Household Words*, December 21, 1850: 306–09.

_____. *Great Expectations.* London: Chapman & Hall, 1861.

_____. "The Lost Arctic Voyagers." *Household Words* December 2 and 9, 1854: 362–65 and 387–93. Available at victorianweb.org/authors/dickens/arctic/pva342.html.

Dippel, John V.H. *To the Ends of the Earth: The Truth Behind the Glory of Polar Exploration.* New York: Prometheus, 2018.

Dobrée, Bonamy, and G.E. Manwaring. *The Floating Republic: An Account of the Mutinies at Spithead and the Nore, 1797.* London: Pelican, 1937.

Dodds, Klaus. "'We Are a Northern Country': Stephen Harper and the Canadian Arctic." *Polar Record* 47 (2011): 371–82.

Dodge, Ernest. *Northwest by Sea.* New York: Oxford University Press, 1961.

_____. *The Polar Rosses: John and James Clark Ross and Their Explorations.* London: Faber & Faber, 1973.

Donnelly, James S., Jr. *The Great Irish Potato Famine.* Stroud: Sutton, 2001.

_____, and Daniel Diehl. *Eat Thy Neighbor: A History of Cannibalism.* Phoenix: Mill Sutton, 2006.

D'Ortenzio, Lori, Michael Inskip, William Manton, and Simon Mays. "The Franklin Expedition: What Sequential Analysis of Hair Reveals about Lead Exposure Prior to Death." *Journal of Archaeological Sciences: Reports* 21 (2018): 401–05.

Drummond, J.C., and W.R. Lewis. "Historical Introduction." In International Tin Research and Development Council, *Historic Tinned Foods.* Greenford: International Tin Research and Development Council Publication No. 85, 1939: 9–33. Available at https://archive.org/stream/in.ernet.dli.2015.211975/2015.211975.-Historic-Tinned_djvu.txt.

Dugan, James. *The Great Mutiny.* London: Andre Deutsch, 1966.

Dunbar, Moira. "The Effect of Sea Ice Conditions on Maritime Arctic Expeditions During the Franklin Era." In Patricia Sutherland, ed., *The Franklin Era in Canadian Arctic History 1845–1859.* Ottawa: National Museums of Canada, Mercury Series, Archaeological Survey of Canada Paper No. 131, 1985, 114–21.

Durey, Michael. "Exploration at the Edge: Reassessing the Fate of Sir John Franklin's Last Arctic Expedition." *The Great Circle: Journal of the Australian Association for Maritime History* 30 (2008): 3–40 and 186–95.

Durnin, J.G.V.A. "Energy Requirements in Health." In J. Watt, E.J. Freeman and W.F. Bynum, eds., *Starving Sailors: The Influence of Nutrition upon Naval and Maritime History.* Greenwich: National Maritime Museum, 1981: 1–8.

Eber, Dorothy. *Encounters on the Passage: Inuit Meet the Explorers.* Toronto: University of Toronto Press, 2008.

_____. *When the Whalers Were Up North: Inuit Memories from the Eastern Arctic.* Montreal: McGill-Queens University Press, 1989.

Eberhard, Adrienne. *Lady Jane Franklin.* Melbourne: Black Pepper, 2004.

Edge, Arabella. *Fields of Ice.* London: Picador, 2011.

Edinger, Ray. *Fury Beach: The Four-Year Odyssey of Captain John Ross and the Victory.* New York: Berkley, 2003.

Edric, Robert. *The Broken Lands.* London: Jonathan Cape, 1992.

Egerton, Francis. "Barrow on the Arctic Voyages." *Quarterly Review* 78 (1846): 45–46.

Elce, Erika Behrisch. "In Between Memory and Discovery: The Franklin Expedition Story Continues to Grow." *The Whig-Standard*, Kingston, Ontario, October 9, 2018. Available at https://

www.thewhig.com/opinion/columnists/in-between-memory-and-discovery-franklin-expedition-story-continues-to-grow.

_____. *Lady Franklin of Russell Square*. Edmonton: Stonehouse, 2018.

_____. "'One of the Bright Objects that Solace Us in These Regions': Labor, Leisure and the Arctic Shipboard Periodical." *Victorian Periodicals Review* 46 (2013): 343–67.

_____. "Voices of Silence, Texts of Truth: Imperial Discourse and Cultural Negotiations in Nineteenth-Century British Arctic Exploration." Ph.D. Dissertation, Queen's University, Kingston, 2002.

_____, "Widows' Men: The Admiralty Board, Precedent, and Pensions for the Widows of the Lost Franklin Expedition." *Journal of Imperial and Commonwealth History* 47 (2019): 28-50.

_____, ed. *As Affecting the Fate of My Absent Husband: Selected Letters of Lady Franklin Concerning the Search for the Lost Franklin Expedition*. Montreal: McGill-Queen's University Press, 2009.

Falconer, Edmund. "Last of the Crew." In Edmund Falconer, *Bequest of My Boyhood: Poems*. London: Tinsley Brothers, 1864, 38–41.

Farrer, K.T.H. "Goldner's Preserved Meats and the Last Franklin Expedition." *Food Science and Technology Today* 15 (2001): 20.

_____. "Lead and the Last Franklin Expedition." *Journal of Archaeological Science* 20 (1993): 399–409.

Feeney, Robert E. *Polar Journeys: The Role of Food and Nutrition in Early Exploration*. Fairbanks: American Chemical Society and University of Alaska Press, 1998.

Fernandez-Jalvo, Y., J.C. Diez, I. Caceres, and J. Rosell. "Human Cannibalism in the Early Pleistocene of Europe (Gran Dolina, Sierra de Atapuerca, Burgos, Spain)." *Journal of Human Evolution* 37 (1999): 591–622.

Finnie, Richard. *The Lure of the North*. Philadelphia: David McKay Co., 1940.

Fisher, Alexander. *Journal of a Voyage of Discovery to the Arctic in His Majesty's Ships* Hecla *and* Griper *in the Years 1819 and 1820*. London: Longman, Hurst, 1821.

Fitzjames, James. "Journal of James Fitzjames Aboard *Erebus*, 1845." *Nautical Magazine & Naval Chronicle* 21 (1852): 158–65 and 195–201.

Fitzpatrick, Kathleen. *Sir John Franklin in Tasmania, 1837–1843*. Melbourne: Melbourne University Press, 1949.

Flanagan, Richard. *Wanting*. Boston: Atlantic Monthly Press, 2009.

Fleming, Fergus. *Barrow's Boys*. London: Granta Books, 1998.

Flinn, Lynn, Christy G. Turner II, and Alan Brew. "Additional Evidence for Cannibalism in the Southwest: The Case of LA-4528." *American Antiquity* 41 (1976): 308–18.

Flood, Josephine. *The Original Australians: Story of the Aboriginal People*. Longford: Allen & Unwin, 2006.

Florio, Margot. *Run Until Dead: A Brief History of the Polar Expedition Dog*. Mercersburg: Polar Press, 2010.

Fluhmann, May. *Second in Command: the Life of Francis R.M. Crozier*. Yellowknife: Department of Information, Government of the Northwest Territories, 1976. Available at https://drive.google.com/file/d/1sW0Fn-p3F4w-xmfDCL1-wj_dERUsM9E/view?fbclid=IwAR1AOlOl q791X3xm23yGiDK4hXyrSHI92PQTgeOI yEP_tNV1gVdqFS5w6_E.

"Footnotes to the Franklin Search." *The Beaver* 285 (Spring 1955): 46–51.

Fordham, Derek. "Lead Poisoning and the Franklin Expedition." *Polar Record* 27 (1991): 371.

Fordyce, A.D. *Outlines of Naval Routine*. London: John Murray, 1826.

Forst, Janine, and Terence A. Brown. "A Case Study: Was Private William Braine of the 1845 Franklin Expedition a Victim of Tuberculosis?" *Arctic* 70 (2017): 381–88.

Fortier, Dominique. *On the Proper Use of Stars*. Translated by Sheila Fischman. Toronto: McClelland & Stewart, 2010.

Fox, Laurel L. "Cannibalism in Natural Populations." *Annual Review of Ecology and Systematics* 6 (1975): 87–106.

Franklin, Sir John. "Letter to Dr. John Richardson." Sent from Disco Bay, July 7, 1845. *Polar Record* 5 (1949): 348–50.

_____. *Narrative of a Journey to the Shores of the Polar Sea in the Years 1819, 20, 21 and 22*. London: John Murray,1823, reprint, Edmonton: M.G. Hurtig, 1969.

_____. *Narrative of a Second Expedition to the Shores of the Polar Sea*. London: John Murray, 1828, reprint, New York: Greenwood Press, 1969.

_____. *Narrative of Some Passages in the History of Van Diemen's Land*. London: Richard and John E. Taylor, 1845, reprint, Cambridge: Cambridge University Press, 2012.

_____. *Thirty Years in the Arctic Regions: The Narrative of a Polar Explorer*. New York: Simon & Schuster for the Explorers Club, 2017.

"Franklin's Journey to the Shores of the Polar Sea." *British Review* 22 (1824): 25.

Fraser, J.K. "Tracing Ross Across Boothia." *Canadian Geographer* 2 (1957): 40–60.

Frey, Joseph. "Inside the *Terror*—Life on Board Sir John Franklin's Lost Ships." *Geographical Magazine* 27 (November 2019). Available at http://geographical.co.uk/people/explorers/item/3525-inside-terror.

Frykman, Niklas. "Connections between Mutinies in European Navies." *International Review of Social History* 58 (2013): 87–107.

_____. "The Mutiny on the *Hermione*: Warfare, Revolution and Treason in the Royal Navy." *Journal of Social History* 44 (2010): 158–87.

_____. "Seamen on Late Eighteenth-Century

European Warships." *International Review of Social History* 54 (2009): 69–93.

Galaburri, Richard. "The Franklin Records: A Problem for Further Investigation." *The Musk-Ox* 32 (1983): 62–65.

_____. *Lost! The Franklin Expedition and the Fate of the Crews of HMS Erebus and Terror*. New York: Black Raven Books, 2011.

Gantt, Paul H. *The Case of Alfred Packer: The Man-Eater*. Denver: University of Colorado Press, 1952.

Garn, Stanley. "The Noneconomic Nature of Eating People." *American Anthropologist* 81 (1979): 92–93.

_____. "Was the Ill-Fated Franklin Expedition a Victim of Lead Poisoning?" *Nutrition Reviews* 47 (1989): 322–23.

_____, and W.D. Block. "The Limited Nutritional Value of Cannibalism." *American Anthropologist* 70 (1970): 106.

Gaul, Ashleigh. "If Any Living Inuk Knew." *UpHere*, December 2014.

Geiger, John, and Alanna Mitchell. *Franklin's Lost Ship: The Historic Discovery of HMS Erebus*. Toronto: HarperCollins, 2015.

_____, and Owen Beattie. *Frozen in Time: The Fate of the Franklin Expedition*. New York: Dutton, 1988.

Gell, Mrs. Eleanor M. *John Franklin's Bride: Eleanor Anne Porden*. London: John Murray, 1930.

Gibbons, Ann. "Archaeologists Rediscover Cannibalism." *Science* 277 no. 5326 (August 1, 1997): 635–37.

Gibson, William. "The Dease and Simpson Cairn." *The Beaver* 264 (1933): 44–45.

_____. "Sir John Franklin's Last Voyage: A Brief History of the Franklin Expedition and an Outline of the Researches Which Established the Facts of Its Tragic Outcome." *The Beaver* 268 (1937): 44–75.

_____. "Some Further Traces of the Franklin Retreat." *Geographical Journal* 79 (1932): 402–08.

_____. "The *Victory* Relics." *The Beaver* 260 (1929): 311–12.

Gilbert, Arthur N. "Buggery and the British Navy, 1700–1861." *Journal of Social History* 10 (1976): 72–98.

Gilder, William H. *Schwatka's Search: Sledging in the Arctic in Search of the Franklin Records*. New York: Scribner's, 1881.

_____. *The Search for Franklin: A Narrative of the American Expedition Under Lieutenant Schwatka*. London: T. Nelson & Sons, 1893.

Gill, Harold B., and Joanne Young, eds. *Searching for the Franklin Expedition: The Arctic Journal of Robert Randolph Carter*. Annapolis: Naval Institute Press, 1998.

Gillin, Edward J. "The Instruments of Expeditionary Science and the Reworking of Nineteenth Century Magnetic Experiment." *Royal Society Publishing*, March 30, 2022. Available at https://royalsocietypublishing.org/doi/10.1098/rsnr.2022.0002

Girouard, Mark. *Return to Camelot: Chivalry and the English Gentleman*. New Haven: Yale University Press, 1981.

Godfrey, Martin. *Mystery in the Frozen Lands*. Toronto: James Lorimer & Co., 2015.

Goetzmann, William. *The Atlas of North American Exploration*. New York: Prentice-Hall, 1992.

Goldman, Lawrence, ed. *The Anthropology of Cannibalism*. Westport: Bergin & Garvey, 1999.

Gontran de Poncins, Jean-Pierre, and Lewis Galantière, *Kabloona*. New York: Reynal and Hitchcock, 1941.

Goodall, Jane. "Infant-Killing and Cannibalism in Free-Ranging Chimpanzees." *Folia Primatologica* 28 (1977): 259–82.

Goodman, Matthew. "Proving Instruments Credible in the Early Nineteenth Century: The British Magnetic Survey and Site-Specific Experimentation." *Notes and Records of the Royal Society* 70 (2016): 251–68.

Goodsir, Robert A. *An Arctic Voyage to Baffin's Bay and Lancaster Sound in Search of Friends with Sir John Franklin*. London: J. Van Voorst, 1850, reprint, Sussex: Arctic Press, 1996.

_____. "The Explorer: A Fragment from the Story of Franklin's Fate by an Arctic Man of Two Voyages." *Australasian* (Melbourne), December 25, 1880: 7.

Grace, Sherrill E. *Canada and the Idea of North*. Montreal: McGill-Queen's University Press, 2001.

_____. "Gendering Northern Narrative." In John Moss, ed., *Echoing Silence: Essays on the Arctic Narrative*. Ottawa: University of Ottawa Press, 1997, 163–81.

_____. "Reconfiguring North: Canadian Identity in the 21st Century." In Marc Maufort and Franca Bellarsi, eds., *Reconfigurations: Canadian Literatures and Postcolonial Identities*. Brussels and Oxford: P.I.E./Peter Lang, 2002, 215–30.

_____. "Re-Inventing Franklin." *Canadian Review of Comparative Literature*. 22 (1995): 707–25.

_____, Eve d'Aeth, and Lisa Chalykoff, eds. *Staging the North: 12 Canadian Plays*. Toronto: Playwrights Canada Press, 1999.

Grann, David. *The Wager: A Tale of Shipwreck, Mutiny and Murder*. New York: Doubleday, 2023.

Griffiths, Franklin, ed. *The Politics of the Northwest Passage*. Montreal: McGill-Queen's University Press, 1987.

Gross, Tom. "The Peter Bayne Story: The Investigation of Inuit Testimony Relating to Sir John Franklin's Last Arctic Expedition." *The Society for Historical Archaeology Newsletter* 4 (2012): 17–78. Available at https://sha.org/assets/documents/Spring2012.pdf.

_____, and Russell S. Taichman. "A Comparative Analysis of the Su-pun-ger and Bayne Testimonies Related to the Franklin Expedition." *Polar Record* 53 (2017): 561–79.

Guillemard, F.H.H. "Franklin and the Arctic." *Blackwood's Magazine* 161 (February 1897): 238–56.

Guly, H.R. "'Polar Anaemia': Cardiac Failure During the Heroic Age of Antarctic Exploration." *Polar Record* 45 (2012): 157–64.

Guttridge, Leonard. *Ghosts of Cape Sabine: The Harrowing True Story of the Greeley Expedition*. New York: Putnam, 2000.

———. *Mutiny: A History of Naval Insurrection*. Annapolis: Naval Institute Press, 1992.

Hacquebard, Louwrens. "Five Early European Winterings in the Atlantic Arctic (1596–1635): A Comparison." *Arctic* 44 (1991): 146–55.

Haggarty, J.M., Z. Cernovsky, M. Husni, P. Kermeen, and H. Merskey. "Seasonal Affective Disorder in an Arctic Community." *Acta Psychiatrica Scandinavica* 105 (2002): 378–84.

Hall, Charles Francis. *Life with the Esquimaux*. London: Sampson Low & Son, 1864, reprint, Rutland: Charles E. Tuttle, 1970.

Hamilton, Charles I. "John Wilson Croker: Patronage and Clientage at the Admiralty, 1809–1857." *The Historical Journal* 43 (2000): 49–77.

———. "Naval Hagiography and the Victorian Hero." *The Historical Journal* 23 (1980): 381–98.

Hansen, Todd. "Additional Documents and Survey on the Franklin Sites of Beechey Island, Nunavut, Canada." *Polar Record* 48 (2012): 195–96.

———. "Documents and Survey re Beechey Island." *Polar Record* 46 (2010): 193–99.

———. "Physical Descriptions of the Beechey Island Headboards." *Polar Record* 53 (2017): 403–12.

Harper, Kenn. *Give Me My Father's Body: The Life of Minik, the New York Eskimo*. Royalton: Steerforth Press, 2001.

———. "The Travels of an Inuit Story." *Nunatsiaq News* online, July 7, 2019. Available at https://nunatsiaq.com/stories/article/taissumani-july-5/.

Harrison, Mark. "Scurvy on Sea and Land: Political Economy and Natural History, c. 1780–c. 1850." *Journal for Maritime Research* 15 (2013): 7–25.

Hayes, Derek. *Historical Atlas of the Arctic*. Seattle: University of Washington Press, 2003.

Headland, R.K., et al. "Transits of the Northwest Passage to End of the 2022 Navigation Season," Scott Polar Research Institute Infosheet, November 18, 2022. Available at https://www.spri.cam.ac.uk/resources/infosheets/northwestpassage.pdf.

Hearne, Samuel. *A Journey from Prince of Wales Fort in Hudson's Bay to the Northern Ocean, 1769, 1770, 1771, 1772*. Edited by Richard Glover. London: A. Strahan & T. Cadell, 1795, reprint, Toronto: Macmillan of Canada, 1958.

Heighton, Stephen. *Afterlands*. Boston: Houghton Mifflin, 2006.

Heywood, Peter. *The Impracticability of a North-West Passage for Ships, Impartially Considered*. London: A.T. Valpy, 1824, reprint, Cambridge: Cambridge University Press, 2011.

Hickey, Clifford G., James M. Savelle, and George B. Hobson. "The Route of Sir John Franklin's Third Arctic Expedition: An Evaluation and Test of an Alternate Hypothesis." *Arctic* 46 (1993): 78–81.

Hill, Jen. "National Bodies: Robert Southey's Life of Nelson and John Franklin's Narrative of a Journey to the Shores of the Polar Sea." *Nineteenth Century Literature* 61 (2007): 417–48.

Hiraiwa-Hasigawa, Mariko. "Cannibalism among Non-Human Primates." In Mark A. Elger and Bernard J. Crespi, eds., *Cannibalism: Ecology and Evolution Among Diverse Taxa*. Oxford: Oxford University Press, 1992: 323–38.

Hoag, Elaine. "Caxtons of the North: Mid-Nineteenth-Century Arctic Shipboard Printing." *Book History* 4 (2001): 81–114.

Hodgetts, Lisa M. "The Rediscovery of HMS *Investigator*: Archaeology, Sovereignty and the Colonial Legacy in Canada's Arctic." *Journal of Social Archaeology* 13 (2012): 80–100.

Holland, Clive. "The Arctic Committee of 1851: A Background Study." *Polar Record* 20 (1980): 117.

———. *Farthest North: A History of Polar Exploration in Eyewitness Accounts*. London: Robinson, 1994.

———. "John Franklin and the Fur Trade, 1819–1822." In Richard C. Davis, ed., *Rupert's Land: A Cultural Tapestry*. Waterloo: Wilfrid Laurier Press, 1988.

———. "William Penny, 1809–92: Arctic Whaling Master." *Polar Record* 15 (1970): 25–43.

———, and James M. Savelle. "My Dear Beaufort: A Personal Letter from John Ross's Arctic Expedition of 1829–33." *Arctic* 40 (1987): 66–77.

Hooper, W.H. *Ten Months Among the Tents of the Tuski: With Incidents of an Arctic Boat Expedition in Search of Sir John Franklin*. London: John Murray, 1853.

Horowitz, B. Zane. "Polar Poisons: Did Botulism Doom the Franklin Expedition?" *Journal of Toxicology: Clinical Toxicology* (2003): 841–47.

Hough, Richard. *Captain Bligh and Mr. Christian*. Annapolis: Naval Institute Press, 2000.

Hudes, Sammy. "Accidents, Not Illness, May Have Killed Many Officers in Franklin Expedition, Says Scottish Study." *Toronto Star*, September 21, 2016. Available at https://www.thestar.com/news/canada/2016/09/21/accidents-not-illness-may-have-killed-many-officers-in-franklin-expedition-says-scottish-study.html1.

Huish, Robert. *The Last Voyage of Captain Sir John Ross, R.N., to the Arctic Regions for the Discovery of a North West Passage*. London: John Saunders, 1835.

Hulan, Renée. *Northern Experience and the*

Myths of Canadian Culture. Montreal: McGill-Queen's University Press, 2002.

Huntford, Roland. *Race to the South Pole: The Expedition Diaries of Scott and* Amundsen. London: Bloomsbury Academic, 2010.

_____. *Scott and Amundsen.* London: Hodder and Stoughton, 1979.

Hutchinson, Gillian. *Sir John Franklin's Erebus and Terror Expedition Lost and Found.* London: Bloomsbury, 2017.

Inglefield, Edward A. *A Summer Search for Sir John Franklin, with a Peep into the Polar Basin.* London: Thomas Harrison, 1853.

Inskeep, Steve. "Remembering John Glenn." *Morning Edition*, NPR, December 9, 2016. Available at https://www.npr.org/2016/12/09/504930256/remembering-john-glenn.

Jackson, Colin Ian. "Three Puzzles from Nineteenth Century Arctic Exploration." *Northern Mariner* 17 (2007): 1–17.

Jacobs, Annalise. "Arctic Circles: The Franklin Family, Networks of Knowledge and Early 19th-Century Arctic Exploration, 1818–1859." Ph.D. Dissertation, University of Illinois at Champagne-Urbana, 2015.

Jacobs, Martina M., and James B. Richardson, eds. *Arctic Life: Challenge to Survive.* Pittsburgh: The Carnegie Institute, 1983.

James, Lawrence. *Mutiny in the British and Commonwealth Forces, 1797–1956.* London: Buchan & Enright, 1987.

James, N. "Franklin's Fate: Discoveries and Prospects," *Antiquity* 91 (2017): 1647–51.

Jenks, Timothy. "Contesting the Hero: The Funeral of Admiral Lord Nelson." *Journal of British Studies* 39 (2000): 422–53.

Johnson, Robert E. "By Want Beleaguered: Sir John Franklin in the Canadian Arctic." In J. Watt, E.J. Freeman, and W.F. Bynum, eds., *Starving Sailors: The Influence of Nutrition upon Naval and Maritime History.* London: National Maritime Museum, 1981, 109–16.

_____. "Doctors Abroad: Medicine and Nineteenth-Century Arctic Exploration." In J. Watt, E.J. Freeman, and W.F. Bynum, eds., *Starving Sailors: The Influence of Nutrition upon Naval and Maritime History.* London: National Maritime Museum, 1981, 101–08.

Johns-Putra, Adeline. "Historicizing the Networks of Ecology and Culture: Eleanor Anne Porden and Nineteenth-Century Climate Change." *ISLE Interdisciplinary Studies in Literature and Environment* 21 (2015): 27–46.

Jones, A.G.E. "Frederick John Hornby." *Mariner's Mirror* 41 (1955): 303–07.

_____. "Henry Peter Peglar, Captain of the Foretop, 1811–48." *Notes & Queries* 31 (1984): 463–68.

_____. *Polar Portraits.* Whitby: Caedmon of Whitby, 1992.

_____. "Protecting the Whaling Ships." *Fram: The Journal of Polar Studies* 2 (1985): 265–69.

_____. "Rear Admiral Sir William Edward Parry: A Different View." *The Musk-Ox* 21 (1978): 3–10.

_____. "Sir James Clark Ross and the Voyage of the Enterprise and Investigator, 1848–49." *Royal Geographical Society Journal*, 137 (1971): 165–79.

_____, and Richard J. Cyriax. "Lt. Edward Griffiths and the Franklin Expedition." *Mariner's Mirror* 39 (1953): 176–86.

Jones, H.C. "The Inuit as Geographers: The Case of Eenoolooapik." *Inuit Studies* 28 (2004): 58–72.

Karpoff, Jonathan M. "Public vs. Private Initiative in Arctic Exploration: The Effects of Incentives and Organizational Structure." *Journal of Political Economy* 109 (2001): 38–78.

Kasten-Mutkus, Kathleen. "Ghosts in the Archive: The Textual Lacunae of the Third Franklin Expedition." *Polar Record* 55 (2019): 417–24.

Kaufman, Matthew H. "Harry Goodsir and the Franklin Expedition of 1845." *Journal of Medical Biography* 12 (2004): 82–89.

Keenleyside, Anne. "The Last Resort: Cannibalism in the Arctic." *The Explorers Journal* 72 (1995): 138–43.

_____, Margaret Bertulli, and Henry C. Fricke. "The Final Days of the Franklin Expedition: New Skeletal Evidence." *Arctic* 50 (1997): 36–46.

_____, X. Song, D.R. Chettle, and C.E. Webber. "The Lead Content of Human Bones from the 1845 Franklin Expedition." *Journal of Archaeological Science* 23 (1996): 461–65.

Kendi, Ibram X. *Stamped from the Beginning: The Definitive History of Racist Ideas in America.* New York: Nation Books, 2016.

Kennedy, William. *A Short Narrative of the 2nd Voyage of the Prince Albert in Search of Sir John Franklin.* London: W.H. Dalton, 1853.

Kerr, R. "Rae's Franklin Relics." *The Beaver* 284 (March 1954): 25–27.

Kinealy, Christine. *The Great Calamity.* Dublin: Gill & Macmillan, 1994

King, Richard. *The Franklin Expedition from First to Last.* London: John Churchill, 1855, reprint, London: Forgotten Books, 2017.

Klutschak, Heinrich. *Overland to Starvation Cove: With the Inuit in Search of Franklin, 1878–1880.* Translated and edited by William Barr. Toronto: University of Toronto Press, 1987.

Knopf, Kerstin. "Exploring for the Empire: Franklin, Rae, Dickens, and the Natives in Australian and Canadian Literature." In Barbara Buchenau and Virginia Richter, eds., *Post-Empire Imaginaries? Anglophone Literature, History and the Demise of Empires.* Leiden: Brill/Rodopi, 2015.

Kofron, John. "Dickens, Collins and the Influence of the Arctic." *Dickens Studies Annual* 40 (2009): 81–93.

Kowal, Walter, Owen Beattie, and Halfdan

Baadsgaard. "Did Solder Kill Franklin's Men?" *Nature* 343 (1990): 319–20. Available at https://doi.org/10.1038/343319a0.

———, Owen Beattie, Halfdan Baadsgaard, and P.M. Krahn. "Source Identification of Lead Found in Tissues of Sailors from the Franklin Arctic Expedition of 1845." *Journal of Archaeological Science* 18 (1991): 193–203.

———, P.M. Krahn, and Owen Beattie. "Lead Levels in Human Tissues from the Franklin Forensic Project." *International Journal of Environmental Analytical Chemistry* 35 (1989): 119–26.

Kriwoken, Lorne K., and John W. Williamson. "Hobart, Tasmania: Antarctic and Southern Ocean Connections." *Polar Record* 29 (1993): 93–102.

Kushner, Ervan F. *Alfred Packer, Cannibal! Victim?* Frederick: Platte 'N Press, 1980.

Lamb, Geoffrey F. *Franklin, Happy Voyager.* London: Ernest Benn, 1956.

Lamb, Jonathan. *Scurvy: The Disease of Discovery.* Princeton: Princeton University Press, 2017.

Lambert, Andrew. *Franklin: Tragic Hero of Polar Navigation.* London: Faber & Faber, 2009.

———. *The Gates of Hell: John Franklin's Tragic Quest for the Northwest Passage.* New Haven: Yale University Press, 2011.

Lambert Richard S. *Franklin of the Arctic.* London: The Bodley Head, 1954.

Larson, Henry. *The North-West Passage, 1940–42 and 1944.* Vancouver: City Archives, 1954.

Latta, Jeffrey B. *The Franklin Conspiracy.* Toronto: Dundurn Press, 2001.

Layman, C.H. *The Wager Disaster: Mayhem, Mutiny and Murder in the South Seas.* London: Unicorn Press, 2015.

Learmouth, Lorenz A. "Notes on Franklin Relics." *Arctic* 1 (1948): 122–23.

Lederman, Leon M., and Christopher T. Hill. *Quantum Physics for Poets.* Amherst: Prometheus, 2011.

Leech, Samuel. *A Voice from the Main Deck.* London: Chatham, 1841.

Lemann, Jennifer. "The Magnetic North Pole Is Rapidly Moving Because of Some Blobs." *Popular Mechanics*, March 7, 2022. Available at https://www.popularmechanics.com/science/environment/a32496561/why-magnetic-north-pole-moving/.

Lemercier-Goddard, Sophie, and Frédéric Regard. "Introduction: The Northwest Passage and the Imperial Project: History, Ideology, Myth." In Frédéric Regard, ed., *The Quest for the Northwest Passage: Knowledge, Nation and Empire, 1576–1806.* London: Pickering & Chatto, 2013, 1–14.

LeMoine, Genevieve. Review of John V.H. Dippel, *To the Ends of the Earth: The Truth Behind the Glory of Polar Exploration* (Amherst NY: Prometheus Books, 2018). *Arctic* 72 (2019): 204–05.

Lentz, John W. "The *Fox* Expedition in Search of Franklin: A Documentary Trail." *Arctic* 56 (2003): 175–84.

Leslie, Edward E. *Desperate Journeys, Abandoned Souls: True Stories of Castaways and Other Survivors.* Boston: Houghton Mifflin, 1988.

Levere, Trevor H. *Science and the Canadian Arctic: A Century of Exploration, 1818–1918.* Cambridge: Cambridge University Press, 1992.

———. "Science and the Canadian Arctic, 1818–76, from Sir John Ross to Sir George Strong Nares." *Arctic* 41 (1988): 127–37.

Levy, Buddy. *Labyrinth of Ice: The Triumphant and Tragic Greeley Polar Expedition.* New York: St. Martin's Press, 2019.

Lewis, I.M. *Religion in Context: Cults and Charisma.* Cambridge: Cambridge University Press, 2001.

Lewis, Michael. *The Navy in Transition, 1814–1864: A Social History.* London: Hodder & Stoughton, 1965.

Lewis-Jones, Huw. *Face to Face: Polar Portraits.* Cambridge: Scott Polar Research Institute, 2009.

———. "'Heroism Displayed': Revisiting the Franklin Gallery at the Royal Naval Exhibition, 1891." *Polar Record* 41 (2005): 185–203.

———. "'Nelsons of Discovery': Notes on the Franklin Monument in Greenwich." *The Trafalgar Chronicle, Yearbook of the 1805 Club*. (2009). Available at http://www.ric.edu/faculty/rpotter/temp/HLJ_Franklin_Monument.pdf.

Lincoln, Margarette. "Shipwreck Narratives of the Eighteenth and Early Nineteenth Century: Indicators of Culture and Identity." *Journal for Eighteenth-Century Studies* 20 (1997): 155–72.

Lind, James. *A Treatise on the Scurvy.* London: A. Millar, 1753.

Livermore, Philip W., Christopher C. Finlay, and Matthew Bayliff. "Recent North Magnetic Pole Acceleration toward Siberia Caused by Flux Lobe Elongation." *Nature GeoScience* 13 (2020): 387–91.

Lloyd, Christopher. *The British Seaman 1200–1860: A Social Survey.* London: Collins, 1968.

———. *Mr. Barrow of the Admiralty: A Life of Sir John Barrow, 1764–1848.* London: Collins, 1970.

———. "Victualling of the Fleet in the Eighteenth and Nineteenth Centuries." In J. Watt, E.J. Freeman and W.F. Bynum, eds., *Starving Sailors: The Influence of Nutrition upon Naval and Maritime History.* Greenwich: National Maritime Museum, 1981, 9–15.

———, and J.S.L. Coulter. *Medicine and the Navy 1200–1900.* vol. IV, 1815–1900. London: E. & S. Livingstone, 1969.

Lloyd-Jones, Ralph. "An Evangelical Christian on Franklin's Last Expedition: Lieutenant John Irving of HMS *Terror*." *Polar Record* 33 (1997): 337–42.

———. "Franklin's Men and Their Families: New

Evidence from the Allotment Books." *Polar Record* 54 (2018): 267–74.
___. "Further Light on Franklin's Men." *Polar Record* 47 (2011); 379–82.
___. "The Men who Sailed with Franklin." *Polar Record* 41 (2005): 311–18.
___. "The Paranormal Arctic: Lady Franklin, Sophia Cracroft and Captain and 'Little Weesy' Coppin." *Polar Record* 37 (2001): 27–34.
___. "The Royal Marines on Franklin's Last Expedition." *Polar Record* 40 (2004): 319–26.
Long, Alison. "A Note Relating to the Birth Date of Captain Francis Rawdon Moira Crozier RN, FRS, FRAS." *Polar Record* 58 (2022). Available at doi:10.1017/S0032247422000225.
Loomis, Chauncey C. *Weird and Tragic Shores: The Story of Charles Francis Hall, Explorer*. New York: Knopf, 1991.
Lyon, David, and Rif Winfield. *The Sail and Steam Navy List*. London: Chatham, 2004.
MacDonald, Janet. *Feeding Nelson's Navy: The True Story of Food at Sea in the Georgian Era*. London: Chatham, 2004.
MacInnis, J.B. "The Breadalbane Project: Implications for the Discovery of Franklin's Ships." In Patricia D. Sutherland, ed., *The Franklin Era in Canadian Arctic History 1845–1859*. Ottawa: National Museums of Canada, Mercury Series, Archaeological Survey of Canada Paper No. 131, 1985, 174–84.
___. *Polar Passage: The Historic First Sail Through the Northwest Passage*. Toronto: Random House Canada, 1989.
Mackaness, George. *Some Private Correspondence of Sir John and Lady Jane Franklin*. Gilberton: D.S. Ford, 1947.
MacKay, Douglas, and W.K. Lamb. "More Light on Thomas Simpson." *The Beaver* 269 (September 1938): 26–31.
Mackinnon, C.S. "The British Man-Hauling Sledging Tradition." In Patricia Sutherland, ed., *The Franklin Era in Canadian Arctic History 1845–1859*. Ottawa: National Museums of Canada, Mercury Series, Archaeological Survey of Canada Paper No. 131, 1985, 129–40.
___. "The Wintering-Over of Royal Navy Ships in the Canadian Arctic, 1819–1876." *The Beaver*, Winter 1984/85: 12–21.
MacLaren, Ian S. "Bones of Empire: Cook and Franklin Reaching to Alaska for a Northwest Passage." In James K. Barnett and Ian C. Hoffman, *Imagining Anchorage: The Making of America's Northernmost Metropolis*. Fairbanks: University of Alaska Press, 2018, 105–26.
___. "From Exploration to Publication: The Evolution of a 19th-Century Arctic Narrative." *Arctic* 47 (1994): 43–53.
___. "John Barrow's Darling Project (1816–46)." In Frédéric Regard, ed., *Arctic Exploration in the Nineteenth Century: Discovering the Northwest Passage*. London: Pickering & Chatto, 2013, 19–36.

___. "John Franklin." In J.M. Heath, ed., *Profiles in Canadian Literature*. Toronto: Dundurn Press, 1986, 25–32.
___. "The Poetry of the 'New Georgia Gazette' or 'Winter Chronicle' 1819–1820." *Canadian Poetry* 30 (1992): 41–73.
___. "Tracing One Discontinuous Line Through Poetry of the Northwest Passage." *Canadian Poetry* 39 (1996): 7–48.
MacPherson, Cluny. "The First Recognition of Beri-Beri in Canada." *Canadian Medical Association Journal* 95 (1966): 278–79.
Madley, Benjamin. "From Terror to Genocide: Britain's Tasmanian Penal Colony and Australia's History Wars." *Journal of British Studies* 47 (2008): 77–106
Malchow, H.L. *Gothic Images of Race in Nineteenth Century Britain*. Palo Alto: Stanford University Press, 1996.
Malin, S.R.C., and D.R. Barraclough. "Humboldt and the Earth's Magnetic Field." *Quarterly Journal of the Royal Astronomical Society* 32 (1991): 279–93.
Mallin, M.G. "In Warm Blood: Some Historical and Procedural Aspects of *Regina v. Dudley and Stephens*." *University of Chicago Law Review* 34 (1967): 387–407.
Mangles, James, ed. *Papers and Dispatches Relating to the Arctic Searching Expeditions of 1850-51-52*, 2nd ed. London: Francis & John Rivington, 1852.
Marchand, Peter. *Life in the Cold: An Introduction to Winter Ecology*. Hanover: University Press of New England, 1987.
Markham, Albert Hastings. "On Sledge Traveling." *Royal Geographical Society Proceedings* (December 1876): 110–20.
___, (with the aid of Sophia Cracroft). *Life of Sir John Franklin and the North-West Passage*. London: George Philip & Son, 1891.
Markham, Clements. "The Expedition of Lieutenant Schwatka to King William's Land." *Royal Geographical Society Proceedings* (November 1880): 657–62.
___. *Franklin's Footsteps*. London: Chaman & Hall, 1853.
___. *The Threshold of the Unknown Regions*. London: Sampson, Low, Marston, Low, & Steele,1873, reprint, Miami: HardPress, 2017.
Marlow, James E. "The Fate of Sir John Franklin: Three Phases of Response in Victorian Periodicals." *Victorian Periodicals Review* 15 (1982): 3–11.
___. "Sir John Franklin, Mr. Charles Dickens and the Solitary Monster." *Dickens Studies Newsletter* 12, no. 4 (December 1981): 97–103.
Martin, Constance. "William Scoresby, Jr. (1789–1857) and the Open Polar Sea—Myth and Reality." *Arctic* 41 (1988): 39–47.
Martin, Ronald, Steven Naftel, Sheila Macfie, Keith Jones, and Andrew Nelson. "Pb Distribution in Bones from the Franklin Expedition: Synchrotron X-ray Fluorescence and Laser

Ablation/Mass Spectroscopy." *Applied Physics A* 111 (2013): 23–29. Available at https://doi.org/10.1007/s00339-013-7579-5.

Maxtone-Graham, John. *Safe Return Doubtful: The Heroic Age of Polar Exploration*. Wellingborough: Patrick Stephens, 1988.

Mays, S., and Owen Beattie, "Evidence for End-Stage Cannibalism on Sir John Franklin's Last Expedition to the Arctic, 1845." *International Journal of Osteoarchaeology* 26 (2016): 778–86.

_____, A. Ogden, J. Montgomery, S. Vincent, W. Battersby, and G.M. Taylor. "New Light on the Personal Identification of a Skeleton of a Member of Sir John Franklin's Last Expedition to the Arctic, 1845." *Journal of Archaeological Science* 38 (2011): 1571–82.

_____, G.J.R. Maat, and H.H. de Boer. "Scurvy as a Factor in the Loss of the 1845 Franklin Expedition to the Arctic: A Reconsideration." *International Journal of Osteoarchaeology* 25 (2015): 334–44.

McClintock, Francis Leopold. "Discoveries of the Late Expedition in Search of Sir John Franklin." *Proceedings of the Royal Geographical Society of London* 4 (1859–1860): 2–13.

_____. "Discoveries by the Late Expedition in Search of Sir John Franklin and His Party." *Journal of the Royal Dublin Society* 1 (1857): 183–250.

_____. "The Expedition in Search of Sir John Franklin." *Journal of the American Geographical and Statistical* Society 1 (1859): 247–52.

_____. "Narrative of the Expedition in Search of Sir John Franklin and His Party." *Royal Geographical Society Journal* 31 (1861): 1–13.

_____. "On Arctic Sledge Travelling." *Proceedings of the Royal Geographical Society of London* 19 (1874–75): 464–79.

_____. *The Voyage of the "Fox" in the Arctic Seas: A Narrative of the Discovery of the Fate of Sir John Franklin and His Companions*. London: John Murray, 1859.

McClure, Robert L.M. *The Discovery of the Northwest Passage by HMS 'Investigator.'* London: Longmans, Brown, Green, Longmans and Roberts, 1857, reprint, Sherrard Osborn, ed. Rutland: Charles E. Tuttle & Co., 1969.

McCorristine, Shane. "Cross-Dressing, Feasts and Fun." *UpHere*, December 2017.

_____. "The Franklin Ghosts" *UpHere*, October-November 2018.

_____. "A Manuscript History of the Franklin Family by Sophia Cracroft (1853)." *Polar Record* 51 (2015): 72–90.

_____. "On the Trail of the *Terror*." *History Today*, September 14, 2016. Available at https://www.historytoday.com/trail-terror.

_____. "Searching for Franklin: A Contemporary Canadian Ghost Story." *British Journal of Canadian Studies* 26 (2013): 39–57.

_____. *The Spectral Arctic: A History of Dreams and Ghosts in Polar Exploration*. London: UCL Press, 2018.

_____, and Victoria Herrmann. "The 'Old Arctics': Notices of Franklin Search Expedition Veterans in the British Press, 1876–1934." *Polar Record* 52 (2016): 215–29.

_____, and Jane S.P. Mocellin. "Christmas at the Poles: Emotion, Food, and Festivities on Polar Expeditions, 1818–1912." *Polar Record* 52 (2016): 562–77.

McDonald, Alexander. *A Narrative of Some Passages in the History of Enoolooapik, a Young Eskimo, Who Was Brought to Britain in 1839, in the Ship "Neptune" of Aberdeen*. Edinburgh: Fraser & Co., 1841.

McDougall, George Frederick. *The Eventful Voyage of H.M. Discovery Ship 'Resolute' to the Arctic Regions*. London: Longmans, Green, Brown, Longmans and Roberts, 1857.

McGoogan, Ken. "Defenders of Arctic Orthodoxy Turn their Backs on Sir John Franklin." *Polar Record* 51 (2015): 220–21.

_____. *Fatal Passage: The Untold Story of Scotsman John Rae, the Arctic Adventurer Who Discovered the Fate of Franklin*. New York: Basic Books, 2002.

_____. *Lady Franklin's Revenge: A True Story of Ambition, Obsession and the Remaking of Arctic History*. London: Bantam, 2006.

_____. *Race to the Polar Sea, the Heroic Adventures of Elisha Kent Kane*. Berkeley: Counterpoint, 2009.

_____, ed. *The Arctic Journals of John Rae*. Victoria: TouchWood Editions, 2012.

_____, ed. *John Rae's Arctic Correspondence 1844–1855*. Victoria: TouchWood Editions, 2014.

McGovern, Thomas H. "The Archeology of the Norse North Atlantic." *Annual Review of Anthropology* 19 (1990): 331–51.

McGreevey, Ronan. "The Role of 'Survivor Cannibalism' During the Great Famine." *The Irish Times*, November 30, 2020. Available at https://www.irishtimes.com/news/ireland/irish-news/role-of-survivor-cannibalism-during-great-famine-detailed-in-new-tv-documentary-1.4423323.

McGregor, Elizabeth. *The Ice Child*. New York: Dutton, 2001.

McGuire, Ian. *The North Water*. New York: Holt, 2016.

McIntyre, Neil, and Nigel N. Stanley. "Cardiac Beriberi: Two Modes of Presentation." *British Medical Journal* 3 (1971): 567–69.

McKee, Alexander. *HMS Bounty*. New York: William Morrow, 1962

McKenzie, W.G. "A Further Clue in the Franklin Mystery." *The Beaver* 299 (Spring 1969): 28–32.

McLaughlin, Joseph B., Jeremy Sobel, Tracey Lynn, Elizabeth Funk, and John P. Middaugh. "Botulism Type E Outbreak Associated with Eating a Beached Whale, Alaska." *Emerging Infectious Diseases* 10 (2004): 1685–87.

McNab, David T. "The Arctic Prescription: Indigenous Knowledge and the Role It Played in the Search for Sir John Franklin's Erebus." *The Aboriginal Business Report*, Issue 3 (February 2016). Available at https://view.imirus.com/1158/document/12046/page/3.

McRobert, David. "Arctic Sovereignty Initiatives in the Canadian North: A Historical Review, 1700-1980." Report for the President's Advisory Committee on Northern Studies, York University, 1982.

Melbye, Jerry, and Scott I. Fairgrieve. "A Massacre and Possible Cannibalism in the Canadian Arctic: New Evidence for the Saunatuk Site (NgTn1)." *Arctic Anthropology* 31 (1994): 57–77.

Melville, Herman. *Moby-Dick: An Authoritative Text, Context, Criticism*. New York: W.W. Norton, 2018 (first published 1851).

_____. *White Jacket, or the World in a Man-of-War*. New York: Harper & Brothers, 1850.

Michener E.A. (Ted). *Ice in the Rigging*. Hobart: National Museum of Tasmania, 2015.

Millar, Keith, and Adrian Bowman. "Cognitive Archaeology: Estimating the Effects of Blood-Lead Concentrations on the Neuropsychological Function of an Officer of the 1845 Franklin Expedition." *Journal of Archaeological Sciences: Reports* 32 (2020). Available at https://doi.org/10.1016/j.jasrep.2020.102449.

_____, and Adrian W. Bowman. "Hartnell's Time Machine Reprise: Further Implications of Zinc, Lead and Copper in the Thumbnail of a Franklin Expedition Crewmember," *Journal of Archaeological Science: Reports* 13 (2017): 286–90.

_____, Adrian W. Bowman, and William Battersby. "The *Erebus*, the *Terror* and the Northwest Passage: Did Lead Really Poison Franklin's Lost Expedition?" *Significance, Journal of the Royal Statistical Society*, April 2014: 20–26.

_____, Adrian W. Bowman, and William Battersby. "A Re-Analysis of the Supposed Role of Lead Poisoning in Sir John Franklin's Last Expedition, 1845–1848." *Polar Record* 51 (2015), 224–38.

_____, Adrian W. Bowman, William Battersby, and R.R. Welbury. "The Health of Nine Royal Naval Arctic Crews, 1848 to 1854: Implications for the Lost Franklin Expedition." *Polar Record* 52 (2016). 423–41.

Mills, William J. *Exploring Polar Frontiers*. Santa Barbara ABC-CLIO, 2003.

Milton-Thompson, G.J. "Two Hundred Years of the Sailor's Diet." In J. Watt, E.J. Freeman and W.F. Bynum, eds., *Starving Sailors: The Influence of Nutrition upon Naval and Maritime History*. Greenwich: National Maritime Museum, 1981, 27–34.

Mocellin, Jane S.P. "A Behavioural Study of Human Responses to the Arctic and Antarctic Environments." Ph.D. Dissertation, University of British Columbia, 1988.

_____. "Levels of Anxiety Aboard Two Expeditionary Ships." *Journal of General Psychology* 122 (1995): 317–24.

_____, and Peter Suedfeld. "Voices from the Ice: Diaries of Polar Explorers." *Environmental Behavior* 23 (1991): 704–22.

Montgomery, H. "An Arctic Mirage." *Fram: The Journal of Polar Studies* (1985): 177–94.

Moss, Sarah. *The Frozen Ship: The Histories and Tales of Polar Exploration*. Katonah: BlueBridge, 2006.

_____. *Scott's Last Biscuit: The Literature of Polar Travel*. Oxford: Signal Books, 2005.

Mowat, Farley. *High Latitudes*. South Royalton: Steerforth Press, 2003.

_____. *Ordeal by Ice*. Boston: Atlantic/Little, Brown, 1960.

_____. *Westviking: The Ancient Norse in Greenland and North America*. Toronto: McClelland & Stewart, 1970.

Murdock, P.E. "The Old Shipwreck." *The Beaver* 284 (March 1954): 42–43.

Murphy, David. *The Arctic Fox: Francis Leopold McClintock*. Toronto: Collins, 2004.

Nadolny, Sten. *The Discovery of Slowness*. New York: Viking, 1987.

Nansen, Fridtjof. *Farthest North: Being the Record of a Voyage of Exploration of the Ship "Fram," 1893–96 and of a Fifteen Months' Sleigh Journey by Dr. Nansen and Lieut Johansen*. London: Harper & Bros., 1897.

Nanton, Paul. *Arctic Breakthrough: Franklin's Expeditions 1819–1847*. Toronto: Clarke Irwin & Co., 1970.

Neale, Jonathan Sayles. "Forecastle and Quarterdeck: Protest, Discipline and Mutiny in the Royal Navy, 1793–1814." Ph.D. Dissertation, University of Warwick, 1990.

Neatby, Leslie. *In Quest of the Northwest Passage*. Toronto: Longmans, Green & Co., 1958.

_____. "Joe and Hannah." *The Beaver* 290 (Autumn 1969): 16–21.

_____. *The Search for Franklin*. New York: Walker, 1970.

_____, ed. *Frozen Ships: the Arctic Diary of Johann Miertsching, 1850–1854*. Toronto: Macmillan of Canada, 1967.

Nelson, L.H. "The Last Voyage of HMS *Investigator*, 1850–1853, and the Discovery of the North-West Passage." *Polar Record* 13 (1967): 757–68.

Newman, Peter C. *Company of Adventurers*. Ann Arbor: University of Michigan Press, 1985.

Nickerson, Sheila. *Midnight to the North: The Untold Story of the Inuit Woman who Saved the Polaris Expedition*. New York: Jeffrey A. Tarcher/Putnam, 2002.

Nordhoff, Charles, and James Norman Hall. *Mutiny on the Bounty*. New York: Grosset and Dunlap, 1932.

Norman, J.N. "A Comparison of the Patterns of Illness and Injury Occurring in Offshore Structures in the Northern North Sea and the Stations of the British Antarctic Survey." *Arctic Medical Research* 50 (1991): 719–21.

Norris, John. "The 'Scurvy Disposition': Heavy

Exertion as an Exacerbating Influence on Scurvy in Modern Times." *Bulletin of the History of Medicine* 53 (1983): 325–38.

Notman, D.N.H., L. Anderson, Owen Beattie, and Roger Amy. "Arctic Paleology: Portable Radiographic Examination of Two Frozen Sailors from the Franklin Expedition (1845–1848)." *American Journal of Roentgenology* 149 (1987): 347–50.

_____, and Owen Beattie. "The Paleoimaging and Forensic Anthropology of Frozen Sailors from the Franklin Arctic Expedition Mass Disaster (1845–1848): A Detailed Presentation of Two Radiological Surveys." In K. Spindler, et al., eds., *Human Mummies: A Global Survey of Their Status and the Techniques of Conservation*. Vienna: Springer-Verlag, 1996: 347–50.

Nourse, Joseph Everett. *American Explorations in the Ice Zones*. Boston: D. Lothrop, 1884.

_____, ed. *Narrative of the Second Arctic Expedition Made by Charles Francis Hall: His Voyage to Repulse Bay, Sledge Journeys to the Straits of Fury and Hecla and to King William's Land, and Residence among the Eskimos during the Years 1864-'69*. Washington, D.C.: Government Printing Office, 1879.

Nugent, Frank. *Seek the Frozen Lands: Irish Polar Explorers 1740–1922*. Cork: The Collins Press, 2003.

O'Dochertaigh, Eavan. *Visual Culture and Arctic Voyages: Personal and Public Art and Literature of a Franklin Search Expedition*. Cambridge: Cambridge University Press, 2022.

Officer, Charles, and Jake Page. *A Fabulous Kingdom: The Exploration of the Arctic*. New York: Oxford University Press, 2001.

O'Loughlin, Ed. *Minds of Winter*. London: Riverrun, 2016.

O'Neill, Patrick. "Theatre in the North: Staging Practices of the British Navy in the Canadian Arctic." *Dalhousie Review* 73 (1994): 356–84.

Opel, Mechtild. "Chocolate in the Arctic." *Trimaris* blog, April 24, 2019. Available at https://www.trimaris.de/?s=Chocolate.

Osborn, Sherrard. *The Career, Last Voyage and Fate of Sir John Franklin*. London: Bradbury & Evans, 1860.

_____. *The Polar Regions*. London: Barnes, 1854.

_____. *Stray Leaves from an Arctic Journal*. New York: G.P. Putnam's Sons, 1852.

_____, ed. *The Discovery of the North-West Passage by H.M.S. "Investigator," Capt. R. M'Clure*. London: Longmans Green, 1857.

_____, and G.F. McDougall, eds. *Facsimile of the Illustrated Arctic News, Published on Board HMS Resolute*. London: Ackermann & Co., 1852.

Owen, Roderic. *The Fate of Franklin*. London: Hutchinson, 1978.

Pack, Stanley W.C. *The Wager Mutiny*. London: Alvin Redman, 1952.

Palin, Michael. *Erebus: One Ship, Two Epic Voyages and the Greatest Naval Mystery of All Time*. Vancouver: Greystone Books, 2018.

Palinkas, Lawrence A. "Going to Extremes: The Cultural Context of Stress, Illness and Coping in Antarctica." *Social Science and Medicine* 35 (1992): 651–64.

_____, and Peter Suedfeld. "Psychological Effects of Polar Expeditions." *The Lancet* 371 (2008): 153–63.

Palmer, Roy. *The Oxford Book of Sea Songs*. Oxford: Oxford University Press, 1986.

Pálsson, Hermann. *The Vinland Sagas: The Norse Discovery of America*. New York: Penguin Classics, 1965.

Paredes, Andrés. "Thomas Blanky: A Life in Hell." *Kabloonas* blog, June 26, 2015. Available at https://kabloonas.blogspot.com/2015/06/-thomas-blanky-live-in-hell.html.

Park, Robert W., and Douglas Stenton. "Use Your Best Endeavours to Discover a Sheltered and Safe Harbour." *Polar Record* 55 (2019): 361–72.

Parkinson, Edward. "'All Well': Narrating the Third Franklin Expedition." In John Moss, ed., *Echoing Silence: Essays on Arctic Narrative*. Ottawa: University of Ottawa Press, 1997, 43–52.

Parrado, Nando. *Miracle in the Andes: 72 Days on the Mountain and My Long Trek Home*. New York: Broadway Books, 2006.

Parry, Ann. *Parry of the Arctic: The Life Story of Admiral Sir Edward Parry*. London: Chatto & Windus, 1963.

Parry, William Edward. *Journal of a Second Voyage of Discovery of a Northwest Passage*. London: John Murray, 1824, reprinted New York: Greenwood Press, 1968.

_____. *Journal of a Third Voyage for the Discovery of a North-West Passage in the Years 1824–25*. London: John Murray, 1826.

_____. *Journal of a Voyage for the Discovery of a Northwest Passage from the Atlantic to the Pacific: Performed in the Years 1819–20 in His Majesty's Ships* Hecla *and* Griper. London: John Murray, 1821.

Payton, Brian. *The Ice Passage*. Toronto: Doubleday Canada, 2016.

Pearson, Mike. "'No Joke in Petticoats': British Polar Expeditions and their Theatrical Presentations." *TDR (The Drama Review)* 48 (2004): 48–59.

Peary, Robert E. *Secrets of Polar Travel*. New York: The Century Co., 1917.

Peri, Alexis. *The War Within: Diaries from the Siege of Leningrad*. Cambridge: Harvard University Press, 2017.

Petrinovich, Lewis. *The Cannibal Within*. New York: Aldine de Gruyter, 2006.

Pharand, Donat. "Canada's Sovereignty over the Northwest Passage." *Michigan Journal of International Law* 10 (1989): 653–78.

_____, and Leonard H. Legault. *The Northwest Passage: Arctic Straits*. Leyden: Brill/Nijhoff, 1984.

Phillips, Caroline. "The Camps, Cairns and Caches of the Franklin and Franklin Search Expeditions." In Patricia Sutherland, ed., *The Franklin Era in Canadian Arctic History 1845–1859*. Ottawa: National Museums of Canada, Mercury Series, Archaeological Survey of Canada Paper No. 131, 1985, 149–73.

Pilø, Lars. "Buried in Ice—The Franklin Expedition Cemetery," *Secrets of the Ice* blog. Available at https://secretsoftheice.com/news/2019/10/28/franklin-expedition/.

Piper, Karen. "Inuit Diasporas: Frankenstein and the Inuit in England." *Romanticism* 13 (2007): 63–75.

Piper, Liza, and John Sandlos. "A Broken Frontier: Ecological Imperialism in the Canadian North." *Environmental History* 12 (2007): 759–95.

Poe, Edgar Allan. *The Narrative of Arthur Gordon Pym of Nantucket*. New York: Hill & Wang, 1960 (originally published 1838).

Pope, Dudley. *The Black Ship*. London: Weidenfeld and Nicolson, 1963.

Potter, Russell. *Arctic Spectacles: The Frozen North in Visual Culture, 1818–1875*. Seattle: University of Washington Press, 2007.

_____. "Celebrating 30 Years of Franklin Fascination." *Canadian Geographic*, October 23, 2017. Available at https://www.canadiangeographic.ca/article/celebrating-30-years-franklin-fascination.

_____. "Introduction: Exploration and Sacrifice: The Cultural Logic of Arctic Discovery." In Frédéric Regard, ed., *The Quest for the Northwest Passage: British Narratives of Arctic Exploration, 1576–1874*. London: Routledge, 2013, 1–17.

_____. *Finding Franklin*. Montreal: McGill-Queen's University Press, 2016.

_____. "The Grave of Lieutenant Irving?" *Visions of the North* blog, January 3, 2010, available at https://visionsnorth.blogspot.com/search?q=John+Irving+Grave.

_____. "A Navigable Northwest Passage." *Visions of the North* blog, September 12, 2012. Available at http://visionsnorth.blogspot.com/2012/09/a-navigable-northwest-passage.html

_____. "The 'Peglar Papers' Revisited." *The Trafalgar Chronicle: The Yearbook of the 1805 Club*, 2014: 202–15. Available at http://www.ric.edu/faculty/rpotter/potter_peglar_trafchron.pdf.

_____, Regina Koellner, Peter Carney, and Mary Williamson, eds. *May We Be Spared to Meet on Earth: Letters of the Lost Franklin Arctic Expedition*. Montreal: McGill-Queen's University Press, 2022.

Powell, Brian D. "The Memorials on Beechey Island, Nunavut, Canada: An Historical and Pictorial Survey." *Polar Record* 42 (2006): 325–33.

Price, Alexa. "'Our Proudest Heritage': Masculinity, Nostalgia, and the Sailing Navy on Display, 1820–1920." Ph.D. Dissertation, The George Washington University, 2019.

Przybylak, Rajmund. "Air Temperature in the Canadian Arctic in the Mid-Nineteenth Century Based on Data from Expeditions." *Prace Geograficzne* 107 (2000): 251–258.

Pyragius, Ramona. "Journey Into White Hell: Canadian Expedition Attempts to Unravel Arctic's Greatest Mystery." *Canadian Motorist* 59 (February 1974): 12–23.

Quinn, David Beers. "The Northwest Passage in Theory and Practice." In John Logan Allen, ed., *North American Exploration*, Vol. 1. Lincoln: University of Nebraska Press, 1997, 292–343.

Rae, John. "Arctic Exploration, with Information Respecting Sir John Franklin's Missing Party." *Journal of the Royal Geographical Society*, 25 (1855): 245–56.

_____. *John Rae Arctic Explorer: The Unfinished Autobiography*. Edited by William Barr. Edmonton: University of Alberta Press, 2019.

_____. *John Rae's Arctic Correspondence 1844–1855*. Victoria: TouchWood Editions, 2014.

_____. "The Lost Arctic Voyagers." *Household Words*, December 23, 1854: 433–37.

_____. *The Melancholy Fate of Sir John Franklin, Together with The Dispatches and Letters of Captain M'Clure*. London: John Bertts, 1854.

_____. *Narrative of an Expedition to the Shores of the Arctic Sea, in 1846 and 1847*. London: T.W. Boone & Co., 1850.

_____. "Rae on the Eskimos." *The Beaver* 284 (March 1954): 38–41.

_____. "Sir John Franklin and His Crew." *Household Words*, February 3, 1855: 12–20.

Ranford, Barry. "Bones of Contention." *Equinox* 74 (1994): 69–87.

_____. "In Franklin's Footsteps." *Equinox* 72 (1994): 46–53.

_____. "More Pieces of the Franklin Puzzle." *UpHere* 2, no. 4 (1995): 36–39.

Rarick, Ethan. *Desperate Passage: The Donner Party's Perilous Journey West*. New York: Oxford University Press, 2008.

Rasmussen, Knud. *Across Arctic America: Narrative of the Fifth Thule Expedition*. Fairbanks: University of Alaska Press, 1927.

_____. "The Fifth Thule Expedition, 1921–1924: The Danish Ethnographical and Geographical Expedition from Greenland to the Pacific." *The Geographical Journal* 67 (1931): 123–38.

_____. *The Netsilik Eskimos: Social Life and Spiritual Culture*. New York: AMS Press, 1976.

Rasor, Eugene. *Reform in the Royal Navy: A Social History of the Lower Deck, 1850 to 1880*. Hamden: Archon Books, 1976.

Read, Piers Paul. *Alive: The Story of the Andes Survivors*. Philadelphia: J.B. Lippincott, 1974.

Reed, H.L., K.R. Reedy and Lawrence A. Palinkas. "Impairment in Cognitive and Exercise Performance During Prolonged Antarctic Residence: Effect of Thyroxine Supplementation in the Polar Triiodothyronine Syndrome." *Journal of Clinical Endocrinology and Metabolism* 86 (2001): 110–16.

Rennie, Donald, Benjamin Covino, Murray Blair & K. Rodahl. "Physical Regulation of Temperature in Eskimos." *Journal of Applied Physiology* 17 (1962): 326–32.

"Review of Captain Franklin's Narrative." *Christian Observer* 24 (1824): 107–18 and 163–75.

Rich, E.E. *Hudson's Bay Company, 1763–1870.* London: Hudson's Bay Record Society, 1959.

———, and A.M. Johnson, eds. *John Rae's Correspondence with the Hudson's Bay Company on Arctic Exploration, 1844–1855.* London: Hudson's Bay Record Society, 1953.

Richards, R.L. "Rae of the Arctic." *Medical History* 19 (1975): 176–93.

Richardson, Sir John. *Arctic Ordeal: The Journal of John Richardson, Surgeon-Naturalist with Franklin, 1820–1822,* Edited by C. Stuart Houston. Montreal: McGill-Queen's University Press, 1984.

———. *Arctic Searching Expedition: A Journal of a Boat Voyage Through Rupert's Land and the Arctic Sea in Search of the Discovery Ships Under Command of Sir John Franklin.* London: Longmans, Brown, Green and Longmans, 1851.

———. *The Polar Regions.* Edinburgh: Adam and Charles Black, 1861.

Richler Mordecai. *Solomon Gursky Was Here.* New York: Knopf, 1990.

Riffenburgh, Beau. *The Myth of the Explorer: The Press, Sensationalism and Geographical Discovery.* London: Belhaven Press, 1993.

Ritchie, J.C. "Northern Fiction, Northern Homage." *Arctic* 31 (1978): 69–74.

Roberts, David. "Dickens and the Arctic." *Horizon,* January 1980: 65–71.

Robinson, J.L. "Conquest of the Northwest Passage by R.C.M.P. Schooner *St. Roch*." *Canadian Geographical Journal* 30 (1945): 52–73.

Robinson, Michael. "Reconsidering the Theory of the Open Polar Sea." In Keith Rodney Benson and Helen M. Rozwadowski, eds., *Extremes: Oceanography's Adventure at the Poles.* Sagamore Beach: Science History Publications/USA, 2007, 15–29.

Rodahl, R., and T. Moore. "The Vitamin A Content and Toxicity of Bear and Seal Liver." *Biochemical Journal* 37 (1943): 166–68.

Rodger, N.A.M. "Commissioned Officers' Careers in the Royal Navy, 1690–1815." *Journal for Maritime Research* 3 (2001): 85–129.

———. "Shipboard Life in the Old Navy: The Decline of the Old Order." In Lewis R. Fischer, et al., eds. *,The North Sea: Twelve Essays on Social History of Maritime Labour.* Stavanger: Maritime Museum, 1992: 29–39.

Rogers, Nicholas. "The Politics of Mutiny: The Pompée at Spithead and Beyond." *International Journal of Maritime History* 33 (2021): 464–88.

Rogers, Stan. "Northwest Passage". 1981. Available at https://www.azlyrics.com/lyrics/stanrogers/northwestpassage.html.

Roland, Charles G. "Saturnism [Lead Poisoning] at Hudson's Bay: The York Factory Complaint of 1833–1836." *Canadian Bulletin of Medical History* 1 (1984): 59–78.

Rondeau, Robin M. "The Wrecks of Franklin's Ships *Erebus* and *Terror*; Their Likely Location and the Cause of Failure of Previous Search Expeditions." *Journal of the Hakluyt Society,* March 2010: 1–11.

Roobol, M. John. *Franklin's Fate: An Investigation into What Happened to the Lost 1845 Expedition of Sir John Franklin.* Canterbury: The Conrad Press, 2019.

———. "Status of the History of the Lost Franklin Expedition of 1845." *Academia Letters* 1461, July 2021. Available at https://www.academia.edu/49826555/Status_of_the_History_of_the_Lost_Franklin_Expedition_of_1845.

———. *Trapped.* Canterbury: The Conrad Press, 2019.

Rosen, L., K.H. Knudson, and P. Fancher. "Prevalence of Seasonal Affective Disorder Among U.S. Army Soldiers in Alaska." *Military Medicine* 167 (2002): 581–84.

Rosenberg, Roger. "The Donner Party: Trapped in the Mountains." *Academia Letters* 2021. Available at https://doi.org/10.20935/AL2482.

Ross, James Clark. *A Voyage of Discovery and Research in the Southern and Antarctic Regions During the Years 1839–1843.* London: John Murray, 1847, reprint, Cambridge: Cambridge University Press, 2011.

———, and James M. Savelle. "Retreat from Boothia: The Original Diary of James Clark Ross, May to October 1832." *Arctic* 45 (1992): 179–94.

———, and James M. Savelle. "'Round Lord Mayor Bay with James Clark Ross: The Original Diaries of 1830." *Arctic* 43 (1990): 66–79.

Ross, John *Narrative of a Second Voyage in Search of the Northwest Passage and of a Residence in the Arctic Regions, 1829–33.* London: A.W. Webster, 1835.

———. *Observations on a Work Entitled 'Voyages of Discovery and Research Within the Arctic Regions,' by Sir John Barrow, Being a Refutation of Numerous Representations Contained in That Volume.* Edinburgh: William Blackwell, 1846.

———. *A Voyage of Discovery, Made Under the Orders of the Admiralty, in His Majesty's Ships Isabella and Alexander, for the Purpose of Exploring Baffin's Bay, and Inquiring into the Probability of a North-west Passage.* London: John Murray, 1819.

Ross, Maurice J. *Polar Pioneers: John Ross and James Clark Ross.* Montreal: McGill-Queen's University Press, 1994.

———. *Ross in the Antarctic: The Voyages of James Clark Ross in Her Majesty's Ships* Erebus

and Terror 1839–1843. Whitby: Caedmon of Whitby, 1982.
Ross, W. Gillies. "The Admiralty and the Franklin Search." *Polar Record* 40 (2004): 289–301.
_____. "The Arctic Council of 1851: Fact or Fancy?" *Polar Record* 40 (2004): 135–41.
_____. *Arctic Whalers, Icy Seas: Narratives of the Davis Strait Whale Fishery*. Toronto: Irwin, 1985.
_____. "Clairvoyants and Mediums Search for Franklin." *Polar Record* 39 (2003): 1–18.
_____. "False Leads in the Franklin Search." *Polar Record* 39 (2003): 131–60.
_____. "The Gloucester Balloon: A Communication from Franklin?" *Polar Record* 38 (2002): 11–22.
_____. *Hunters on the Track: William Penny and the Search for Franklin*. Montreal: McGill-Queen's University Press, 2019.
_____. "Nineteenth-Century Explorations of the Arctic." In John Logan Allen, ed., *North American Explorations*, Vol. 3. Lincoln: University of Nebraska Press, 1997, 244–331.
_____. "The Type and Number of Expeditions in the Franklin Search, 1847–1859." *Arctic* 65 (2002): 57–69.
_____. "William Penny (1809–1892)." *Arctic* 36 (1983): 380–81.
Rowbotham, Sheila. "Canned Food Sealed Sailors' Fate." *History Today* 37 (1987). Available at https://www.historytoday.com/archive/canned-food-sealed-icemens-fate.
Ruddock, Alvyn A. "Columbus and Iceland: New Light on an Old Problem." *The Geographical Journal* 136 (1970): 177–89.
Ruggles, Richard I. *A Country So Interesting: The Hudson's Bay Company and Two Centuries of Mapping, 1670–1870*. Montreal: McGill-Queen's University Press, 1991.
Rule, B.J. *Polar Knight: The Mystery of Sir John Franklin*. New Smyrna Beach: Luthers Publishing, 1998.
Russell, Penny. "Wife Stories: Narrating Marriage and Self in the Life of Jane Franklin." *Victorian Studies* 48 (2005): 35–57.
_____, ed. *This Errant Lady: Jane Franklin's Overland Journey to Port Philip and Sydney, 1839*. Canberra: National Library of Australia, 2002.
Rutter, Owen, ed. *The Court Martial of the Bounty Mutineers*. London: Hodge, 1931
Ryan, Karen. *Death in the Ice: The Mystery of the Franklin Expedition*. Gatineau: Canadian Museum of History, 2018.
St. Maur, Gerald. *Odyssey Northwest*. Edmonton: Boreal Institute for Northern Studies, 1983.
Saladié, P., and A. Rodriguez-Hidalgo. "Archaeological Evidence for Cannibalism in Pre-Historic Western Europe from Homo Antecessor to the Bronze Age." *Journal of Archaeological Method and Theory* (2016). Available at doi: 10.1007/s10816-016-9306-y.
Sanday, Peggy Reeves. *Divine Hunger: Cannibalism as a Cultural System*. New York: Cambridge University Press, 1986.
Sandler, Martin. *Resolute: The Epic Search for the Northwest Passage and Sir John Franklin, and the Discovery of the Queen's Ghost Ship*. New York: Sterling, 2006.
Savelle, James M. "Effects of Nineteenth Century European Exploration on the Development of the Netsilik Inuit Culture." In Patricia D. Sutherland, ed., *The Franklin Era in Canadian Arctic History 1845–1859*. Ottawa: National Museums of Canada, Mercury Series, Archaeological Survey of Canada Paper No. 131, 1985, 192–214.
_____, and Clive Holland. "John Ross and Bellot Strait: Personality versus Discovery." *Polar Record* 23 (1987): 411–17.
Savitt, Ronald. "Frederick Schwatka and the Search for the Franklin Expedition Records, 1878–1880." *Polar Record* 44 (2008): 193–210.
Savours, Ann. "The British Admiralty and the Arctic 1773–1876." *Pole Nord*, 1983: 153–67.
_____. *The Northwest Passage in the Nineteenth Century: Perils and Pastimes of a Winter in the Ice*. London: Hakluyt Society, 2003.
_____. *The Search for the Northwest Passage*. New York: St. Martin's Press, 1999.
_____, and Margaret Deacon. "Nutritional Aspects of the British Arctic (Nares) Expedition of 1875–76 and Its Predecessors." In J. Watt, E.J. Freeman and W.F. Bynum, eds., *Starving Sailors: The Influence of Nutrition upon Naval and Maritime History*. Greenwich: National Maritime Museum, 1981, 131–62.
Schaeffer, Jonathan. *Toward No Earthly Pole: Letters from John Franklin's Last Expedition*. Amazon, 2019.
Schulz, Kathryn. "Literature's Arctic Obsession." *New Yorker*, April 24, 2017. Available at https://www.newyorker.com/magazine/2017/04/24/-literatures-arctic-obsession4-26.
Schutt, Bill. *Cannibalism: A Perfectly Natural History*. Chapel Hill: Algonquin Books, 2018.
Schwartz, B.S., and W.F. Stewart. "Lead and Cognitive Function in Adults: A Question and Answers Approach to a Review of the Evidence for Cause, Treatment and Prevention." *International Review of Psychiatry* 19 (2007): 671–92.
Schwatka, Frederick. *The Long Arctic Search: The Narrative of Lt. Frederick Schwatka, USA, 1878–1880, Seeking the Records of the Lost Franklin Expedition*. Edited by Edouard Stackpole. Mystic: Marine Historical Association, 1965.
Schweger, Barbara. "Documentation and Analysis of the Clothing Worn by Non-Native Men in the Canadian Arctic Prior to 1920, with an Emphasis on Footwear." Ph.D. Dissertation, University of Alberta, 1983.
Scoresby, William. *The Arctic Whaling Journals of William Scoresby*. London: Hakluyt Society, 2003.

_____. *The Franklin Expedition, or, Considerations on Measures for the Discovery and Relief of Our Absent Adventurers in the Arctic Regions.* London: Longman, Brown, Green & Longmans, 1850.

_____. "On the Greenland or Polar Ice." *Memoirs of the Wernerian* Society 2 (1815): 268–338, reprint, Whitby: Caedmon of Whitby, 1918.

Scott, G. Richard, and Sean McMurry. "The Delicate Question: Cannibalism in Prehistoric and Historic Times." In Kelly Dixon, Julie Schablitsky and Shannon Novak, eds., *An Archaeology of Desperation: Exploring the Donner Party's Alder Creek Camp.* Norman: University of Oklahoma Press, 2011, 211–43.

"Sea Ice: Terminology, Formation and Movement." *Polar Record* 4 (1944): 126–33.

Seaver, Kirsten A. *The Last Vikings: The Epic Story of the Great Norse Voyages.* London: Bloomsbury Academic, 2021.

Seeman, Berthold. *Narrative of the Voyage of HMS Herald During the Years 1854–51 Under the Command of Captain Henry Kellett R.N.* London: Reeve & Co., 1853.

Shapton, Leanne. "A Record of Oblivion." *New York Times Magazine,* March 20, 2016: 72–79.

Shartman, Ivan M. "Vitamin Requirements of the Human Body." In J. Watt, E.J. Freeman and W.F. Bynum, eds., *Starving Sailors: The Influence of Nutrition upon Naval and Maritime History.* Greenwich: National Maritime Museum, 1981, 17–26.

Shaw, R.D. "Three-Day Visitors: The Samo Response to Colonialism in Western Province, Papua New Guinea." In N.M. McPherson, ed., *Colonial New Guinea: Anthropological Perspectives.* Pittsburgh: University of Pittsburgh Press, 2001, Ch. 9.

Shelley, Mary Wollstonecraft. *Frankenstein, or the Modern Prometheus.* London: Penguin Classics, 2018 (originally published 1818).

Shillinglaw, John. *Narrative of Arctic Discovery.* London: William Shoberl, 1851.

Shirley, Ann. "Lead Poisoning and the Franklin Expedition." *Polar Record* 28 (1992): 73.

Sides, Hampton. *In the Kingdom of Ice.* New York: Doubleday, 2014.

Simmonds, Peter L. *Sir John Franklin and the Arctic Regions,* 6th ed. Buffalo: George H. Derby & Co., 1852.

Simmons, Dan. *The Terror.* Boston: Little, Brown, 2007.

Simpson, Alexander. *The Life and Travels of Thomas Simpson, Arctic Traveler.* London: Richard Bentley, 1845.

Simpson, A.W. Brian. *Cannibalism and the Common Law: The Story of the Tragic Last Voyage of the* Mignonette *and the Strange Legal Proceedings to Which It Gave Rise.* Chicago: University of Chicago Press, 1984.

_____. "Regina v. Archer and Muller (1875): The Leading Case That Never Was." *Oxford Journal of Legal Studies* 2 (1982): 181–96.

Simpson, Lindsay. *The Curer of Souls.* Sydney: Random House Australia, 2006.

Simpson, Thomas. *Narrative of the Discoveries on the North Coast of America.* London: Richard Bentley, 1843.

Smith, Kiona. "Strands of Hair Shed Light on Doomed 19th-century Arctic Expedition." *Ars Technica online,* September 30, 2018. Available at https://arstechnica.com/science/2018/09/-did-lead-poisoning-finish-off-a-doomed-arctic-expedition/.

Smith, Michael. *Captain Francis Crozier: Last Man Standing?* Cork: Collins, 2006.

_____. *Icebound in the Arctic: The Mystery of Captain Francis Crozier and the Franklin Expedition.* Dublin: The O'Brien Press, 2021.

_____, *Polar Crusader: Sir James Wordle—Exploring the Arctic and Antarctic.* Edinburgh: Birlinn, 2004.

Smucker, Samuel, ed. *Arctic Explorations and Discoveries during the Nineteenth Century.* New York: C.M. Sexton, 1859.

Smyth, William Henry, and Sir Edward Belcher. *The Sailor's Word-book.* London: Blackie & Son, 1867.

Snaebsjornsdottir, Bryndis, and Mark Wilson. *Nanoq: Flat Out and Bluesome, A Cultural Life of Polar Bears.* London: Black Dog, 2006.

Snelling, William J. *The Polar Regions of the Western Continent Explored.* Boston: W.W. Read, 1831.

Snow, William Parker. *Voyage of the Prince Albert in Search of Sir John Franklin: A Narrative of Every-day Life in the Arctic Seas.* London: Longman, Green, Brown & Longman, 1851.

Solomon, Evan. "What Really Happened to the Franklin Expedition?" *Outpost Magazine,* August 13, 2014. Available at https://outpostmagazine.com/the-lost-expedition-the-fall-of-franklin-s-men/.

Solway, David. *Franklin's Passage.* Montreal: McGill-Queen's University Press, 2003.

Sorensen, Chris. "HMS *Terror:* How the Final Franklin Ship was Found." *Maclean's,* September 20, 2016. Available at https://www.macleans.ca/news/canada/how-trust-led-to-hms-terror/.

Sorokin, Pitirim. *Hunger as a Factor in Human Affairs.* Gainesville: University Presses of Florida, 1975.

_____. *Man and Society in Calamity: The Effects of War, Revolution, Famine, Pestilence Upon the Human Mind, Behavior, Social Organization and Cultural Life.* New York: E.P. Dutton, 1942.

Spenneman, Dirk H.R. "Cannibalism in Fiji: The Analysis of Butchering Marks on Human Bones and the Historical Record." *Domodomo* 2 (1997): 29–46

Spufford, Francis. *I May Be Some Time: Ice and the English Imagination.* New York: St. Martin's Press, 1996.

Stackpole, Edouard A., ed. *The Long Arctic*

Search: The Narrative of Lieutenant Frederick Schwatka, U.S.A., 1878–1880, Seeking the Records of the Lost Franklin Expedition. Chester: Pequot Press, 1977.

Stam, David H., and Deirdre Stam. *Adventures in Polar Reading: The Book Cultures of the High Latitudes.* New York: The Grolier Club, 2019.

_____. *Books on Ice: British and American Literature of Polar Exploration.* New York: The Grolier Club, 2005.

_____. "Silent Friends: Books and Reading on Polar Expeditions." In K. Caning and V.S. Jakobsen, eds., *Poles Apart—Poles Online: Proceedings of the 19th Polar Libraries Colloquy.* Copenhagen: Danish Polar Centre, 2002, 113–19.

Stamp, Tom, and Cordelia Stamp. *William Scoresby, Arctic Scientist.* Whitby: Caedmon of Whitby, 1976.

Stark, Peter. *Driving to Greenland.* New York: Lyons & Burford, 1994.

Stark, Suzanne. *Female Tars: Women Aboard Ship in the Age of Sail.* Annapolis: Naval Institute Press, 2017.

Steadman, David W., Susan C. Antón, and Patrick V. Kirch. "Ana Manuku: A Prehistoric Ritual Site on Mangaia, Cook Islands." *Antiquity* 74 (2000): 873–83.

Steele, Peter. *The Man Who Mapped the Arctic: The Intrepid Life of George Back, Franklin's Lieutenant.* Vancouver: Raincoast Books, 2003.

Steffansson, Vilhjalmur. "Arctic Controversy: The Letters of John Rae." *Geographical Journal* 120 (1954): 486–93.

_____. *The Friendly Arctic.* New York: Macmillan, 1921.

_____. *My Life with the Eskimo.* London: Macmillan, 1913.

_____. "Rae's Arctic Correspondence." *The Beaver* 248 (March 1954): 36–37.

_____. *Unsolved Mysteries of the Arctic.* New York: Macmillan, 1939.

Stein, Glenn M. *Discovering the Northwest Passage: The Four-Year Arctic Odyssey of H.M.S. Investigator and the McClure Expedition.* Jefferson: McFarland, 2015.

_____. "'A Most Inhospitable Coast: The Report of Lieutenant William Hobson's 1859 Search for the Franklin Expedition on King William Island," *Arctic* 67 (2014), 511–22.

_____. "Scattered Memories and Frozen Bones: Revealing a Sailor of the Franklin Expedition, 1845–1848." *Journal of the Orders and Medals Research Society* 46 (2007): 224–32.

Stenton, Douglas R. "'Finding Harry Peglar': Re-Examining the Discovery of a Franklin Expedition Sailor's Skeleton by the 1859 McClintock Search Expedition." *Polar Record* 59 (2022). Available at doi:10.1017/S0032247422000237.

_____. "Finding the Dead: Bodies, Bones and Burials from the 1845 Franklin Northwest Passage Expedition." *Polar Record* 54 (2018): 197–212.

_____, Stephen Fratpietro, Anne Keenleyside, and Robert W. Park, "DNA Identification of a Sailor from the 1845 Franklin Northwest Passage Expedition." *Polar Record* 57 (2021): e14. Available at doi:10.1017/S0032247421000061.

_____, Anne Keenleyside, Stephen Fratpietro, and Robert Park. "DNA Analysis of Human Skeletal Remains from the 1845 Franklin Expedition." *Journal of Archaeological Science: Reports* 16 (2017): 409–17.

_____, Anne Keenleyside, Phillippe Froesch, and Robert W. Park. "A Franklin Expedition Officer's Burial at Two Graves Bay, King William Island, Nunavut." *Journal of Archaeological Science: Reports* 35 (2021): 102687.

_____, Anne Keenleyside, and Robert W. Park. "The 'Boat Place' Burial: Skeletal Evidence from the 1845 Franklin Expedition." *Arctic* 68 (2015): 32–44.

_____, Anne Keenleyside, Diana Trepkov, and Robert W. Park. "Faces from the Franklin Expedition? Craniofacial Reconstructions of Two Members of the 1845 Northwest Passage Expedition." *Polar Record* 52 (2016): 76–81.

_____, and Robert W. Park. "The 'Cast Iron Site'—A Tale of Four Stoves from the 1845 Franklin Northwest Passage Expedition." *Arctic* 73 (2020): 1–12.

_____, and Robert W. Park. "History, Oral History and Archaeology: Reinterpreting the "Boat Places" of Erebus Bay." *Arctic* 70 (2017): 203–18.

Stephens, Simon. *Heisenberg.* London: Bloomsbury Methuen Drama, 2015.

Stern, Pamela. *Daily Life of the Inuit.* Santa Barbara: Greenwood/ABC-CLIO, 2010.

Stevenson, Alex. "Feminine Fortitude—Part 1 of the Story of Jane Lady Franklin." *North* 15 (March-April 1968): 34–39.

Stewart, David J. "'Rocks and Stones I'll Fear No More': Anglo-American Maritime Memorialization 1700–1940." Ph.D. Dissertation, Texas A&M University, 2004.

Stewart, W.F., and B.S. Schwartz. "Effects of Lead on the Adult Brain: a 15-Year Exploration." *American Journal of Industrial Medicine* 50 (2007): 729–39.

Stirling, Ian. *Polar Bears.* Ann Arbor: University of Michigan Press, 1998.

Stone, Harry. *The Night Side of Dickens: Cannibalism, Passion, Necessity.* Columbus: Ohio State University Press, 1994.

Stone, Ian R. "The Arctic Portraits of Stephen Pearce." *Polar Record* 24 (1988): 55–58.

_____. "'The Contents of the Kettles': Charles Dickens, John Rae and Cannibalism on the 1845 Franklin Expedition." *The Dickensian,* 83 (1987): 7–16.

_____. "An Episode in the Franklin Search: The Prince Albert Expedition 1850." Part 1, *Polar Record* 29 (1993): 127–42, and Part 2, *Polar Record* 29 (1993): 197–208.

———. "Fairholme v Fairholme's Trustees: Presumption of Death on the 1845 Franklin Expedition." *Polar Record* 32 (1996): 239–41.

———. "The Franklin Search in Parliament." *Polar Record* 32 (1996): 217–28.

Strachan, Graham. "The Terror: Could Hair Strands from Fife Doctor Harry Goodsir Solve Skeleton Riddle?" *The Courier*, March 4, 2021. Available at https://www.thecourier.co.uk/fp/nostalgia/2020351/the-terror-could-hair-strands-from-fife-doctor-harry-goodsir-solve-skeleton-riddle/.

Struzik, Edward. *Northwest Passage*. Toronto: Key Porter, 1991.

Stuster, Jack W. "Bold Endeavors: Behavioral Lessons from Polar and Space Exploration." *Gravitational and Space Biology Bulletin* 13, no. 2 (2000): 49–58.

———. *Bold Endeavors: Lessons from Polar and Space Exploration*. Annapolis: Naval Institute Press, 1996.

Surtees, Robert Smith. *Mr. Sponge's Sporting Tour*. London: Bradbury & Evans, 1853.

Sutherland, Patricia D., ed. *The Franklin Era in Canadian Arctic History 1845–1859*. Ottawa: National Museums of Canada, Mercury Series, Archaeological Survey of Canada Paper No. 131, 1985.

Sutherland, Peter C. *Journal of a Voyage in Baffin's Bay and Barrow Straits*. London: Longmans, Green, Brown & Longmans, 1852.

Swanston, Treena, et al. "Franklin Expedition Lead Exposure: New Insights from High Resolution Confocal X-ray Fluorescence Imaging of Skeletal Microstructure." *PLOS One* 13, no. 8 (2018). Available at https://journals.plos.org/plosone/article?id=10.1371/journal.pone.0202983.

Sweet, Jessie M. "Robert Jameson and the Explorers: The Search for the Northwest Passage." *Annals of Science* 31 (1974): 21–47.

Swinburne, Algernon Charles. *The Death of Sir John Franklin*. In E. Gosse and T.J. Wise, eds., *The Posthumous Poems of Algernon Charles Swinburne*. London: William Heineman, 1917, 79.

Taichman, Russell H., Tom Gross, and Mark P. MacEachern. "A Critical Assessment of the Oral Condition of the Crew of the Franklin Expedition." *Arctic* 70 (2017): 25–36.

Talbot, Mary Ann. *The Life and Surprising Adventures of Mary Ann Talbot, in the Name of John Taylor*. London: R.S. Kirby, 1809, reprint, Paul Royster, ed., Lincoln: UNL Libraries Faculty Publication No. 32, 2006.

Tanaka, Yuki. *Hidden Horrors: Japanese War Crimes in World War II*, 2nd ed. Lanham: Rowman & Littlefield, 2017.

Tapley, Caroline. *John Come Down the Backstay*. New York: Atheneum, 1974.

Taylor, Andrew. *Arctic Blue Books: Parliamentary Papers on Exploration in the Canadian North*. Washington, D.C.: Government Printing Office, 1959. Index available online at https://www.umanitoba.ca/faculties/arts/anthropology/bluebooks/.

Taylor, Charles. "Foucault on Freedom and Truth." *Political Theory* 12 (1984): 152–83.

Taylor, Stephen. *Sons of the Waves: The Common Seaman in the Heroic Age of Sail*. New Haven: Yale University Press, 2020.

Thacher, Dana. "Salvaging on the Coast of Erebus Bay, King William Island: An Analysis of Inuit Interactions with Material from the Franklin Expedition." *Arctic* 71 (2018): 431–43.

Thiess, Derek J. "Dan Simmons's The Terror, Inuit 'Legend,' and the Embodied Horrors of History." *Journal of the Fantastic in the Arts* 29 (2018): 222–41.

Thomas, David N. *Frozen Oceans: The Floating World of Pack Ice*. London: Firefly, 2004.

Thompson, E.P. *The Making of the English Working Class*. London: Victor Gollancz, 1963.

Thorén, Ragnar. *Picture Atlas of the Arctic*. Amsterdam: Elsevier, 1969.

Todd, A.L. *Abandoned: The Story of the Greeley Arctic Expedition 1881–1884*. New York: McGraw-Hill, 1961.

Tomlinson, Barbara. "The Explorers of the North-West Passage: Claims and Commemoration." *Church Monuments* 22 (2007): 111–32.

Trafton, Stephen. "Did Lead Poisoning Contribute to the Deaths of Franklin Expedition Members?" *Information North* 15, No. 9 (Nov. 1989).

———. "The Location and Mapping of the Camps and Cairns Associated with Sir John Franklin's Last Arctic Expedition on the Northern Shores of King William Island." *The Mariner's Mirror* 77 (1991): 407–15.

Traill, H.D. *Life of Sir John Franklin, RN*. London: John Murray, 1896.

Travis-Henikoff, Carole A. *Dinner with a Cannibal*. Santa Monica: Santa Monica Press, 2008.

Trimmer, John D. "The Present Situation in Quantum Mechanics: A Translation of Schrödinger's 'Cat Paradox' Paper." *Proceedings of the American Philosophical Society* 124 (1980): 323–38.

Trocki, Carl A. *Opium, Empire and the Global Political Economy: A Study of the Asian Opium Trade, 1750–1950*. London: Routledge, 1999.

Troubetzkoy, Alexis S. *Arctic Obsession: The Lure of the Far North*. New York: St. Martin's Press, 2011.

Turner, Christy G., II, and Jacqueline Turner. *Man Corn: Cannibalism and Violence in the Pre-Historic American Southwest*. Salt Lake City: University of Utah Press, 1999.

———. "Cannibalism in Chaco Canyon: The Charnel Pit Excavated in 1926 at Small House Ruin by Frank H.H. Roberts, Jr." *American Journal of Physical Anthropology* 91 (1993): 421–39.

———. "Cannibalism in the Prehistoric American Southwest: Occurrence, Taphonomy, Explanation and Suggestions for Standardized World Definition." *Anthropological Science* 103 (1995): 1–22

Turner, Joseph Addison. *The Discovery of Sir John Franklin and Other Poems*. Mobile, 1858.
Tyrrell, J.B. "A Story of a Franklin Search Expedition." *Transactions of the Canadian Institute* 8 (1908–09): 393–402.
Uusma, Bea. *The Expedition: A Love Story*. Translated by Agnes Broom. London: Head of Zeus, 2014.
Vaughan, Richard. *The Arctic: A History*. Dover: Sutton, 1994.
Verne, Jules. *The Voyages and Adventures of Captain Hatteras*. Boston: James R. Osgood & Co., 1874.
Victory Point Record. Available at https://commons.wikimedia.org/wiki/File:Franklin expeditionnote.jpg#/media/File:Franklin expeditionnote.jpg.
Vidal, Owen Alexander. *A Poem Upon the Life and Character of Sir John Franklin*. Oxford: T. & G. Shrimpton, 1860.
Villa, Paola. "Cannibalism in Prehistoric Europe." *Evolutionary Anthropology* 1 (1992): 93–104.
Vollmann, William T. *The Rifles*. New York: Viking, 1994.
Wadhams, Peter. *A Farewell to the Ice: A Report from the Arctic*. New York: Oxford University Press, 2017.
_____, and Maria Pia Casarini, "Signs of Life." *Geographical Magazine* 66(4) (1994): 26–27.
Wahlgren, Eric. *The Vikings and America*. New York: Thames and Hudson, 1986.
Wallace, Hugh N. *The Navy, the Company and Richard King: British Exploration in the Canadian Arctic, 1829–1860*. Montreal: McGill-Queen's University Press, 1980.
_____. "Private Expeditions in the Search for Sir John Franklin." In Patricia D. Sutherland, ed., *The Franklin Era in Canadian Arctic History, 1845–1859*. Ottawa: National Museums of Canada, Mercury Series, Archaeological Survey of Canada Paper No. 131, 1985, 42–53.
Wallace, R.C. "Rae of the Arctic." *The Beaver* 284 (March 1954): 28–33.
Walpole, Garth. *Relics of the Franklin Expedition*. Edited by Russell Potter. Jefferson: McFarland, 2017.
Wamsley, Douglas, and William Barr. "Early Photographers of the Arctic." *Polar Record* 32 (1996): 295–316.
Ware, Chris. *The Bomb Vessel: Shore Bombardment Ships of the Age of Sail*. Annapolis: Naval Institute Press, 1994.
Warkentin, Germaine. Introduction. In Germaine Warkentin, ed., *Canadian Exploration Literature*. Toronto: Oxford University Press, 1990, ix–xxi.
Waterman, Jonathan. *Arctic Crossing: A Journey Through the Northwest Passage and Inuit Culture*. Lanham: Lyons Press, 2001.
Waterman, Laura. *Starvation Shore*. Madison: University of Wisconsin Press, 2019.
Watson, Paul. *Ice Ghosts: The Epic Hunt for the Lost Franklin Expedition*. New York: W.W. Norton, 2017.
Watt, J. "Some Consequences of Nutritional Disorders in Eighteenth-Century British Circumnavigations." In J. Watt, E.J. Freeman and W.F. Bynum, *Starving Sailors: The Influence of Nutrition upon Naval and Maritime History*. Greenwich: National Maritime Museum, 1981, 51–71.
Weiss, Peter. *The Aesthetics of Resistance*, Vol. 2. Durham: Duke University Press, 2020 (originally published in German 1978).
Welky, David. *A Wretched and Precarious Situation: in Search of the Last Arctic Frontier*. New York: Norton, 2017.
Wetherell, David. "Accounts of Fighting and Cannibalism in Eastern New Guinea During the Missionary Contact Period, 1877–1888, as Told to Charles Abel." *Pacific Studies* 26 (2003): 37–52.
Wheatley, Kim. "The Arctic in the *Quarterly Review*." *European Romantic Review* 20 (2009): 465–90.
Wheeler, Sara. *The Magnetic North: Notes from the Arctic Circle*. New York: North Point Press, 2009.
Whidden, John D. *Ocean Life in the Old Sailing Days*. Boston: Little, Brown, 1908.
White, Rosanna. "Ceremonies of Possession: Performing Sovereignty in the Canadian Arctic." Ph.D. Dissertation, University of London, 2019.
White, Tim D. *Prehistoric Cannibalism at Mancos—5MTUMR 2346*. Princeton: Princeton University Press, 1992.
Whitfield, Jerome T., Wandagi H. Pako, John Collinge, and Michael P. Alpers. "Mortuary Rites of the South Fore and Kuru." *Philosophical Transactions of the Royal Society B: Biological Sciences* 363 (2008): 3721–24.
Whitridge, Peter. "Landscapes, Houses, Bodies, Things: 'Place' and the Archaeology of Inuit Imaginaries." *Journal of Archaeological Method and Theory* 11 (2004): 213–48.
Wiebe, Rudy. *A Discovery of Strangers*. Toronto: Alfred A. Knopf Canada, 1994.
_____. *Playing Dead: A Contemplation Concerning the Arctic*. Edmonton: NeWest, 1989.
Wilkinson, Douglas. *Arctic Fever: The Search for the Northwest Passage*. Toronto: Clarke Irwin & Co., 1971.
Wilkinson-Latham, Robert. *The Royal Navy 1790–1970*. London: Osprey, 1977.
Williams, Glyn. *Arctic Labyrinth: The Quest for the Northwest Passage*. London: Allen Lane, 2009.
_____. *Voyages of Delusion: The Search for the Northwest Passage in the Age of Reason*. New York: HarperCollins, 2002.
Williams, Guy. *The Age of Miracles: 19th Century Medicine*. Chicago: Academy Chicago, 1987.
Williams, Naomi. *Landfalls*. New York: Farrar, Straus & Giroux, 2015.

Wilson, Eric. *The Spiritual History of Ice: Romanticism, Science and the Imagination*. New York: Palgrave Macmillan, 2003.

Wilson, Evan. *A Social History of British Naval Officers, 1775–1815*. Martlesham: Boydell Press, 2017.

Wilson, Harry. "Sir John Franklin's HMS Terror Believed Found in Arctic." *Canadian Geographic*, September 12, 2016. Available at https://www.canadiangeographic.ca/article/-sir-john-franklins-hms-terror-believed-found-arctic.

Wilson, John. *Graves of Ice: The Lost Franklin Expedition*. Toronto: Scholastic Canada, 2014.

———. *Lost (Orca Currents)*. Vancouver: Orca Book, 2016.

———. *North with Franklin: The Lost Journals of James Fitzjames*. Markham: Fitzhenry and Whiteside, 1999.

Wilson, Malcolm. "Sir John Ross's Last Expedition, in Search of Sir John Franklin." *The Musk-Ox* 1973, 3–11.

Winton, John. *Hurrah for the Life of a Sailor: Life on the Lower Deck of the Victorian Navy*. London: Michael Joseph, 1977.

Witzenburg, Frankie. "The Lost Franklin Expedition." *Naval History Magazine*, October 2020. Available at https://www.usni.org/magazines/naval-history-magazine/2020/october/lost-franklin-expedition.

Wolchover, Natalie. "A Different Kind of Theory of Everything." *New Yorker*, February 19, 2019. Available at https://www.newyorker.com/science/elements/a-different-kind-of-theory-of-everything.

Wonders, William C. "Project Franklin." *Polar Record* 14 (1968): 333–36.

———. "Search for Franklin." *Canadian Geographic Journal* 5 (1968): 116–27.

Woodcock, G. "The Search for Franklin." *History Today* 20 (1970): 686–94.

Woodman, David C. "Inuit Accounts and the Franklin Mystery." In John Moss, ed., *Echoing Silence: Essays on Arctic Narrative*. Ottawa: University of Ottawa Press, 1997, 53–59.

———. "Probing the Franklin Mystery." *Canadian Geographic*, 1995. Available at https://www.canadiangeographic.ca/sites/cgcorp/files/images/web_articles/2014-victoria-strait-expedition/docs/ 1995_marapr_canadian_geographic.pdf.

———. *Strangers Among Us*. Montreal: McGill-Queen's University Press, 1996.

———. *Unravelling the Franklin Mystery: Inuit Testimony*, 2d ed. Montreal: McGill-Queen's University Press, 2015.

Woodward, Frances J. "The Franklin Search in 1850." *Polar Record* 5 (1950): 532–42.

———. "Joseph René Bellot, 1826–53." *Polar Record* 5 (1950): 398–407.

———. *Portrait of Jane: A Life of Lady Franklin*. London: Hodder & Stoughton, 1951.

Wordie, J.M., and Richard J. Cyriax. "Centenary of the Sailing of Sir John Franklin with the *Erebus* and *Terror*." *Geographical Journal* 106 (1945): 169–95.

Wright, John K. "The Open Polar Sea." *Geographical Review* 43 (1953): 338–65.

Wright, Noel. *New Light on Franklin*. Ipswich: W.J. Cowell, 1949.

———. *The Quest for Franklin*. London: Heinemann, 1959.

Wylie, Herb. *Speculative Fictions: Contemporary Canadian Novelists and the Writing of History*. Montreal: McGill-Queen's University Press, 2002.

Yang, Ji-Sheng. *Tombstone: The Great Chinese Famine, 1958–1962*. New York: Farrar, Straus & Giroux, 2012.

Young, Delbert. "Was There an Unsuspected Killer Aboard the *Unicorn*?" *The Beaver* 304, no. 3 (1973): 4–15.

Zweig, Paul. *The Adventurer*. London: Dent, 1974.

Films and TV Series

Arctic Ghost Ship (NOVA, 2015).
Arctic Passage—Prisoners of the Ice (NOVA, 2005).
Arctic Tomb (History Channel, 2001).
Buried in Ice—The Franklin Expedition (NOVA, 1988).
Franklin's Lost Expedition (Crossing the Line Pictures, 2005).
Franklin's Lost Ships (Lion TV/90th Parallel Productions-CBC, 2015).
The Mysterious Franklin Disappearance (CBC, 1994).
The North Water (BBC/CBC, 2021).
The Northwest Passage—The Last Great Frontier (Mill Creek, 2014).
Passage (John Walker, 2008).
Scott of the Antarctic (Ealing Studios, 1948).
The Terror (AMC, 2018–19).

Websites

Arctic.mysticseaport.org.
Arcticnorthwestpassage.blogspot.com.
https://battersby-archive.wixsite.com/website/blog?fbclid=IwAR2yHeg7Fwiv7TBl5i1qdMAJjT7DsSDJ4BstEd5_QglWY5nW3zwvmOL0mso.
Canada's Digital Collections—John Franklin. http://collections.ig.ca/arctic/explorer/franklin.htm.
Erebusandterrorfiles.blogspot.com.
Franklin in the Public Eye, www.ric.edu/rpotter/publiceye.html.
The Franklin Trail, www.franklintrail.com.
Franklinexpedition.blogspot.com

Franklin-Expedition.fandom.com
Franklinsghost.blogspot.com.
Illuminator.blog (Logan Zachary). www.illuminator.blog.
Irish Antarctic Explorers, http://www.iol.ie/~south-aris/irishexp.htm.
Kabloonas.blogspot.com.
KenMcGoogan.blogspot.com.
Kieranmulvaney.com.
National Maritime Museum, www.nmm.ac.uk/education/fact_franklin.html.
Parks Canada archaeology: https://www.pc.gc.ca/en/lhn-nhs/nu/epaveswrecks/culture/archeologie-archeology/explore/subaquatique-underwater.
Parks Canada tour of HMS *Terror*: https://www.youtube.com/watch?v=OxyTZ3F7mkA.
Parson's World of Sir John Franklin, http://homepages.enterprise.net/rogerp/franklin.html.
Remembering the Franklin Expedition—Facebook group.
Sir John Franklin, www.cronab.demon.co.uk/frank.htm.
Sir John Franklin Was Here! https://web.archive.org/web/20130606014858/http://www.sirjohnfranklin.com/pages/framea.html.
Starvationcove.blogspot.com.
verticalarctic.blogspot.com/.
Visionsofthenorth.blogspot.com.

Web-Based Resources

Ananquaq, Jean-Marie, and Guy Mary-Rousseliere. "The Schwatka Expedition as Seen by the Inuit." n.d. Available at https://lookaside.fbsbx.com/file/The%20Schwatka%20Expedition%20as%20seen%20by%20the%20Inuit%20-%20Eskimo%2C%20n.s.%-2038-39%2C%201990%2C%20pp%205-11%2C%20.pdf?token=AWzpLpNdDIW_nu00I1tR3wBgHbddA0fL7V0IpmuVUlQMcAN7xesn1ruIsYh2ex42kiFuB-HSO_g7U8UiuYHTzPMuPTjRsNmhYNVPexZlz4zZdZZ1_XGafvct3QPPpdf78sazRpVFA-W3kY1FFRKMUcHI.
"The Life Story of William Braine RM." Available on the *Remembering the Franklin Expedition* group on Facebook. https://www.facebook.com/groups/11434844549/permalink/10157616864449550.
Carney, Peter. "Are You Friendly?—We Are!" *Erebus & Terror Files*, December 4, 2011.
_____. "Black Men, Welsh Wigs and the Knights Templar." *Erebus & Terror Files*, December 17, 2011.
_____. "J'Accuse!—The Case of Stephan Goldner—Britain's Dreyfus." *Erebus & Terror Files*, April 30, 2020. Available at https://erebusandterrorfiles.blogspot.com/2020/04/jaccuse-case-of-stephan-goldner.html13-20.
Coleman, Ernest C. "The 1845 Franklin Expedition: Cannibalism? Lead Poisoning?" 1998. Retrieved from the Roger Parsons Lincolnshire World website, http://www.rogerparsons.info/franklin.html.
Cunningham, Sgt. W.K. *Memorandum Book of the 1839–43 Antarctic Expedition.* Available at https://www.hakluyt.com/PDF/Campbell_Part2_Journal.pdf?fbclid=IwAR1443h7oZPFZnr4kL0Yqba0ulPvvg1_LvWsmgeo-SdnrJfALHo34EawT2Aoon.
Dagneau, Charles. "Interpretative Essay." *The Franklin Mystery: Life and Death in the Arctic* website, https://www.canadianmysteries.ca/sites/franklin/interpretation/experts/interpretationDagneau_en.htm.
Dickens, Charles. "The Noble Savage." n.d. Copied from www.ric.edu/faculty/rpotter/temp/noblesav.html.
Freebairn, Alison. "Lost and Found: the Beechey Island Papers." *VisionsNorth* blog, https://visionsnorth.blogspot.com/2019/09/lost-and-found-beechey-island-papers.html. Posted September 10, 2019.
Frey, Joseph. "Inside the Terror—Life on Board Sir John Franklin's Lost Ships." *Royal Geographical Society* blog, http://geographical.co.uk.
Gehrmann, Kristina. *Im Eisland*. English version available at https://tapas.io/episode/963713.
Hatfield, Philip. "The Search for the Northwest Passage." *History Today* 67 (2017). Available at https://www.historytoday.com/archive/search-northwest-passage.
Hayes, Alfred. "I Dreamed I Saw Joe Hill Last Night." 1930. Available at http://www.protestsonglyrics.net/Labor_Union_Songs/Joe-Hill.phtml.
Hoekstra, Kyle. "The Adventurous Seafaring Women of the Age of Sail, in Their Own Words." Atlas Obscura website, November 25, 2019. https://www.atlasobscura.com/articles/women-age-of-sail?fbclid=IwAR2uLc9GF_PzVDpCbAG83i6GOmqb-OPnSXH5tx6zJiogrZsHNWzn6BRnebc.3-14.
Hong, Jackie, and Jesse Winter. "HMS Terror, Second Ship from Doomed Franklin Expedition, Found in Terror Bay." TheStar.com, September 12, 2016.
Keller, Wolfgang, Ben Li, and Carol Shiue. "China's Soaring Foreign Trade: Made in Britain c. 1840?" Available at https://voxeu.org/article/china-s-soaring-foreign-trade-made-britain-c-1840.
Kogvik, Sammy. "The Franklin Mystery: HMS Terror." Available at https://www.arcticfocus.org/stories/franklin-mystery-hms-terror/.
McGoogan, Ken. "Mystery Solved!!! Polar Bears Explain Fate of the Franklin Expedition." *Ken McGoogan* blog, September 7, 2018. http://kenmcgoogan.blogspot.com/2018/09/mystery-solved-polar-bears-explain-fate.html.
McNab, David, and Paul-Emile McNab.

"William Kennedy and the Search for Franklin." *Arctic Focus*. https://www.arcticfocus.org/stories/william-kennedy-and-search-franklin/.

Moore, Fiona. "Comparing Colonialisms in Dan Simmons' Novel, *The Terror* and its AMC Adaptation." Available at https://www.academia.edu/40129260/Comparing_colonialisms_in_Dan_Simmons_novel_The_Terror_and_its_AMC_adaptation.

Opel, Mechtild. "Chocolate in the Arctic." Available at http://www.trimaris.de/2019/04/24/-chocolate-in-the-arctic/?fbclid=IwAR0QBFeIPt1BCb8ZM__W58mWbNojYXEkz2IlIzrwDENkgHqRinDZlAz__Xo.

Pilø, Lars. "Buried in the Ice: The Franklin Expedition Cemetery." Available at https://secretsoftheice.com/news/2019/10/28/franklin-expedition/.

Potter, Russell. "The Fate of Franklin's Ships: What We Know Now." 2016. Available on the Remembering the Franklin Expedition Facebook group.

———. "The Grave of Lieutenant Irving?" *Visions North blog*, January 3, 2010. Available at https://visionsnorth.blogspot.com/2010/01/-grave-of-lieutenant-irving.html16-16.

———. 'The Library of the Erebus and Terror." *Visions of the North* blog, April 26, 2009. Available at https://visionsnorth.blogspot.com/2009/04/library-of-erebus-and-terror.html.

———. "A Navigable Northwest Passage." *Visions of the North* blog, September 12, 2012. Available at http://visionsnorth.blogspot.com/2012/09/a-navigable-northwest-passage.html.

Pryor, Jonathan. "Interment Without Earth: A Study of Sea Burials During the Age of Sail." *New England Burials at Sea* blog, November 26, 2010. Available at https://www.newenglandburialsatsea.com/interment-without-earth-a-study-of-sea-burials-during-the-age-of-sail/.

Roobol, M. John. "Status of the History of the Lost Franklin Expedition of 1845." Academia Letters 1461, July 2021. Available at htpps://www.academia.edu/49826555/Status_of_the_History_of_the_Lost_Franklin_Expedition_of_1845.

Rosenberg, Roger. "The Donner Party: Trapped in the Mountains." *Academia Letters*, Article 2482 (2021). Available at https://doi.org/10.20935/AL2482.

Smith, Kiona. "Strands of Hair Shed Light on Doomed 19th-Century Arctic Expedition." *Ars Technica Online,* September 30, 2018. Available at https://arstechnica.com/science/2018/09/did-lead-poisoning-finish-off-a-doomed-arctic-expedition/.

Watson, Paul. "The Wreck of HMS Erebus: How a Landmark Discovery Triggered as Fight for Canada's History." *Buzzfeed*, September 14, 2015. Available at https://www.buzzfeed.com/paulwatson/the-wreck-of-erebus.

Index

aboriginals, Tasmanian 32
Adelaide Peninsula 74, 77, 87, 89, 111, 119, 122–23
Aden 39
Admiral Makarov 65
Admiralty 4–5, 7–8, 10–11, 13, 20, 24–25, 30, 32–33, 35, 37–38, 41, 44, 47, 49, 52, 57–59, 63, 66, 74–77, 79, 93–94, 96, 98, 102, 108, 124, 133, 135–36
Advice 43
Africa 4, 24, 26, 40, 47, 107–08, 116, 118
"Aglooka" 121–22
Alaska 6, 31, 59, 65, 77, 94–96
HMS *Albert* 40
HMS *Albion* 39
alcohol 108–10
HMS *Alexander* 5, 103
Alexandria, Egypt 44
Algiers 22, 27
allotments 44–45
Amazon 116
Amundsen, Roald 1, 10, 65–67
Amundsen Gulf 65
Anasazi 115, 120
Anchorage, Alaska 51
Anderson, James 85
Andes 116, 119
Andrée expedition 99
Anson, George 108
Antarctic 7–9, 23–26, 28, 35, 38, 45, 98, 105, 107, 130
Anthropocene 118
anthropophagia 120; *see also* cannibalism
Antigua 31
anti-semitism 102
Arctic (Canadian) Archipelago 3, 5, 8, 17, 52
Arctic Circle 10, 12
"Arctic Fever" 4
Arctic Ocean 3, 30
Armitage, Thomas 47
Articles of War 89–90
Asia 3–4, 65, 107
HMS *Assistance* 51
Athenaeum Club 30
Athens 39, 42
Atlantic Ocean 3, 25, 28, 33, 38, 76, 96, 114

aurora borealis 57
Austin, Horatio 11, 51, 60
Australia 25, 30, 39, 42, 94, 117

Back, George 7–8, 28, 35, 39, 76, 83–86
Back's Fish River 12–15, 77, 82–85, 88, 121, 134, 136
Baffin Bay 3, 6, 10, 17–18, 56, 97, 103
Baffin Island 5, 33, 97
Baghdad 38
Baker Lake 121
"Balthazar" (television series) 113
Baltic Sea 96
Baltimore, Maryland 27
Banbridge, Northern Ireland 32–33
Banks Island 10, 59–61, 63, 94
Barretto Junior 10, 26, 44, 52–53
Barrow, George 38
Barrow, John 4–8, 10, 30, 37–38, 40, 44, 52, 59, 61, 65, 67, 77, 81, 98, 133, 136
Barrow, John, Jr. 38
Barry, Lady Lucy 31, 42, 131
Bartlett, Bob 96
Basra, Iraq 38
Bathurst Island 57
Battersby, William 37–38
Bavaria 31, 39
Bayne, Peter 124
HMS *Beagle* 26, 39
Beardsley, Martyn 129
Beattie, Owen 53, 99, 106, 115, 119–20
Beaufort, Francis 35
Beaufort Sea 12, 53, 57, 61, 65, 78
Beck, Adam 18
HMS *Bedford* 27, 30
Beechey, Frederick William 9
Beechey Island 11–13, 16–18, 20, 46, 49–53, 56–57, 59, 62–65, 70–74, 76, 98–99, 101–104, 106, 112, 123–25, 134
HMS *Beelzebub* 26
Beirut 38
HMS *Bellerophon* 30
Bellot, Joseph-René 94
HMS *Belvidere* 42–43

Bengal Cavalry 29
Bergen-Belsen 117
beriberi 99, 101
Bering Sea 57
Bering Strait 4, 10, 57, 94–96
Berton, Pierre 1, 56, 130
Blanky, Thomas 8, 43–44, 48, 69, 72
Bligh, William 90
"boat place" 42, 84, 92, 110
bomb vessels 22–23, 26, 34
Boothia Peninsula 7, 10, 20, 28, 34, 44, 58, 61, 66–67, 71, 75, 78, 81
botulism 20, 49, 99, 101–03
HMS *Bounty* 33, 90
Braine, William 46, 49, 99, 103
Brazil 37
HMS *Briton* 33
broadsheets 53, 80
Browne, William 60, 63
Buchan, Buchan, David 4–5, 30
Burwash, Lachlan 87
Bylot, Robert 3

cairns 39, 60, 62, 66, 125, 135
Calais 22
California 94
HMS *Calliope* 39, 105
Campbell, R.J. 33
campsites 15, 71, 84–85, 99, 112
Canada 16, 40
cannibalism 6, 15, 19–20, 41, 46, 85, 99, 104–105, 114–120
Cape Felix 61, 66, 68, 71, 78, 86, 124
Cape Herschel 12, 77, 79
Cape Town, South Africa 23
Cape Walker 10, 57, 59, 62, 66, 68
Caracas 91
Caribbean Sea 31, 65
Cavell, Janice 12, 131
Chaco Canyon 115, 120
Chambers, George 45
Channel Islands 39–40
Chantrey Inlet 12, 46, 77, 82–84, 111
Chatham shipyard 24, 28
Cheap, David 90
Cherokee class frigates 26
Chesapeake Bay 27

197

Index

Chesterfield Inlet 97
Chile 90, 94
China 7–8, 38–39, 105
chocolate 49, 109–10
Christmas 24, 53, 72, 74, 80
Chukchi Sea 96
Church of England 43
Churchill, Manitoba 97, 112, 122
Clerke, Charles 38
HMS *Clio* 39–40
Clipper Adventurer 67
clothing 50, 72–73, 104, 131
Collinson, Richard 94
Colorado 119
Columbus, Christopher 4, 107
Coningham family 37
Cook, James 25, 38
Cook Islands 116
Cookman, Scott 102
Copenhagen, Battle of 22, 30, 39
Copper Inuit 63
Coppermine River 3, 6, 31, 83
HMS *Cordelia* 42
Cork, Ireland 33, 117
HMS *Cornwallis* 38
Cornwallis Island 11, 57, 62, 65
Coronation Gulf 65
Couch, Edward 38
HMS *Cove* 35
Cracroft, Sophia 25, 32, 35, 96
Crimean War 98
Croker, John Wilson 5
"Croker Mountains" 5, 68
Crozier, Francis 1, 7–10, 12, 14–17, 24–25, 27–29, 32–39, 42, 44–46, 51–52, 61, 69, 71–72, 75–77, 79–82, 84, 86, 88–89, 91, 98, 103, 112, 115, 119, 121–23, 125, 130–31, 134, 136
Crozier, George 33
Crozier's Landing 14, 46, 112
Cyriax, Richard 47, 54, 109, 124–25

Da Gama, Vasco 107–08
daguerreotypes 8, 26–27, 36, 42
Dahmer, Jeffrey 116
Daly, James 45
Darwin, Charles 26, 39
Davis, John 3
Davis Strait 5–6, 12, 65, 98
Davy, Robert 27
Dealy Island 94
Dean Cemetery 43
Dease, Peter 7, 12, 77, 79
Deptford 45
Des Voeux, Charles 71
Devon Island 5, 11–12, 17, 27, 50
Devonport 28
Dickens, Charles 114–17, 120
Diggle, John 45
dogs, use of 54–56, 119
Donkin & Gamble 103
Donner Party 104, 114, 116, 119
HMS *Dorothea* 4
HMS *Dotterel* 33, 38
Dutch East India Company 107

East India Company (UK) 58
Eber, Dorothy 18, 134
HMS *Edinburgh* 42
Edinburgh, Scotland 41–43, 87, 127–28
Edward III, King 22
Egypt 117
Elce, Erika Behrisch 17, 20, 47, 131
Ellesmere Island 104
England 17, 23–24, 26, 41
English Channel 33
Enoolooapik 43
HMS *Enterprise* 10, 68, 81, 93–94
HMS *Erebus* 1, 7–10, 13, 15, 17–28, 34–49, 51, 53, 57, 59–61, 64, 66–68, 70–72, 74–75, 79, 81, 83–85, 87–89, 92, 96, 98–100, 102, 105, 107, 109, 111, 114–15, 119–22, 124–26, 128, 130, 133–35
Essex 118
Euphrates River 7, 38
Evans, Thomas 45
HMS *Excellent* 38
exhumation 47, 53
Experimental Squadron 37

Facebook 17, 134
Fairholme, Adam 41
Fairholme, James 8, 36, 40–41
Falconer, Edmund 121
Falkland Islands 23, 25, 35
Falmouth 118
Fiji 116
Fitzjames, James 7–8, 14, 26–27, 29, 36–40, 44–45, 49, 52, 57, 68, 71–72, 84, 89, 95, 103, 105, 131
Flexible Flyer 54
Flinders, Matthew 30–31, 39
Fluhmann, May 32, 103
Forbes, Lord Walter 40
Fort McHenry 22, 27
Fort Providence 6
HMS *Forth* 30
Fortnum & Mason 53
Foxe, Luke 3
Foxe Basin 33
Fram 96
France 31, 90, 113
Frances Mary 118
Francis Spaight 118
Frankenstein 4, 31
Franklin, James 29
Franklin, Lady Jane 7, 12, 31–32, 47, 56, 59, 77, 93–94, 96, 98, 124, 126
Franklin, Sir John 1, 3–14, 16–21, 23–68, 70, 72–88, 90–114, 136; grave 124–25, 127; naval career 29–32
Franklin, Willingham 29
Franklin, Willingham, Jr. 29
Franklin Strait 65
Franklinites 12, 15, 89, 96, 115, 134

Fraser patent stove 24, 50
Freebairn, Alison 49, 57
Frobisher, Martin 3
frostbite 42, 105
HMS *Fury* 9, 34, 75–76, 103
Fury and Hecla Strait 34
Fury Beach 20, 34, 44, 57, 67–69, 71, 81–84, 86

Galway, Ireland 117
Gambier, Sir James 37
Gambier, Robert
HMS *Ganges* 38, 40
Geiger, John 18
Géricault, Théodore 118
Gibraltar 126
Gibson, William (Hudson's Bay Company) 85
Gibson, William (*Terror* steward) 47
Gjoa 66–67
Glenn, John 100
Golding, Robert 45
Goldner, Stephen 9, 72, 100–03, 106, 113
Goodsir, Archibald 41
Goodsir, Harry 27, 36, 40–41, 43, 47, 52, 106, 120, 126
Goodsir, Robert 41, 43
Gore, Graham 8, 36, 40–41, 70–71, 74–79
Gore, John 38
Grace, Sherrill 130
Great Barrier Reef 30
Great Expectations 116
Great Leap Forward (China) 117
Great Slave Lake 6, 14–15, 82–83, 85
Greece 31, 38–39, 117
Greeley expedition 104–05, 118
Greenhithe 9–10, 32, 35, 43, 63, 110
Greenland 3–6, 8, 10, 12, 17–19, 24, 26, 35–36, 40, 43–45, 52–53, 73, 76, 88, 100, 103, 130, 134
Greenland Sea 4
Greenwich 23, 47, 74
Gregory, John 47, 128
Griffin, Jane *see* Franklin, Lady Jane
HMS *Griper* 6, 44, 61
Guelphic Order of Hanover 31, 35, 126
Gursky, Ephraim 18, 43, 121

Hall, Charles Francis 18, 40, 46, 84, 92–93, 96–97, 111, 115, 121–24, 126, 134
HMS *Hamadryad* 33
Hanseatic 67
Hardy, William 45
Harper, Stephen 130, 133, 135
Hartnell, John 46, 49, 99, 112, 123
Hawaii 11, 38, 94–95; *see also* Sandwich Islands
Hayes, Rutherford B. 59

Index

HMS *Hecla* 6, 9, 22, 34, 61, 76, 81
HMS *Hermione* 91
Hertfordshire 37
Hickey, Clifford 62
Hickey, Cornelius 45, 62, 89, 120
Hobart, Tasmania (Van Diemen's land) 23, 25, 35, 130
Hobson, George 62
Hobson, Lt. William 14, 71, 84–85, 91, 109, 122
Hodgetts, Lisa 133
Holodomor 117
Horie, Kenichi 65
Hornby, Frederick 75
Houndsditch, London 101
HMS *Howe* 27
Hudson, Henry 3
Hudson Strait 28
Hudson's Bay 3, 6–7, 28, 30–31, 33, 64, 82, 93, 97, 121
Hudson's Bay Company 3, 6–7, 12, 14–15, 56, 77, 82–83, 85, 94, 97, 100, 136
Huguenots 22
Huntford, Roland 1

Ice Blink 102
Iceland 3–4
Igloolik 34, 55
Iglulingmiut 55
Illustrated Londons News 9, 36
In-nook-poo-zhe-juk 115, 122
India 38
Indian (Native American) 54, 131, 133
Indian Ocean 38
Indonesia 5
Inglefield, Edward 124
HMS *Intrepid* 63, 74
Inuit 2, 8, 12, 15, 18, 34, 41, 43, 46, 54–56, 63, 72, 75–76, 81–82, 85, 87, 92–93, 95–97, 100, 104, 109, 111, 114–15, 120–24, 126, 129–31, 133–35
Inuktitut 3, 82, 121
HMS *Investigator* 30, 59, 63, 68, 81, 93–94, 96, 109
Iraq 7–8, 38, 105
Irbis 65
Ireland 28, 33, 35, 39, 76, 117
Iron Maiden 53
Irving, Lt. John 42–43, 46, 71, 87–88, 115, 127, 137
Isabel 94
Isabella 5, 71, 103
Iwillik 122

James Ross Strait 66–68
Japan 65
Jews 43–44, 102; *see also* anti-semitism
Johnson, Lyndon 1
Johnson, Thomas 45
Johnston, Sir Harry 116
Jones, A.G.E. 47
Jopson, Thomas 45

Kamchatka 95
Kamookak, Louis 87
Karluk 96
Kasten-Mutkus, Kathleen 16
Keenleyside, Anne 115, 119–20
Kellett, Henry 53, 63, 94
Kennedy, William 94
Kerguelen Island 25
Kerry (Irish county) 117
Key, Francis Scott 27
King, Richard 130
King William Island 10, 12–13, 15, 18, 20, 28, 33, 39, 42–43, 46, 60–61, 66–67, 69–71, 81, 84, 86–87, 89, 100, 104, 106, 190–10, 112, 114–15, 117, 119–20, 122, 124–26, 131, 134
Klempenfelt, Richard 90
Kolyma River 96
Korea 65
Kuru 116

Labrador 3, 101
HMS *Lady Franklin* 43
Lake Borgne 27
Lambert, Andrew 8, 52, 129
Lancaster Sound 5–6, 10–12, 17, 28, 33, 40, 44, 60–61, 65, 67–68, 81, 84, 93–94, 131
Larsen, Henry 61, 65, 86
lead poisoning 20, 49, 51, 89, 99, 106–07, 112–13
Leningrad 117
Le Vesconte, Henry 15, 36, 38–40, 46–47, 115, 118, 128
Lincolnshire 29
Lind, James 108
Lindblad Explorer 65
Lisbon 27
Little, Edward 75
Lloyd-Jones, Ralph 44
locomotives 9, 26, 47, 79, 87
London 4, 6–9, 26, 31–32, 35–38, 49, 53, 95, 101–02, 114, 119–20, 125, 128
Long, Alison 33
Ludwig, King of Bavaria 39
Lyon, George 44, 55

Mackenzie River 3, 6, 31, 94, 122
MacLaren, Ian 133
Madras 29
Magellan, Ferdinand 108
magnetism 7, 51–52
Magwich, Abel 116
Maine 118
Malory, Thomas 133
SS *Manhattan* 65
Manitoba 30, 112, 122
Maori 25, 116–17
Marchand, Peter 104
Markham, Albert 37
Markham, Clements 37
Matty Island 68
Mayo 117
Mays, Simon 109
Mazatlán 95
McClintock, Francis Leopold 12–15, 20, 35, 43, 46–47, 62–63, 71, 84–85, 93, 109, 122–23
McClintock Channel 10, 58–61, 66, 70, 84
McClure, Robert 35, 59–61, 63, 65, 94–96, 109
McClure Strait 61
McDonald, Alexander 43, 49
McGoogan, Ken 67, 111, 129
HMS *Meander* 33
Mediterranean Sea 23, 27, 31, 37–40, 42–43, 105, 126
Médusa 118
Melville, Herman 114, 118
Melville Island 6, 33, 54–55, 57, 61–62, 65, 74, 94
Melville Peninsula 34, 55, 96–97, 121–23
Mercy Bay 63
Methodist Church 31, 42, 131
Mexico 95
Middlesex 45
Midlands 45
Mignonette 118
Millar, Keith 100, 106
Moby-Dick 118
Moldavia 102
Montreal Island 15, 85–86
Monty Python and the Holy Grail 22, 133
Morin, O.J. 63
Mount Etna 42
Mowat, Farley 121
Munk, Jens 111–12
Murray, John 4, 30, 35, 133
Musin, François Étienne 23
mutiny 87–92

Nadolny, Sten 29
Nansen, Fridtjof 96
Napoleonic Wars 3–4, 22, 26, 30, 33, 39, 90–91
Navarino, Battle of 31, 39
Neanderthals 116
Nelson, Horatio 4, 22, 39, 126
Neptune (dog) 122
Netherlands 65
New Georgia Gazette 72
New Orleans, Battle of 27, 30
New South Wales, Australia 31, 90
New Zealand 25, 116 17
Newfoundland 3, 60, 101
Nicholas I, Tsar 31
Niger River 40
The Nore (mutiny at) 90–91
Norsemen 4
North Magnetic Pole 52
North Pole 4, 22, 30, 34, 44, 55
Northwest Passage 3–5, 7, 11, 22, 27–28, 30, 38–39, 44, 52, 55, 57, 59, 61, 65–67
Northwest Territories, Canada 65, 115
Norton Sound 96
Norway 34
Norwegian Sea 4
Nottingham Galley 118

Okhotsk, Russia 95
Ommanney, Erasmus 102
Ootjoolik 121, 134
open Polar Sea 4–5, 7, 10–11, 30, 52, 57, 96
Opium Wars 8, 38–40
O'Reilly Island 20
Orkneys 10, 45
Otho, King of Greece 31, 38–39, 42
Ottawa, Canada 134
Ottoman Empire 31, 39
Oxford University 31

Pacific Ocean 33, 38, 43, 58, 94–95, 116
Packer, Alfred 119
Pakeha Maori 117
Papua New Guinea 1, 116
Park, Robert 85, 92
Parker, Richard 118
Parks Canada 15, 20, 59, 128, 133
Parliament (UK) 94, 102, 121, 123
Parry, William Edward 5–6, 9, 28, 33–35, 38, 44, 49, 54–57, 60–62, 65, 72, 75, 81, 94, 97, 103, 122, 125
Parry Bay 97
Peake, Henry 26
Peary, Robert 108
Peddie, John Smart 43
Peel Sound 13, 58–63, 65–66, 68
Peggy 118
Peglar, Harry 16, 47
Peglar Papers 16
Pelly Bay 81, 121–22
Pembroke, Wales 23
pemmican 56
Peninsular War 37
Penny, William 43, 93
Persian Gulf 38
Perth, Scotland 40
Phipps, Constantine 4
Pilkington, William 45
Pim, Lt. Bedford 95–96
Pitcairn Island 33
HMS *Plover* 95
Plymouth, England 23
"Poctes Bay" 66–67
Poe, Edgar Allan 114
Point Barrow, Alaska 77
Point Beechey, Alaska 31
Point Le Vesconte 15
polar bear 111–12
HMS *Polyphemus* 30
Pond Inlet 95–97, 122
Porden, Eleanor Anne 30–31, 131
Port Leopold 68, 81
Port Louis, Falkland Islands 25
Portsmouth 27, 39, 42, 90
Portugal 3, 34, 37
Potter, Russell 41–42, 47, 127
Presbyterian Church 33, 35, 42
Price, Alexa 123

Prince Albert 94
Prince of Wales 10, 12
Prince of Wales Island 13, 57, 59–63, 66, 68, 79, 94
Prince of Wales Strait 59–61, 94
Prince Regent Inlet 6, 34, 57, 68, 71, 95
propellers 9, 26, 87
HMS *Pyramus* 37

Quarterly Review 4, 30
Québec 44
HMS *Queen Charlotte* 33
Queen Maud Gulf 75
The Queen v Dudley & Stevens 118
Queensland 39

Rae, John 12, 18, 41, 56, 64–68, 82–83, 92–93, 96–97, 100, 109–10, 114, 120, 122, 126, 129
Rae Strait 66–68
HMS *Rainbow* 31, 38
HMS *Rattler* 10
Reid, James 8, 36, 48, 69, 72
Repulse Bay 28, 39, 82, 97, 122
HMS *Resolute* 53, 59, 65, 74, 94
Resolute, Cornwallis island 65
Richardson, John 93, 125
Richler, Mordecai 18, 43, 121
Riga, Latvia 44
Rio de Janeiro 23, 26, 37
Rogers, Stan 49, 53
Rondeau, Robin 85
Roobol, John 122
Ross, James Clark 7, 9, 13, 22, 24–28, 34–36, 38, 44, 55, 58, 62, 66, 68, 71, 75–76, 78–79, 81, 83, 86, 93, 98, 127, 130
Ross, John 5–6, 8, 28, 44–45, 54, 57, 68, 71, 76, 81, 103, 108, 122
Ross, W. Gillies 93
Ross Ice Shelf 25, 35
Rossbank 25
Royal College of Surgeons, Edinburgh 41, 43
Royal Geographical Society 31
Royal Geological Society 31
Royal Marines 44–46, 49, 99, 103
Royal Naval College 39, 41–43, 46, 88, 108, 127
Royal Navy 4, 7, 22, 27, 33–34, 44–45, 89–91
Royal Society 4, 7, 30, 52
"Rum Rebellion" 90
Rumsfeld, Donald 2
Russell Island 57
Russia 4, 31, 65, 79, 95–96, 117

Sabine, Edward 7, 9, 52, 125
Saint Andrew 129
Saint Brendan 3
Saint Helena 33, 103
St. Mary's, Georgia 27
St. Petersburg, Russia 44, 96
HMS *St. Vincent* 37

Sandwich Islands (Hawaii) 94–95
Sasketchewan 30
Saunatuk (cannibalism site) 115
Savelle, James 62
Schaeffer, Jonathan 78
Schrödinger, Erwin 18–20
Schwatka, Lt. Frederick 14–15, 42–43, 46, 84, 88, 93, 96, 100, 127
Scoresby, William 3, 5
Scotland 41, 43, 88
Scott, Robert Falcon 1, 8, 123, 130
Scott Polar Research Institute 33
scurvy 15, 20, 23, 34, 59, 73, 89–90, 94, 96, 99, 101, 107–09, 112–13, 120, 122, 136
Seeger, Pete 1
Setumenin 46
Seward Peninsula 96
Shelley, Mary 4, 31
Siberia 95–96
Sicily 42
Sierra Nevada 104, 116, 119
Simmons, Dan 2, 18, 20, 41, 76, 79, 89, 120, 123
Simonstown, South Africa 24, 26
Simpson, Thomas 7, 12, 56, 77, 79, 99
Simpson Strait 46, 67, 75, 77, 85, 123
Singapore 38
sledging 14, 20, 44, 53–55, 57, 62–64, 70–71, 74, 77–78, 96, 99, 101, 106, 131
Smith, Michael 37
Somerset Island 10, 13, 34, 57–58, 68, 81, 94
Sophia 43
South Africa 47
South Pole 123
Southampton 28, 121
Southampton Island 28, 121
Southern Ocean 24
Spain 27, 107
Spilsbury 29
Spithead 90–91
Spitzbergen 4, 30, 34
Stalin, Josef 117
Stanley, Stephen 41
Starvation Cove 15, 20, 46, 111, 115, 119
Steffansson, Vilhjamur 110
Stein, Glenn 47
Stenton, Doug 85, 92, 99
Stokes, John Lort 39
Stonington, Connecticut 27
Straits of Magellan 84
Stromness 41
Su-pung-er 124
Surgeon's Hall Museum 41
Surrey 45
Sydney, Australia 25, 118
Syria 38

Taichman, Russ 76

Index

Tambure volcano 5
Tasmania *see* Van Diemen's land
Tennyson, Alfred 124
Terra Nova 55
HMS *Terror* 1–2, 7–10, 13, 15, 17–28, 34–36, 39–49, 53, 57, 59–61, 64, 66–68, 70–72, 74–76, 79, 81, 83, 85, 87–89, 96, 98–100, 102, 105, 107, 109, 111–12, 114–15, 119–23, 127, 133–34, 137
Terror Bay 15, 20, 46, 85, 87, 89, 111–12, 115, 119, 134
Thames River 8–10, 26, 33, 90
Thompson, James 44, 46, 103
Thorne, Robert 4
RMS *Titanic* 17
Todd Islets 15, 46, 85, 119
Tomsk 95
Tooshooarthariu 121–22
Topham shipyard 27
Torrington, John 46, 49, 53, 99
Trafalgar, Battle of 4, 30, 39, 126
Traill, H.D. 129
Treaty of Nanking 38
Treblinka 117
Trent 4, 30
trichinosis 99, 111–12
tuberculosis 20, 31, 49, 99, 112–13
Tuktoyaktuk 65
Tulloch Point 85
Turnagain Point 75
Turner, Christy 120

Turner, Jacqueline 120
Two Graves Bay 15

Ukraine 117
United States 27, 65, 93, 115
Uruguay 116, 119
Utah 119

Valparaiso 94
Van Diemen's land (Tasmania) 7, 23, 31–32, 52, 72, 79, 94
Verne, Jules 4
HMS *Vesuvius* 26
Victoria, Queen 41
Victoria Island 10, 12, 59, 61–63, 66, 75–76, 78, 94
Victoria Strait 10, 12, 20, 58, 61–63, 65–70, 75, 98, 105, 125, 128, 134
HMS *Victory* 22
Victory Point, King William Island 14, 39, 46, 66, 72, 77, 82, 85, 101, 119, 124, 126
Victory Point Record 10–11, 13–14, 16, 20, 39, 57, 60, 62, 64, 66, 69, 73–77, 79–80, 84, 88, 98, 101, 111, 123, 127
Vietnam 1
Vikings 3
Viscount Melville Sound 6, 59–60, 63
vitamins 101, 105–6, 108
Vollmann, William 18, 121

Wagenborg shipping line 65, 90

HMS *Wager* 90
Wall, Richard 45
War of 1812 22, 27
Washington Bay 15, 122
Washington, D.C. 27
Waterloo Place, London 125–26
Weddell, James 25
Weddell Sea 26
Wellington Channel 57, 62
Wertheimer, John 102
Wesley, John 131
West Indies 27, 91
Westminster Abbey 124–26
Whalefish Islands, Greenland 10, 43–44
Whitby 44
White House 27, 59
William IV, King 31
Winter Harbor (Melville Island) 54, 62
Wollaston Land 10, 59
Wood, John 4
Woodman, David 18, 42, 82, 92, 96–97, 121–23, 134
Woolwich 8–9, 26, 41, 45
Wright, Oliver 26

Yakutsk 95
York Factory 64, 97
Yorkshire 44
Young, David 45

zinc 20, 49, 99, 101, 112

www.ingramcontent.com/pod-product-compliance
Ingram Content Group UK Ltd.
Pitfield, Milton Keynes, MK11 3LW, UK
UKHW050701160426
5217IPUK00038B/1814